Wildlife of the
Tibetan Steppe

Wildlife of the

George B. Schaller

The University of Chicago Press
Chicago and London

Tibetan Steppe

The University of Chicago Press, Chicago 60637
The University of Chicago Press, Ltd., London

Paperback edition 2000
Printed in the United States of America

07 06 05 04 03 02 01 00   5 4 3 2

ISBN: 0-226-73652-0 (cloth)
ISBN: 0-226-73653-9 (paperback)

Library of Congress Cataloging-in-Publication Data

Schaller, George B.
Wildlife of the Tibetan steppe / George B. Schaller.
p. cm.
Includes bibliographical references and index.
ISBN 0-226-73652-0 (alk. paper)
1. Ungulata—China—Chang Tang Plateau. 2. Mammals—China—Chang Tang Plateau. I. Title.
QL737.U4S34 1998
599′.0951″5—dc21   97-36053
CIP

♾ The paper used in this publication meets the minimum requirements of the American National Standard for Information Sciences—Permanence of Paper for Printed Library Materials, ANSI Z39.48-1992.

Contents

Acknowledgments

MANY INDIVIDUALS and organizations assisted or participated in the project. It was initiated as a collaborative venture between the Wildlife Conservation Society in New York and the Ministry of Forestry in Beijing together with the China Wildlife Conservation Association. Wang Menghu, Qing Jianhua, Jiang Hong, Wang Wei, Meng Sha, and Fan Zhiyong provided help and encouragement.

Organizational support in Xinjiang was provided by the Forest Bureau in Urumqi, and I remember with pleasure the cooperation provided by Liang Guodeng, Li Hong, Talipu, and Lu Hua. In Qinghai, I also collaborated with the Forest Bureau, whose staff, especially Guo Gieting and Zheng Jie, offered much assistance. The Northwest Institute of Endangered Animals in Xian, Shaanxi, participated in some of the research in Xinjiang and Qinghai and provided an assistant, Ren Junrang.

The National Environmental Protection Agency, with the help of Jin Jiangming and Xue Dayuan, initially sponsored the project in Tibet, and this enabled me to begin several years of fruitful collaboration with two local organizations: the Tibet Plateau Institute of Biology and the Tibet Forest Bureau (now Tibet Forestry Department). The logistic, financial, and scientific support helped make it possible to conduct the project in Tibet. Three expeditions were made with the institute, five with the Forest Bureau, and one with both. At the institute, Gu Binyuan was my main colleague, but many others, including Ni Zhicheng, Chi Doa, Chanjue Zhuoma, Dunzhu, Xiong Jigu, and Dawa, also participated. At the Forest Bureau, the help and cooperation of Liu Wulin were invaluable, as was the support of Yin Binggao, Yang Jianxiang, Lu Wei, Danzhen, and others. The Tibet Bureau of Science and Technology generously gave permission to conduct the project.

Wang Haibin accompanied me on two trips in China as coworker and interpreter, and Qiu Minjiang on five trips. They later went to the United States for advanced degrees in biology. Wang Xiaoming of East China Normal University in Shanghai, who had worked with me on a panda project and subsequently went to France for an advanced degree, again joined me, this time for one survey in Tibet and the one in Inner Mongolia. My son Mark Schaller came on one journey in Xinjiang. All contributed much to the project. I owe special gratitude to Ding Li, who volunteered to join us on five expeditions in Tibet, and with cheerful efficiency helped ensure the success of each venture.

Because my knowledge of range ecology is scant, I asked Daniel Miller, now with the International Centre for Integrated Mountain Development, Kathmandu, to accompany me on two trips to the Chang Tang. His research contributed importantly to the project and his company was a pleasure. He also made helpful comments on chapter 15.

The project in Mongolia represented a joint effort with the Mongolian Association for the Conservation of Nature and Environment and the Ministry for Nature and Environment. I am especially indebted to former Minister Z. Batjargal and to J. Tserendeleg for support. R. Tulgat, G. Amarsanaa, S. Chuluunbaatar, and S. Amgalanbaatar ably participated in the fieldwork, and G. Dembereldorj, O. Byambaa, and others were also of assistance. Two of my trips to Mongolia were under the auspices of the United Nations Development Program—Global Environmental Facility (UNDP-GEF) Biodiversity Programme in Ulan Bator, and I am particularly indebted to J. Griffin and D. Batbold for their hospitality, help, and companionship in the field. Thomas McCarthy assisted me during one winter's work on snow leopards and then took on that project for a graduate degree. That winter my son Eric Schaller also helped with the research, and Joel Bennett and Luisa Bennett filmed it. These individuals, together with J. Tserendeleg, made a congenial team in our mountain camp.

At the Wildlife Conservation Society, I am indebted to William Conway and John Robinson for generous support throughout this long-term project. Various staff members assisted me, including Robert Cook and William Karesh (veterinary advice), Ellen Dierenfeld (nutritional analyses), Tracy McNamara (pathology), and Judy Kramer (parasites). Peter Walsh advised on statistics.

George Amato analyzed mitochondrial DNA of various ungulates, and his efforts, together with those of John Gatesy of the University of Arizona, provided me with critical data and insights concerning the evolutionary relationships of my study animals. Chapter 13 could not have been written without their contribution; indeed, George Amato is coauthor of the chapter.

Ni Zhicheng of the Tibet Plateau Institute of Biology identified the plants. Microhistological analyses of feces for the determination of food habits were done by the Composition Analysis Laboratory, Fort Collins, Colorado, and I thank Teresea Foppe for this work. Matson's Laboratory, Milltown, Montana, aged chiru incisors. Mineral analyses of plants were done by the Animal Health Diagnostic Laboratory, Lansing, Michigan.

Samuel McNaughton, Syracuse University of New York, advised on mineral content of soils.

I am greatly indebted to Elisabeth Vrba, Yale University, for her insightful comments on the relationship between morphology, molecular biology, and behavior in the phylogeny of bovids. Chapters 13 and 14 reflect her generous involvement. John W. Olsen and P. Jeffrey Brantingham, University of Arizona, analyzed the stone tools from the Chang Tang and placed them into the context of central Asian archeology, a contribution for which I am most grateful.

The scientific contributions of individuals involved in the field and laboratory work are reflected in the joint publications based on the project (see the reference list in this volume), and I wish to express my deep appreciation to them all for their collaboration.

A number of individuals provided me with information about wildlife and other topics, among them Richard Harris, Daniel Miller, Peter Molnar, Gary McCue, John Bellezza, Galen Rowell, Rodney Jackson, and Melvyn Goldstein. Others who provided assistance in the course of the project include Wang Zhongyi, Li Bosheng, Wang Kewei, Maria Boyd, Brad Simpson, Le Jen Chen, and Alec le Sueur. Nancy Nash helped to produce a snow leopard poster and other conservation-education material for use by the forest departments, and this assisted the project.

Richard Keane once again contributed his artistic talent to one of my reports by drawing the chiru sketches. Sharon Wirt prepared the illustrations of stone tools. The maps and graphs were done by Thomas Gamble.

The project was made possible by generous funding from several donors, especially from the Liz Claiborne–Art Ortenberg Foundation, the Sacharuna Foundation, and the Robert C. and Helen C. Kleberg Foundation, as well as from the Pattee Family, the Armand G. Erpf Fund, Kenhelm Stott, and others. The donation of a Toyota Land Cruiser and an Isuzu Trooper from the respective companies greatly assisted the work.

My wife, Kay, was deeply involved in all aspects of the work, both in the field and in the preparation of this volume. She loves camp life in the central Asian highlands as much as I do, and she accompanied me on seven trips to Xinjiang, Tibet, and Mongolia, during which she recorded wildlife, made vegetation transects, and otherwise participated in the research. She contributed more than any other person to this project, and I express to her my deepest gratitude for her interest, loyalty, and dedication.

1

Introduction

Travel and Research in China's Highlands

Roads! There are no other paths there than those beaten out by wild yaks, wild asses and antelopes. We made, literally made, our way, while I charted the country and captured for the pages of my sketch-book as many views as possible of glorious mountain giants with snow-capped peaks and labyrinths of winding valleys.

We penetrated deeper and deeper into the unknown, putting one mountain chain after the other behind us. And from every pass a new landscape unfolded its wild, desolate vistas towards a new and mysterious horizon, a new outline of rounded or pyramidal snow-capped peaks.

Those who imagine that such a journey in vast solitude and desolation is tedious and trying are mistaken. No spectacle can be more sublime. Every day's march, every league brings discoveries of unimagined beauty.

Sven Hedin (1909)

A BLIZZARD IN OCTOBER 1985 had covered the eastern edge of the great uplands known as the Chang Tang with 30 cm of snow. Here at 4500 m on the high Tibetan Plateau it was crushingly cold as I trudged across the infinity of white toward a range of hills suspended above the earth by a low layer of cloud. My goal was a herd of Tibetan wild asses, or kiangs,[1] which, my binoculars had revealed, were pawing craters in the snow to reach the meager grass. Distances are deceptive in these uplands, and I was still not near the kiangs when cloud and land merged into one. I was adrift in a white void, the silence astral, able to see only a few meters. As I stood considering my route, a herd of Tibetan antelope, or chirus, appeared, plodding in a tawny file through the knee-deep snow. There were males, handsome with black faces and erect horns, and females with youngsters. A second herd came, and another, all heading mutely toward the northeast as if floating through space. They passed by, seemingly oblivious to my presence, and vanished like dream spirits into cloud. Where did they come from? Where were they going? Little was known about the biology of kiangs, chirus, and other species of the unique large-mammal

1. Appendix A gives the scientific names of wild mammal species mentioned in the text, and Appendix B lists the bird and reptile species observed in the Chang Tang Reserve.

community in the Chang Tang. Was their future secure in these remote uplands? The raw bite of cold seeped into me, but I waited for more herds, seeking to extend my covenant with species I had come so far and waited so many years to observe. This was my first visit to the Chang Tang and I silently resolved to return.

During the early 1960s, I had an opportunity to study wildlife in India. I had to decide whether to concentrate on the large mammals of peninsular India, of the Himalaya, or of the western part of the Tibetan Plateau in Ladakh. Each of these regions has its unique and fascinating faunal assemblage, none of which had been studied. Tigers, axis deer, blackbuck antelope, and gaurs, to name just four, inhabit the lowlands. The alpine pastures of the Himalaya support snow leopard and several species of wild sheep and goats. But the Tibetan Plateau, remote and mysterious, the home of wild yaks, chirus, and kiangs, intrigued me most. I had read the travel accounts of Sven Hedin and other explorers. Their descriptions of the harsh, wind-swept landscape, with its tawny, treeless plains and mountains dusted with snow, lured me on. And especially remote and desolate was the Chang Tang (also spelled Jhang Thang, Qian Tang, and Qiang Tang), which extends across northwestern Tibet, the southwestern corner of Qinghai Province, and north into Xinjiang to the arid Kunlun Range. The Chang Tang gave me a great sense of unknown possibilities.

The political realities of the time dictated that I concentrate on the lowlands, and my book *The Deer and the Tiger* (1967) was the result. Later, during the 1970s, I roamed widely through the high mountains of Pakistan, Nepal, and, to a lesser extent, India to provide a baseline of information about the large mammals, as described in the book *Mountain Monarchs: Wild Sheep and Goats of the Himalaya* (Schaller 1977b). Starting in late 1980 and until early 1985, I continued research on the mountain fauna, this time along the forested eastern margin of the Tibetan Plateau in the Sichuan Province of China. There my Chinese colleagues and I concentrated on giant pandas, red pandas, Asiatic black bears, and takins, as described in *The Giant Pandas of Wolong* (Schaller et al. 1985) and other publications. However, the longing for a project in the Chang Tang never left me. It was reinforced by a visit to Lhasa and other parts of southern Tibet in 1980 and by a two-month wildlife survey in eastern Qinghai in 1984. This survey led to a five-year agreement with Beijing's Ministry of Forestry to collaborate in a project on the large mountain mammals of western China. Our joint work had three major aims: to determine the current status and distribution of species, with particular focus on snow leopards; to study selected wildlife populations in detail; and to locate areas that deserved to become reserves or otherwise receive protection.

Within the purview of our project were the two autonomous regions

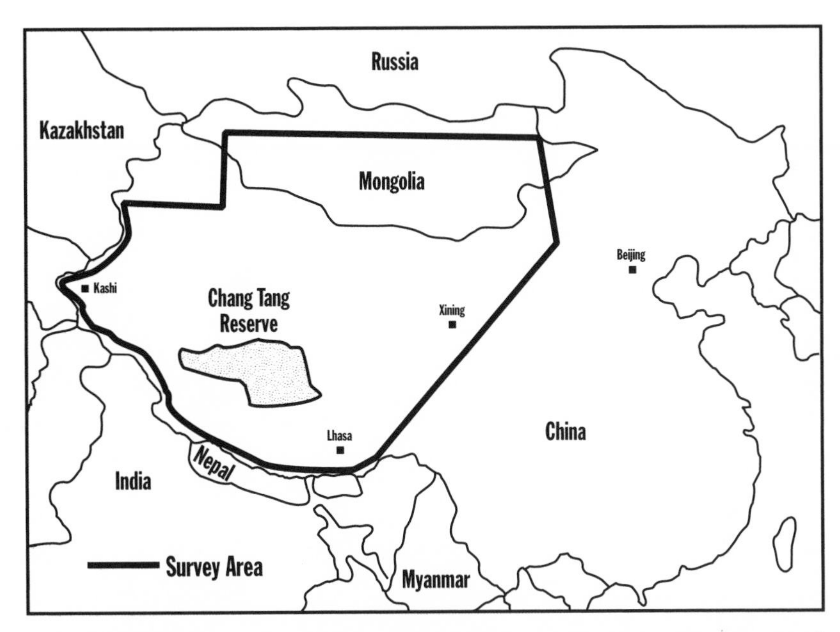

1.1. The general survey region of the project in China and Mongolia and the location of the Chang Tang Reserve, the main study site. Figures 1.2, 1.3, and 1.4 show the areas in greater detail. (Scale: 1 cm = 300 km)

of Xinjiang and Tibet, Qinghai Province, and the western edge of Gansu Province, a total area of about 3,500,000 km^2 (figs. 1.1, 1.2, and 1.3). When speaking of Tibet, I refer solely to the autonomous region, whereas when speaking of the Tibetan Plateau, I refer to the geographical feature, the huge uplift that is nearly half the size of the United States excluding Alaska. Although I spent several months in Xinjiang to conduct wildlife surveys in the Tian Shan (*shan* means "mountain" in Chinese), Karakoram, and Pamirs, I concentrated on the Tibetan Plateau. Specifically, I worked mainly on the great rangelands that cover about two-thirds of the plateau, most of them above 4300 m, above tree line and agriculture. Of 15 major field trips to western China, 12 included the plateau. The region we had to survey was so vast that we could only sample it (fig. 1.3), and even this took much longer than anticipated. I spent months in the field every year from 1984 to 1996 and the task has barely begun (table 1.1).

The original project plans were modified over the years. Since my local collaborators were not prepared to remain afield for more than two to three months at a time, intensive investigations at particular sites were not possible. Instead our field trips consisted of surveys with occasional halts at certain places for as long as three weeks. The emphasis of the project

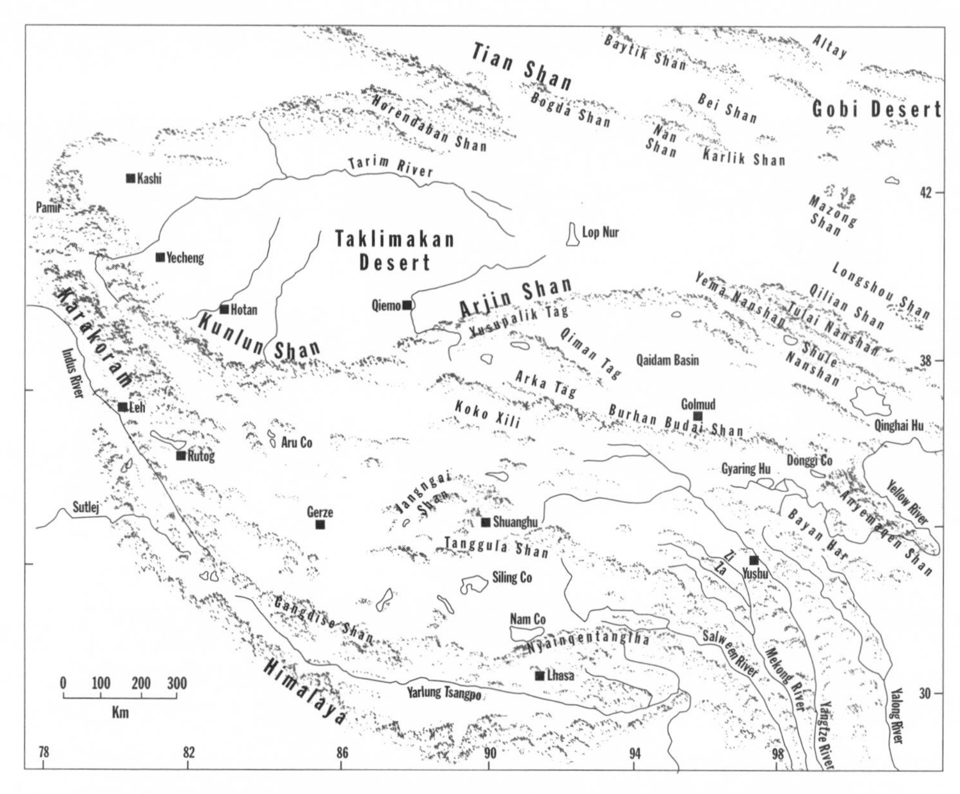

1.2. The principal mountain ranges and river systems on and adjoining the Tibetan Plateau.

also shifted from snow leopards to the wildlife of the Chang Tang, particularly to the plains ungulates—the chiru, Tibetan gazelle, Tibetan argali, wild yak, and kiang—and this book concentrates on the natural history of these species (table 1.2). The chiru became the special focus of our work. Not only is it the most abundant plains ungulate in the Chang Tang, but it also defines the ecosystem by its migrations, much as the wildebeest does in Tanzania's Serengeti. Although an antelope in appearance, the chiru's morphology hints at a relationship to sheep and goats (Pilgrim 1939). Where do its affinities lie? I was also intrigued by how the various ungulate species have adapted to and share the sparse forage of the Chang Tang. Studies on the African savannahs, such as in Serengeti National Park, have shown that the members of the ungulate community partition the resources to some extent on the basis of plant species, growth stage, plant part, and other criteria (Sinclair and Arcese 1995). Compared to the Serengeti, the Chang Tang is desolate, with few plant and ungulate species.

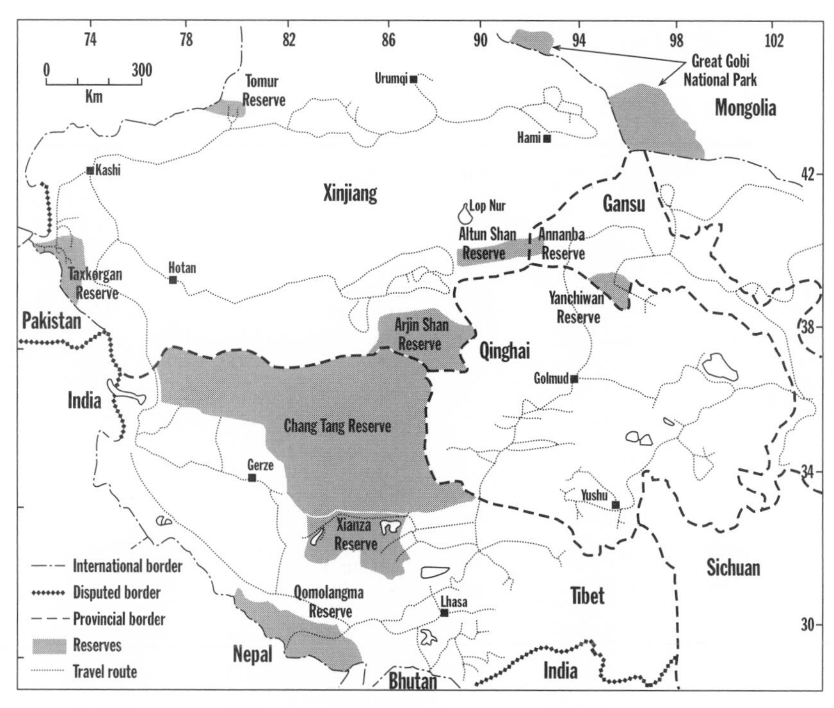

1.3. My travel routes in western China during the project (excluding the Chang Tang Reserve) and the major nature reserves in the region.

But that a cold, dry steppe can support a diverse ungulate assemblage was already shown during the Pleistocene in Alaska, where mammoth *(Mammuthus)*, camel *(Camelops)*, musk ox *(Symbos)*, *Bison*, horse *(Equus)*, and caribou *(Rangifer)*, to name a few, shared an ecosystem (Guthrie 1990). We can only infer how they did so. The Chang Tang offers a last opportunity to study a high, cold steppe with its fauna still intact.

Conservation initiatives are usually made in response to crises after wildlife has been decimated and habitat modified. Plains ungulates are particularly vulnerable, as the slaughter of pronghorns and bison on the American prairies during the late nineteenth century has shown. The northern Chang Tang, unpopulated except for a few thousand pastoralists around the periphery, provided an exciting opportunity to study a whole faunal assemblage and protect an ecosystem before it was seriously damaged. Consequently, I worked primarily in the northern Chang Tang of Tibet from 1988 to 1994 (Schaller 1997).

The research in the Chang Tang provided me with both personal satisfaction and new information. But in today's age of environmental destruction, any project must also consider what it leaves behind. The primary

Table 1.1 Dates and general locations of main fieldwork on the project

	Year	Month	Location
China			
Qinghai	1984	7–8	Anyemaqen Shan and southeast Qinghai
	1985	9–11	Qilian Shan and southwest Qinghai
	1986	8–9	Anyemaqen Shan and southeast Qinghai
		11–12	Burhan Budai Shan and southwest Qinghai
	1993	11	Southwest Qinghai
Xinjiang	1985	5–6	Taxkorgan Reserve
	1986	5–7	Taxkorgan Reserve and central Tian Shan
	1987	4–6	Arjin Shan and east and west Tian Shan
	1988	7–8	Arjin Shan and Aksai Chin
Tibet	1988	8–10	Chang Tang and northeast Tibet
	1989	11	Lhasa
	1990	7–10	Chang Tang
	1991	6–8	Chang Tang
		11–12	Chang Tang
	1992	5–8	Chang Tang
	1993	10–11	Chang Tang
	1994	5–6	Chang Tang
	1995	10	Southeast Tibet
Inner Mongolia	1996	6–7	Western half of Inner Mongolia
Mongolia	1989	8–9	Gobi Desert and eastern steppes
	1989–90	12–1	Gobi Desert and central and west Altay
	1990	5–6	Gobi Desert
		11–12	Central Altay
	1992	10–11	Central Altay and Gobi Desert
	1993	7–8	Eastern steppes and Gobi Desert
	1994	8	Gobi Desert

aim of our work was to gain the kind of knowledge about wildlife, grasslands, and pastoralists that might help them coexist. There are certain places on earth that are so unique that they must remain undeveloped and unaffected by greed, that must be preserved without compromise as part of a country's natural and cultural heritage. The northern Chang Tang is one such place. The Tibet Autonomous Region's government recognized this, and in November 1992 the establishment of the Chang Tang Reserve was announced. The reserve consists officially of about 247,000 km^2, but with another 37,000 km^2 of protected area contiguous with it to the west, the total is 284,000 km^2. This area was legally established by an administrative permit issued by the Tibet Autonomous Region government dated 19 July 1993. The Chang Tang Reserve is the second-largest reserve in the world, exceeded only by Greenland National Park, which consists mostly of ice cap. Actually, the reserve is larger than the official figures suggest. Independent recent calculations by Daniel Miller and myself indicate that

Table 1.2 Main carnivores and ungulates of the Tibetan Plateau discussed in this report

Common English name	Scientific name	Tibetan name	Chinese name
CARNIVORES			
Lynx	*Felis (Lynx) lynx*	*yi, amu, dezay, dbyi*	*sheli*
Snow leopard	*Uncia (Panthera) uncia*	*saa, sha, samo, zik*	*xuebao*
Tibetan brown bear	*Ursus arctos pruinosus*	*dom, demo, mi-de*	*zongxiong*
Tibetan sand fox	*Vulpes ferrilata*	*wa, wamo*	*shahuli*
Red fox	*Vulpes vulpes*	*wa, wamo*	*huli*
Wolf	*Canis lupus*	*changku*	*lang*
UNGULATES			
Asiatic ibex	*Capra ibex sibirica*	*sakm, skin* (in Ladakh)	*baishanyang*
Blue sheep (bharal)	*Pseudois nayaur*	*na, nawa, napo, napik*	*shiyang, yanyang*
Chiru (Tibetan antelope, orongo)	*Pantholops hodgsoni*	*tso, tsi, tsod*	*zangling*
Goitered gazelle	*Gazella subgutturosa*	—	*ehouling*
Kiang (Tibetan wild ass)	*Equus kiang*	*kiang*	*yelu*
Tibetan argali (nyan)	*Ovis ammon hodgsoni*	*nyan*	*datouyang, panyang*
Tibetan gazelle	*Procapra picticaudata*	*goa, gowa*	*zangyuanling*
White-lipped deer	*Cervus albirostris*	*sawa, sha*	*baichunlu*
Wild Bactrian camel	*Camelus (bactrianus) ferus*	*amu, amon*	*yeluotuo*
Wild yak	*Bos grunniens*	*drong, dong, yaa*	*yemaoniu*

the reserve is roughly 334,000 km^2 (128,590 square miles), about 50,000 km^2 larger than the official size, and this figure will be used here. Whatever the precise size, the reserve is huge, larger than the state of New Mexico, or almost as large as Germany.

This book focuses mainly on the results of our work in what is now the Chang Tang Reserve. In addition, it includes data from other parts of the Tibetan Plateau and, where pertinent, from other parts of western China and even Mongolia. A few hundred wild Bactrian camels persist in China and some wander seasonally up onto the northern rim of the Tibetan Plateau. For political reasons, most of these animals were not accessible to me at the time. Instead I studied camels briefly in 1989 in Great Gobi National Park in Mongolia. During six subsequent visits to Mongolia (table 1.1), I did research on snow leopards in the Altay Mountains and on brown bears in the Gobi, and I observed Mongolian gazelles, goitered gazelles, Gobi argalis, and Mongolian wild asses (kulans), obtaining valuable material to compare with the same or similar species on the Tibetan Plateau. One extensive survey in western Inner Mongolia filled the geographical gap between Mongolia and the Tibetan Plateau.

The project encompassed thirteen years, but the actual amount of time spent in the field, as measured in number of days between leaving from and returning to town, was a meager 6.0 months in Qinghai, 7.2 months in Xinjiang, and 13.6 months in Tibet, of which about one year was spent in the Chang Tang Reserve. Total field time was about 28 months, including two brief periods in western Gansu and one in Inner Mongolia. However, my total time in China was 42 months, the difference between the two representing waits for permits, preparations, and so forth. Visits to Mongolia totaled 13.5 months, of which 9.1 months were spent in the field. Thus this report is based on 3.1 years of fieldwork.

It is a sad fact that, because of hunters, wildlife in western China persists mainly in remote or rugged terrain that is often difficult of access. We covered great distances by car and occasionally by horse, camel, or yak to reach potential study sites. Sometimes we selected a site because we had been told that much wildlife persisted and that it might qualify as a reserve, and at other times we merely traveled, observing, talking to local people, evaluating. Often we saw so little that even a hare was worthy of note. For instance, we traveled 15,250 km by car around the desolate rim of the Taklimakan Desert and tallied just 18 goitered gazelles. But then, on occasion, we came upon an aggregation of animals that provided a glimpse into the past as it was a hundred years ago.

To place our far-flung research program in China into perspective and to provide some historical depth to the obscure and exotic places we visited, I give here a brief overview of our travels (fig. 1.3; see also the distribution maps of species and other maps for specific localities). These travels were always done in collaboration with local biologists and officials and often also with my wife, Kay. They included all months from April into January, from spring into the height of winter (table 1.1).

Inner Mongolia

Inner Mongolia was once famous for the number and variety of its large mammals, with argalis abundant in the hills, and kulans, goitered gazelles, and especially Mongolian gazelles thronging the steppe and semidesert. In summer 1996, Wang Xiaoming of East China Normal University and I crisscrossed the western half of Inner Mongolia, west of about 112° E (the area shown in fig. 5.3). Large tracts consist of stony desert, where wild camels once roamed, and of shifting sand dunes, those of the Badain Jaran Desert alone covering 40,000 km^2. A number of ranges—the Helan, Yabrai, Lang, Daqing, and others—rise above the plain to an elevation of 2300 m and more, and a few of these have stands of conifers and deciduous broad-leaved trees. Several of the ranges have, or until recent years had, unusual assemblages of large mammals, including red deer, musk deer, and goral, species that one would not expect isolated in desert. This presented us with intriguing aspects of animal distribution and, by inference, past changes in vegetation and climate (Wang and Schaller 1996). However, the scarcity of wildlife in the region was distressing, with Przewalski's gazelles extinct and snow leopards almost so, camels, kulans, and Mongolian gazelles mainly visitors at the Mongolian border, and the others scarce and fragmented. Unrestricted hunting since 1960 has almost eliminated one of the world's great wildlife populations.

Xinjiang

In Xinjiang we worked mainly in the snow-capped Tian Shan, Kunlun Shan, and other ranges that ring the Taklimakan Desert on three sides (figs. 1.2 and 1.3). The sand dunes and rock-littered plains of the Taklimakan are almost lifeless except where glacier-fed streams tumble from the mountains. Such streams soon disappear underground, with perhaps some *Tamarix* shrubs to trace their routes. Ancient history still pervades these desert tracts. There are fortified oases with crumbling walls surrounded by irrigated fields. And buried beneath the sands of the Taklimakan, a Turkic word meaning "enter and you will not come out," are forgotten civilizations. For over two millennia caravans, pilgrims, and armies carefully skirted this fearsome sea of sand as they traveled to and from China along the Silk Road, as this route became known in the nineteenth century. The Silk Road began in Xian and Lanzhou and then turned northwest up the Heixi Corridor of Gansu, following the Great Wall and along an even greater obstruction, the eastern margin of the Tibetan Plateau. Dunhuang, the last outpost, was already an important town in the first century B.C.; it was also a treasure-house of Buddhist frescoes and ancient manuscripts in Sanskrit, Sogdian, Tibetan, and other languages and became the focus of archeological raids by foreigners early this century.

North of Dunhuang and Hami is the eastern end of the Tian Shan. We drove to Barkol, once a fortified town, and from there investigated three ranges—the Bei Shan, Nan Shan, and Karlik Shan—that rise from the desert floor at about 1500 m to over 4000 m. Little wildlife remained. During the Cultural Revolution (1966–1976) commercial hunting was widespread. We were told that local governments organized communal hunts during which so many ibexes and argalis were killed that trainloads of carcasses were shipped east to market.

Driving west along the Tian Shan, we reached the Turfan depression, at 154 m below sea level, the lowest point in central Asia. Here too is the ancient town of Yarkhoto; although abandoned a thousand years ago, the adobe walls of its homes, watchtower, and monastery still stand. In the central Tian Shan we censused ibex and argali for two weeks in a 750 km^2

block of mountains named the Horendaban Shan. There, as in other parts of the Tian Shan, we traveled on horseback. The Horendaban Shan is much less arid than the mountains to the east. Lush alpine meadows extend up to 3300 m, and spruce *(Picea)* and birch *(Betula)* occur in patches.

The highest parts of the Tian Shan, with several peaks exceeding 6500 m, were incorporated into the 3000 km^2 Tomur Feng Reserve in 1986. The reserve adjoins the border of Kyrgystan. The lower slopes, below 3000 m, have alpine grassland broken by stands of spruce and birch, and cottonwoods *(Populus)* grow along streams. Over half of the reserve consists of barren peaks and rubble-covered glaciers. We found spoor of snow leopard and a moderate number of ibex during our 12-day survey.

At the western edge of the Taklimakan lies the medieval city of Kashgar (Kashi), a famous trading center, a crossroads from which caravans laden with textiles, ivory, glass, fur, and iron once departed in all directions. In the thirteenth century, Marco Polo found there "very fine orchards and vineyards and flourishing estates." Now only the covered market with its many stalls and tea shops and its throng of Kazaks, Uygurs, Tajiks, and others in traditional clothes still has that ancient aura.

The Karakoram highway leads south from Kashgar and generally follows one branch of the Silk Road. The highway passes the glacier peaks of Kongur and Muztag Ata, both over 7500 m high, skirts the ruins of a fort at Taxkorgan, and continues up and over the Khunjerab Pass into the Hunza Valley of Pakistan. There, where the frontiers of China, Pakistan, Afghanistan, and Tajikistan meet, is the 14,000 km^2 Taxkorgan Reserve. We spent over two months in the western half of the reserve to survey wildlife in the remotest tracts. The surveys included the rolling Pamirs, home of the last Marco Polo sheep in China; the rugged Karakoram around the Mintaka Pass, meaning "the place of a thousand ibex," a renowned place of hardship for caravans struggling south on the Silk Road; and the desert canyons of the Kunlun Shan bordering the Yarkant River, where we especially sought blue sheep.

From Kashgar eastward the road hugs the base of the Kunlun Shan for over 1000 km as it passes through the oases towns of Shache, Yecheng, Hetian, Yutian, Minfeng, Qiemo, and Ruoqiang. These towns, listed in the same order, were once called Yarkant, Kargilik, Khotan, Keriya, Niya, Cherchen, and Chargalik, names that are part of history. In 1865 the British surveyor William Johnson reached Khotan, the first European for centuries to enter that region, which under the Ming Dynasty (1368–1644) retreated into obscurity and isolated itself from the West. Nearly a hundred years ago, Sven Hedin wandered the Taklimakan around Keriya and Cherchen, and Aurel Stein sought lost cities beneath the shifting sands of Niya. Nikolai Przewalski passed through Chargalik in 1877, making that same year the scientific discovery of wild Bactrian camels in the nearby Lop Nur area.

The Kunlun Shan and its eastern branch, the Arjin Shan (known as Altun Tag in Uygur), delineate the northern border of the Tibetan Plateau. South of Ruoqiang lies a large basin between two ranges, the Qiman Tag and Arka Tag. This area, 45,000 km^2 in size, was designated as the Arjin Shan Reserve in 1983. It is contiguous with Tibet's Chang Tang Reserve. The National Environmental Protection Agency, which administers Arjin Shan Reserve, did not allow us to enter, but we made extensive surveys west and north of the reserve. In 1987, we headed south toward Muztak, an ice dome 6973 m high, on the Xinjiang-Tibet border. For nearly three weeks we had the atavistic pleasure of traveling by camel caravan in the manner of the Russian explorers Nikolai Przewalski, Pyotr Kozlov, and V. Roborovsky, who had crisscrossed this region a century or more before. After surmounting the Kunlun Shan, here so desolate that the sight of a bird elicited comment, our caravan reached the high plateau, where chirus and kiangs occurred in modest numbers but wild yaks were rare. A year later we surveyed an area north of the reserve, the Akoto Shan, or Yüsüpalik Tag, especially for Tibetan argalis.

Qinghai

West of Xining, the capital of Qinghai Province, the road soon winds up onto the grasslands of the Tibetan Plateau. This great pastoral area is intensely green in summer, and the blossoms of buttercups, forget-me-nots, louseworts, gentians, polygonums, asters, poppies, and potentillas, to name a few, add vivid color. There in a huge basin lies Qinghai Hu, the Blue Sea Lake, as the Chinese call it; the Mongolians, who have lived here for centuries with the Tibetans, know the lake as Koko Nur, also Blue Lake. Twice, in 1984 and 1986, we drove south through eastern Qinghai (figs. 1.2 and 1.3). Along the way we camped near Hei Hai Hu (Black Sea Lake), also known as Donggi Co or Tossum Nur. This lake lies at the edge of the Anyemaqen Shan, as this most eastern extension of the Kunlun Shan is known. The highest peak in this range, though a mere 6282 m, was once thought by several Western travelers to be the highest mountain on earth. We worked for a total of about three weeks in this range, both in Qinghai and the part that extends into Gansu, principally on blue sheep and snow leopards.

The Anyemaqen region is the heartland of the Golok, a Tibetan people who were once known for their harassment of all outsiders. Przewalski's expedition of 1879–1880 was assaulted: "They rushed to the attack with yells. The hoofs of their steeds sounded hollow in the damp soil, their long spears bristled and glistened, their long cloth robes and black, floating locks streamed behind them in the wind. Like a cloud, this savage, bloodthirsty horde dashed upon us" (letter dated 8 August 1884, quoted in Rockhill 1891). And as recently as 1948, Clark (1954) had a similar encounter.

The rich grasslands south and west of Anyemaqen Shan, around the source of the Yellow River and the twin lakes of Gyaring and Ngoring, were according to Rockhill "the most wonderful hunting ground in Asia" (1891). Even in 1947, Migot found "a tremendous lot of wildlife in this region, which is in effect a sort of sanctuary undisturbed by man. Herds of yaks, wild asses, and gazelles were all quite easy to get near" (1957). But roads were built through the area during the 1950s, and today the wild yak is gone and one encounters only a few Tibetan gazelles and occasional kiangs. Instead there is much livestock, at an average density of about 40 head per square kilometer of pasture.

Yushu, once called Jyekundo, lies in southeast Qinghai. The town clusters around a hill that was once crowned by a large monastery. Destroyed during the Cultural Revolution in the late 1960s, like so many others, the monastery was a melancholy shell, though it was being rebuilt. From Yushu we drove west to the county town of Zhidoi. "Road workers, officials, and the military have shot all wildlife near the roads," we were told. "You must go away from the roads." We did. We sought distant valleys but found few animals except herds of blue sheep on small isolated ranges. We often met herdsmen, rifles slung over shoulders. On a subsequent trip we visited Zadoi, an area with many rugged limestone massifs, most no higher than 5500 m. Lush pastures surround these massifs, and gnarled juniper *(Sabina)* forests grow on some slopes. The landscape was so pleasing and the pastoralists so hospitable that we spent a month on horseback exploring north and south of Zadoi, mainly in the drainage of the Za River, a tributary to the Mekong. We found a moderate number of blue sheep, some white-lipped deer, and widespread sign of snow leopard.

To reach the Qilian Shan in the northeast corner of Qinghai, we drove up the Heixi Corridor of Gansu and then ascended the range on a track that leads to a sulfur mine. The Qilian Shan and the ranges that parallel it—the Tulai, Yema, Shule, and Danghe Nanshan—are haggard, a maze of eroded slopes and stony defiles, gray, sienna, and ocher in hue. There is little vegetation at low elevations, but beginning at about 3300 m, *Ephedra* and *Caragana* shrubs appear, and above 3800 m are alpine meadows. We selected a mountain block of 610 km^2 between the Shule and Tulai Nanshan for intensive work. By camel and on foot we censused blue sheep, white-lipped deer, and other wildlife for three weeks.

From the Qilian Shan we drove to Dunhuang in Xinjiang and from there south across a low pass of the Arjin Shan back to Qinghai into the desolate Qaidam Basin and on to the town of Golmud. Over a century ago, Przewalski noted that "large animals are, however, scarce in Tsaidam" (Prejevalsky 1876). South of Golmud the Burhan Budai Shan, as this part of the Kunlun Shan is called, presents a barrier to the high plateau beyond. We spent several days in the Burhan Budai Shan in 1985 and 1986, but our main goal each year was the Chang Tang. High, bleak, and lonely, the Chang Tang extends from about the Golmud-Lhasa road westward for over 1200 km to western Tibet. Most of the area lies above 4500 m. The vegetation consists of a sparse cover of grasses, sedges, forbs, and low to procumbent shrubs. The Golmud-Lhasa road follows an old caravan trail for much of its route. Over a century ago, William Rockhill, Nikolai Przewalski, and several other explorers used this route in their unsuccessful quest for Lhasa, the holy sanctuary of Tibetan Buddhism and the home of the Dalai Lama.

During the winters of 1985 and 1986, along some 300 km of road, we censused chiru, kiang, and Tibetan gazelle between the Kunlun Pass in the north and Tanggula Pass in the south. With the communities of Wudaoliang and Tuotuohe as a base, we also traveled cross-country east and west away from the road. The first year we arrived just after an October blizzard had blanketed the region, followed by temperatures that dropped to −40°C and lower. Much wildlife and livestock died of starvation. We made a nine-day trip east of the road during this period and surveyed about 400 km from a tractor-pulled wagon. The responses of animals to conditions of particular severity were instructive to observe, but the trip was more an exercise in hardship than research. On a visit to this area, Przewalski reported that "in February 1870 a caravan which left Lhassa [*sic*] 300 strong, with 1,000 beasts of burden, in a violent snow-storm followed by severe cold, lost all the animals and fifty men besides" (Prejevalsky 1876). The following year, 1986, was snow-free but the wind was so strong and the cold so biting that gusts whipped my scope and tripod away unless I held them firmly while watching chirus, and my eyeballs began to freeze.

In 1993, we returned for a brief visit to investigate the wildlife in the upper Tuotuohe Valley. There are plans to place the southwest corner of Qinghai under the administration of Tibet, and we hope that the Chang Tang Reserve will be extended eastward to include the Tuotuohe Valley and adjoining region.

Tibet

The forested maze of mountains in eastern Tibet was not part of my program and I only penetrated this area along the western margin in 1988 and 1995 (fig. 1.3). The valley of the Yarlung Tsangpo in southern Tibet, from the longitude of Lhasa westward, is heavily populated and there are many fields of barley, wheat, and rape. Wildlife persists locally between the Himalaya and Gangdise Shan to the north, but I did no research in these mountains. In 1990, we drove west up the Yarlung Tsangpo and beyond the headwaters into the valley of the Indus. Mount Kailas (Kangrinboqe), sacred to both Hindus and Buddhists, was in cloud. However, Gurla Mandhata rose with crystal clarity to 7694 m beyond the waters of holy Mapam Yumco (Manasarowar), the "Lake of Supreme Consciousness." At intervals along our route to Shiquanhe, the administrative center for western Tibet, we stopped to query pastoralists about wildlife in the nearby mountains.

North of the Gangdise and Nyainqentanglha Ranges and west of about 95° E lies the Chang Tang, covering over half of Tibet. I visited various parts of the Chang Tang, from the eastern edge, where it grades into forest; west to the marshes around Xianza, known for their black-necked cranes; and on to the holy lake Tangra Yumco and the high pastures near Coqen. But I concentrated on the most remote and sparsely populated part, north of about 32° N, north of a road that crosses the Chang Tang from Nagqu west past Siling Co (*co* means "lake" in Tibetan) and the county town of Gerze to Shiquanhe. Much of the area north of this road is now within the Chang Tang Reserve (fig. 1.4).

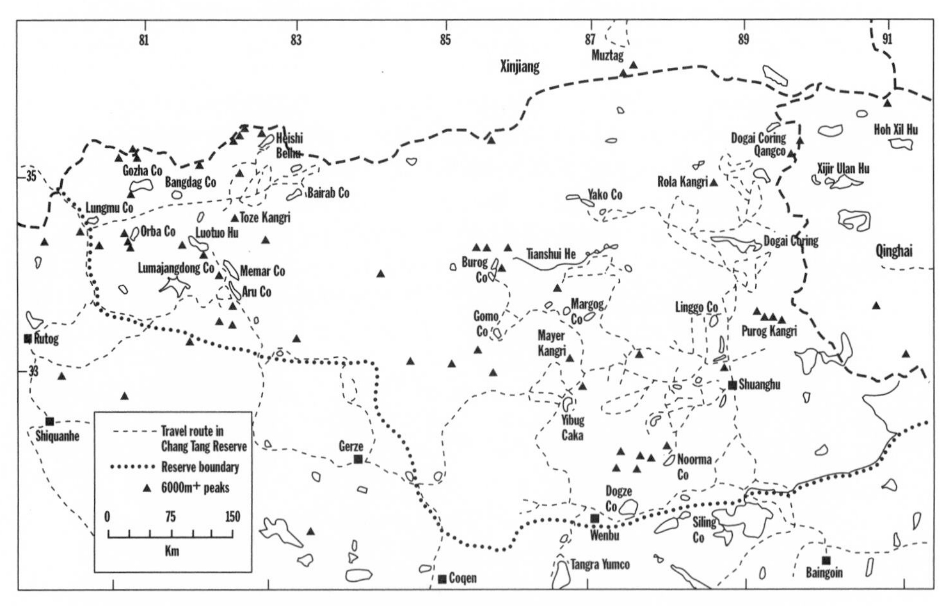

1.4. My travel routes within the Chang Tang Reserve (1988–1994), showing areas surveyed for wildlife. Important lakes and peaks are also indicated.

I first entered the Chang Tang from the west. We had driven from Kashgar in Xinjiang to Yecheng and then into the mountains where a road ascends a desolate valley hemmed in by the Kunlun on one side and the Karakoram on the other. Suddenly, on cresting a ridge, the Tibetan uplands spread before us, rumpled brown hills and flats extending eastward to the horizon. Snatches of white cloud streamed overhead. This was the Aksai Chin, a disputed tract which China had annexed from India in 1962. Beyond the Aksai Chin the road turns south to Rutog in Tibet, an oasis of fields among sterile hills. Nearby was Panggong Co, fjordlike and 150 km long, with a small island upon which hundreds of brown-headed gulls nested. Heading northeast of Rutog, we drove cross-country past Lumajangdong Co to Aru Co (fig. 1.5). On 20 July 1891, Hamilton Bower and W. Thorold, two British army officers on a spying mission, were the first westerners to behold the lake:

> Over a pass 17,876 feet, and then down a long narrow valley which suddenly debouches on Lake Aru Cho (17,150 feet),—a fine sheet of water running north and south, salt like nearly all the Tibetan lakes, and of a deep blue colour. To the south-west and north-west some fine snowy mountains rise up into the blue sky, while on the east low undulating barren-looking hills are seen. In every direction antelope and yak in incredible numbers were seen, some grazing, some lying down. No trees, no signs of man, and this peaceful-looking lake, never before seen by a European eye, seemingly given over as a happy grazing ground to the wild animals. (Bower 1894)

Their expedition continued east across a narrow isthmus between Aru Co and Memar Co. Two subsequent expeditions, one led by H. Deasy in 1897 and the other by C. Rawling in 1903, visited the lake (see Deasy 1901; Rawling 1905), and after that no other westerners did so until our arrival in 1988. To my delight "the happy grazing ground" that Bower had observed 97 years earlier still existed. We spent a mere three days in the Aru Basin, but I was determined to return. Fortunately we were able to revisit the basin in 1990 and 1992 to spend a total of 39 days there (fig. 1.6). During our surveys in the Chang Tang we never found a site to equal the Aru Basin. It offered the tranquil beauty of its snow-capped peaks and turquoise waters and, above all, the wildlife, particularly the many wild yaks that became, for me, the totems of this wilderness.

Cross-country travel is possible in most parts of the northern Chang Tang. In winter, when the ground is dry and streams are frozen, such travel is easy, whereas in summer heavy precipitation may leave the soil so sodden that a vehicle all too often sinks to the axles. A day or two may then be spent digging, an onerous exercise at 5000 m. Everything we might need for two months or so had to be brought with us, including gasoline and food, though we occasionally supplemented our diet by buying a sheep from a herdsman. We usually required two cars—we relied mostly on Toyota Land Cruisers—and a large truck to carry supplies (fig. 1.7).

While traveling we constantly scanned the terrain for large mammals and recorded all those we saw. However, visibility of species differs markedly. A wild yak on a hillside can readily be spotted at 5 km, whereas a Tibetan gazelle may be overlooked at a few hundred meters. To derive a measure by which to compare species density in different areas, we tallied all ungulates in strips of 300 m on each side of our travel route, a transect

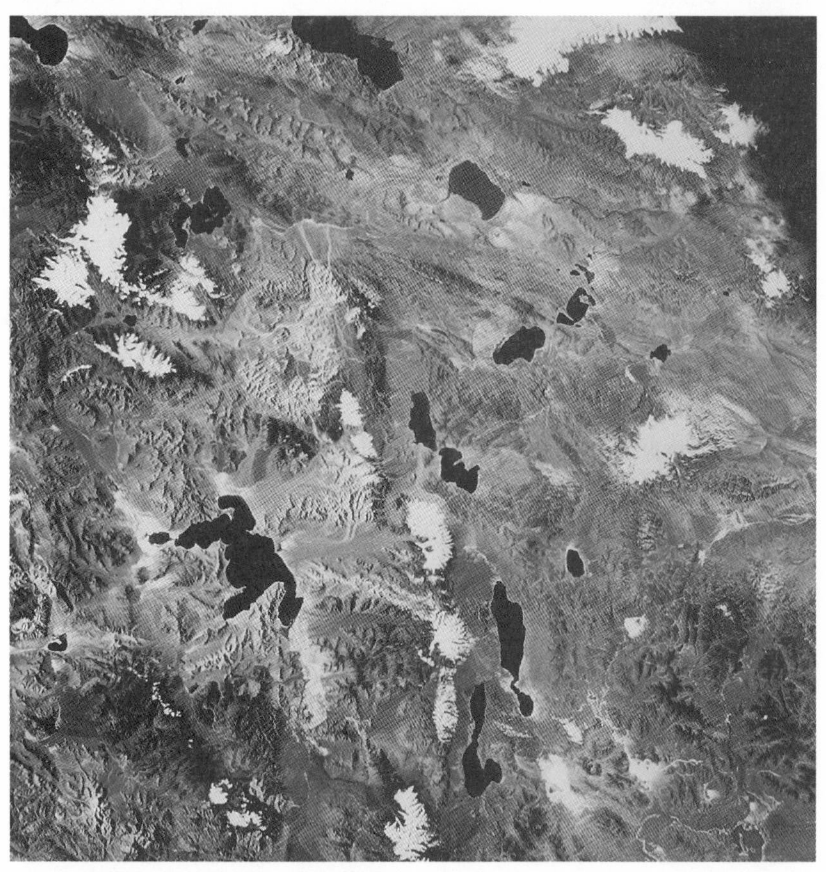

1.5. A satellite photograph (viewed southeast to northwest) covering about 55,000 km^2 in the western part of the Chang Tang Reserve. The four-fingered lake is Lumajangdong Co. The Aru Basin is right of center near the bottom, with Aru Co just south of Memar Co. The glacier-capped Aru Range forms a barrier between Lumajangdong Co and the Aru Basin. Just northwest of the Aru Basin is Luotuo Hu, and the glaciated massif to the northeast is Toze Kangri. The large Gozha Co is in the upper central part at the edge of the photograph.

1.6. Our camp in the Aru Basin, looking east across Aru Co. (July 1992)

1.7. We are camped in the eastern part of the Chang Tang Reserve with a view south toward the Purog Kangri massif. (June 1994)

width of 600 m. Animals were checked with a 20-power scope for age, sex, and other information. They tended to be shy even in the most remote areas because motorized hunters penetrate deeply into the Chang Tang; indeed, between 1988 and 1993 wildlife in general seemed to become ever more skittish. Chirus and wild yaks may flee from a vehicle at a distance of 1 km or more. We often worked on foot to watch undisturbed animals as well as to study vegetation and census wildlife in precipitous terrain.

We traveled by vehicles rather than by pack animals in the Chang Tang in part because we could in a few months survey terrain that would have taken a nineteenth-century explorer several years. Even so, our ventures retained an aura of exploration. When I walked the shore of Heishi Beihu or gazed upon Burog Co, its waters hidden high among the peaks of the Zangser massif, I knew that no westerner had been there before me. Few westerners have penetrated what is now the Chang Tang Reserve: just 14 expeditions since Henri, Prince d'Orléans, and Gabriel Bonvalot (1892) crossed from north to south in 1889 (fig. 1.8). Most of these travelers, including Sven Hedin, who made three journeys, sought the Holy Grail of Asian explorers, the city of Lhasa. But Tibetan troops deflected them all near Siling Co, Nam Co, or Nagqu, and this era of exploration ceased in 1908 for political reasons. In 1950, the American Frank Bessac fled from the Communist army in Xinjiang south across the Chang Tang accompanied by another American and three White Russians. Attacked by Tibetan border guards north of Siling Co, only Bessac and one White Russian survived to reach Lhasa (Donner 1985). Because several of these expeditions provided valuable historical information about the wildlife in that region, I have indicated their approximate routes in figure 1.8. In recent years a number of Chinese expeditions, especially from the Academy of Sciences and the Tibet Forest Bureau, have traveled in the Chang Tang, and some collected data on the fauna and flora (*Research on Flora and Fauna* 1979; Feng, Cai, and Zheng 1986; Feng 1991a, 1991b; Zhen and Liu 1994).

Several more points need comment. I mention many mountains, rivers, towns, and other geographical features with obscure names. These are made even more obscure by the fact that each may have more than one spelling and even two or three different names, depending on whether the Tibetan, Uygur, Chinese, or Mongolian language is used. Maps of western China characteristically have a hodgepodge of old and new ethnic names, as well as names whose origins are difficult to fathom. Explorers had a proclivity for naming places in honor of someone—the Arka Tag was changed to the Columbus Range, Gozha Co to Lake Lighten, and Bairab Co to Lake Markham—but fortunately such names had only a fleeting life on some foreign maps. In general, I use current names with the simplest spellings, and most localities are shown on the maps in this report.

Research in China and Mongolia is always done in full cooperation with local institutions and biologists. Many individuals contributed to the research on which this book is based. Those who assisted with gathering data are joint authors of previous scientific papers or have been acknowledged in those papers (see the references section). In Tibet, the research was done under the auspices of the Tibet Forest Bureau and the Tibet Plateau Institute of Biology. Gu Binyuan of the Tibet Plateau Institute of Biology and Liu Wulin of the Tibet Forest Bureau were my principal coworkers and their contribution is evident from our joint reports (Schaller and Gu 1994; Schaller and Liu 1996). I usually worked with local biologists only, but in 1993 and 1994 our team was joined by Daniel Miller, an American range ecologist now with the International Centre for Integrated Mountain Development in Nepal, to look specifically at the impact of livestock on the rangelands of the Chang Tang Reserve (Miller and Schaller 1996, 1997).

An estimated 12 million yaks and 30 million sheep and goats inhabit the pastures of the Tibetan Plateau (Clarke 1987). Livestock and wildlife have adapted to each other and coexisted there for several thousand years

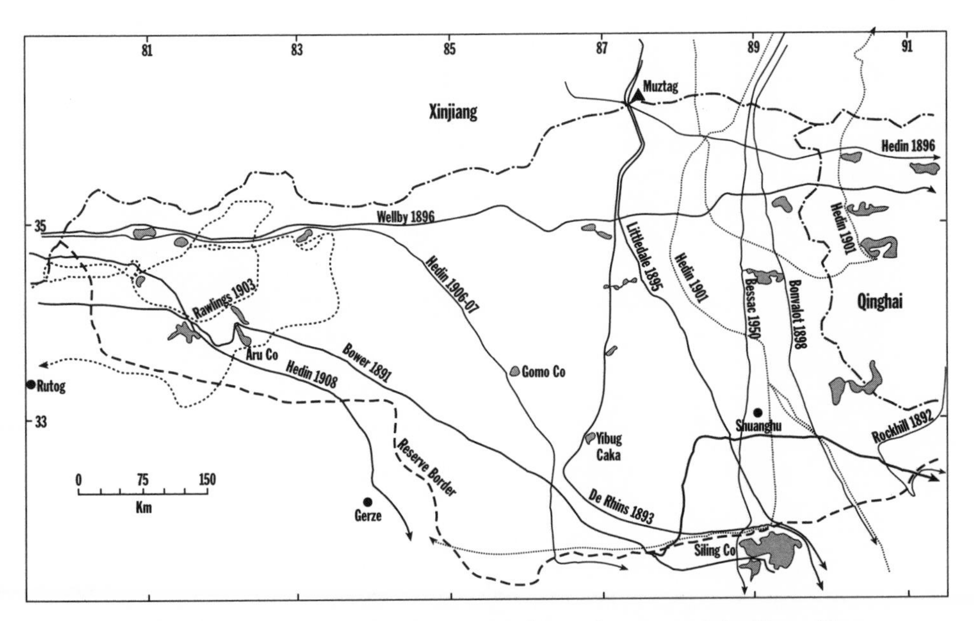

1.8. The approximate routes of all Western expeditions that traversed the Chang Tang Reserve (1892–1908, 1950). Three journeys by H. Deasy (1896–1900) in the western part are not indicated because his routes were similar to those of C. Rawling.

in a more or less sustainable interaction that remains little studied. The fact that both have subsisted together for so long in the Chang Tang testifies not only to the resilience of the wildlife and the ecosystem but also to the ecologically appropriate grazing practices, based on a complex system of seasonal rotation and pasture allocation, that were developed by the nomads (Goldstein and Beall 1990). However, traditional grazing strategies have in recent decades been affected by population growth, government reform policies in land tenure, an emphasis on increased livestock production, and development such as the construction of fences. As a consequence, some rangelands outside the reserve have been degraded and the rest are coming under even greater pressure. Wild ungulates are now increasingly viewed as competing with domestic ones for forage. But the past has shown that, if properly managed, livestock and wildlife can coexist. The challenge is to maintain a well-adapted system of grazing practices together with the unique wildlife on a sustained basis (see chapter 15).

Over three decades have passed since I first thought of a wildlife study on the high Tibetan Plateau, years during which I did research on three distinct large-mammal faunas in the adjoining regions—but not in the one that had for so long been my private quest. With persistence, luck, and generous support from China, I was finally able to conduct that project also. The species accounts in this book are fragmentary and in places have an aura of nineteenth-century natural history, a product of the sporadic and superficial way the project had to be conducted. But they do provide new information and an overview of the unique assemblage of large mammals on the great rangelands of the Tibetan Plateau, biologically still one of the least-known regions on this planet.

The existence of the Chang Tang Reserve indicates that China is concerned about its natural heritage. And it is my fervent hope that our research in the Chang Tang and this book will contribute in a small way so that the chiru and kiang and wild yak can continue to roam in security and freedom across the roof of the world.

2

The Tibetan Plateau

> This center of heaven,
> This core of the earth,
> This heart of the world,
> Fenced round with snow,
> The headland of all rivers,
> Where the mountains are high and
> The land is pure.
>
> Anonymous Tibetan poet (eighth to ninth century)

WITH 85% OF ITS landmass lying above 3000 m and 50% above 4500 m, the Tibetan Plateau is the highest and most imposing such area on earth, higher than any part of the United States except for a few peaks in Alaska. Some 2.5 million km^2 in size, it has the shape of a gigantic teardrop lying on its side, the larger end in the east (fig. 1.2). This plateau, remote and high and hemmed in by mountain chains, has long been a barrier between the civilizations on the Indian subcontinent and those in China and central Asia. The Himalaya guards the southern rim of the plateau in one continuous sweep of 2250 km, each end marked by a massive mountain, Nanga Parbat on the Indus in the west and Namjagbarwa at the great bend of the Yarlung Tsangpo in the east. Beyond Namjagbarwa the mountains fold south and east through northern Myanmar (Burma) and the deeply eroded river gorge country of Yunnan. The northern boundary of the plateau is formed by the ramparts of the Kunlun and Arjin Ranges and, farther east, by the Qilian Shan. The western tip vanishes into a knot of bristling peaks where the Himalaya, Karakoram, Kunlun, and the more gentle Pamirs meet. The eastern boundary extends from the western edge of Gansu southward into Sichuan, where the plateau consists of a tossed sea of ridges and deep valleys.

The plateau has a distinctive geologic history, an understanding of which has relevance to today's distribution of plant and animal communities. Asia grew from various collisions of drifting fragments of the earth's crust. Over 300 million years ago the crustal block that is now the Taklimakan Desert collided with Asia, and the two were sutured together, the impact forming a mountain range where the Tian Shan now rises (Molnar 1989). Other such blocks, from north to south, are the Kunlun block, with a suture that runs near Wudaoliang just south of the Burhan Budai Shan in Qinghai; the Chang Tang block, with a suture near Amdo south of the Tanggula Shan; and the Lhasa block, with a suture in the valley of the Yarlung Tsangpo. At one time this valley was the southern edge of the Asian continent (Chang et al. 1986). The plateau was thus cobbled together with a number of crustal fragments. The area was then above sea level but of low relief, except perhaps for a mountain range at the southern rim where now the Gangdise Shan is located (Harrison et al. 1992).

In the mid-Cretaceous, around 130 million years ago, the Indian subcontinent broke away from the southern continent, Gondwanaland, and drifted northward. As it approached the southern edge of Asia, the deep-sea ocean crust slipped under the edge of the Asian continent. Intrusions of granite forced upward from a molten crustal material and volcanic activity raised the precursor of the Gangdise Shan. The Tethys Sea, hemmed between Asia and the inexorably advancing Indian subcontinent, contracted. About 40–50 million years ago the two landmasses collided. The Tethys Sea vanished. The northern edge of India slid a short distance beneath the southern edge of Asia and the two welded along the valleys of the Yarlung Tsangpo and Indus. The impact buckled and folded and stacked up the northern edge of India to form the Himalaya (Molnar 1989). To the south the impact formed a huge depression that over the eons filled with sediments to create the Gangetic plain. The high elevation of the plateau is the main legacy of this cataclysmic event. Compression led to folding and thrust faulting, and this elevated the area until the earth's crust beneath the plateau averaged 70 km in thickness, twice that of most other parts of the world. In the north, the impact caused the Taklimakan block to underthrust the plate to the south to help create the Kunlun and Arjin Shan at the northern rim of the plateau (Molnar 1989).

After the initial rise of the plateau, there apparently was a period of quiescence, but this was followed by rapid uplift in the mid-Miocene about 20 million years ago. It was thought until recently that the plateau rose significantly during a final and major uplift to reach its present height during the late Pliocene and Pleistocene (Wadia 1966; R. Xu 1981; S. Xu 1981). But further geologic, climatic, and other investigations have revised this idea. Harrison et al. stated that "the present elevation of much of the Tibetan plateau was attained by about 8 million years ago" (1992) and that by the late Miocene the geography was similar to that of today. However, Molnar has suggested that "an uplift of 1,000 to 2,000 m in the last 2 to 10 million years is plausible," primarily in the Pliocene (1989). Since atmosphere cools about 6.5°C for every 1000 m in elevation, a high plateau by the late Miocene rather than only since the Pleistocene would have had a major impact on the region's fauna and flora. About 2.5 million years ago the warm, damp climate that was characteristic of the Pliocene was replaced by the cold periods of the Pleistocene with their ice ages. The presence of the plateau helped cause this diversification of climate, and it became an important regulator of the Asian monsoon (Raymo and Ruddiman 1992).

Jet streams in the innermost layer of the atmosphere up to about 15 km in elevation travel rapidly and move the lower air across the landscape, bringing rain, snow, and clear weather. Terrestrial features such as the plateau affect these jet streams both by their bulk and by variations in their temperature. Deflected by the barrier of the plateau, jet streams flow around it, causing them to meander. In winter, cold air settles on the plateau, whereas in spring and summer, the plateau heats up and the warm air gives rise to a high-pressure system that south of the Himalaya establishes an eastward-flowing jet stream which brings the June monsoon to the subcontinent. If, however, a rare snow cover blankets the plateau, the surface reflects, rather than absorbs, the sun's warmth and heats up slowly. With the high-pressure system of spring now weak, the monsoon may be late and the rains sporadic, affecting the lives of millions of people in the lowlands (Reiter 1981; Reiter and Gao 1982; Chen, Reiter, and Feng 1985; Shen, Reiter, and Bresch 1986). The Himalaya shields the plateau from the full force of the monsoon except in deep valleys that breach the rampart.

A monsoon-related climate change occurred 7–8 million years ago (Harrison et al. 1992), further evidence of the presence of a high plateau at that time. Studies of fossil pollen and the presence of such species as *Hipparion* in the Gyirong Basin at an elevation of 4200 m just north of the Himalayan peak Xixabangma have been taken as evidence of a warm moist climate during the Pliocene followed by an uplift of the plateau (Chen 1981). But as Molnar (1989) pointed out, most such fossil material has been collected along the northern margin of the Himalaya, whose pattern of uplift differed markedly from that of the plateau itself. The crust of the Indian subcontinent continues to slide beneath the Himalaya at 10–20 mm per year. As noted by Molnar, "in the last few million years, horizontal compression and crustal thickening have elevated the edges of the Tibetan Plateau, while the interior has not undergone the same processes" (1989). Indeed the interior has been spreading east and west for 2–6 million years at a rate of 5–15 mm per year and has actually been collapsing in height (Harrison et al. 1992). This spreading, together with erosion, has tended to flatten the landscape (fig. 2.1).

The climatic history of the Pleistocene during the past 2 million years or so has been highly complicated. There were either three or four major glacial periods in Europe and Asia, depending on area, separated by warmer, interglacial periods. In addition, warm and cold oscillations oc-

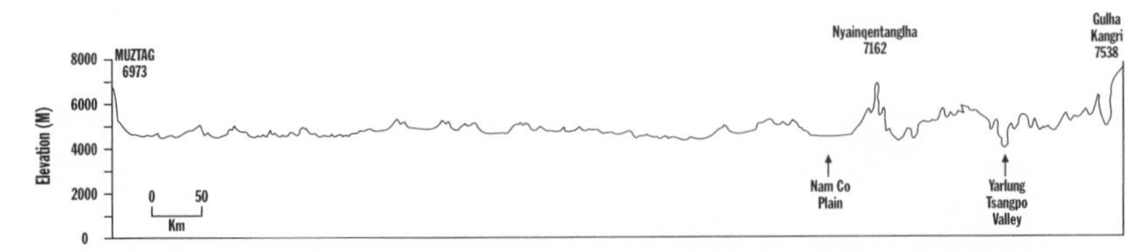

2.1. A cross section of Tibet from the crest of the Himalaya on the Bhutan border north across the Chang Tang to the Xinjiang border, showing topographic relief. (Adapted from Yang et al. 1983.)

curred within each period (Kurtén 1968; Frenzel 1968). The peak of the most recent ice age was about 18,000 years ago (Ruddiman and Kutzbach 1991), and the last interglacial, before the current one, was about 30,000 years ago (Olschak, Gansser, and Bührer 1987). Kuhle (1987b) hypothesized that the whole plateau except the Qaidam Basin and the Yarlung Tsangpo Valley was covered by an ice cap up to 1000 m thick. There is no question that during the last ice age—the Würm in Europe and Tali in China—glaciers were extensive on the plateau, judging by moraines, hanging valleys, and other evidence. Glaciers persist on most peaks that exceed 6000 m in height and tongues may descend below 5000 m. The snow line varies from about 4800 to 5900 m, depending on exposure and latitude, and it was at least 1000 m lower during the last glacial peak (Kuhle 1987a). However, the Chang Tang shows no indication of a vast ice sheet. The level ridge lines with the occasional emergent peaks were formed by erosion (Chang et al. 1986) and spreading of the plateau (Molnar 1989), not by the scouring of glaciers. Moraine deposits made up a mere 2% of the terrain in the Chang Tang of Qinghai ("No Ancient Ice Sheet on Tibetan Plateau" 1990). Similarly, in northern Tibet we saw moraines only near the base of major mountains ranges. S. Xu (1981) estimated that in the Tanggula Shan the glaciers covered a surface area 11 times greater during the late Pleistocene than today. Ice cover on the plateau and adjoining ranges is today only about 4%, compared to as much as 20% at the height of glaciations (Derbyshire et al. 1991).

The change from a glacial to an interglacial period is evident in the water levels of lakes. In early postglacial times warm summers and reduced snowfall during cold winters caused the ice to retreat and lake basins to fill. Subsequently the climate cooled and became more arid, and, exposed to intense solar radiation and high wind, lake levels dropped. Demidoff (1900) observed long ago that most lakes have ancient beach lines well above current water levels and that some basins are now completely dry, and Hedin (1903) mentioned beach lines over 100 m above some lake levels. At Tangra Yumco in the central Chang Tang the highest beach line was 215 m above lake level according to my altimeter, and at Lungma Co

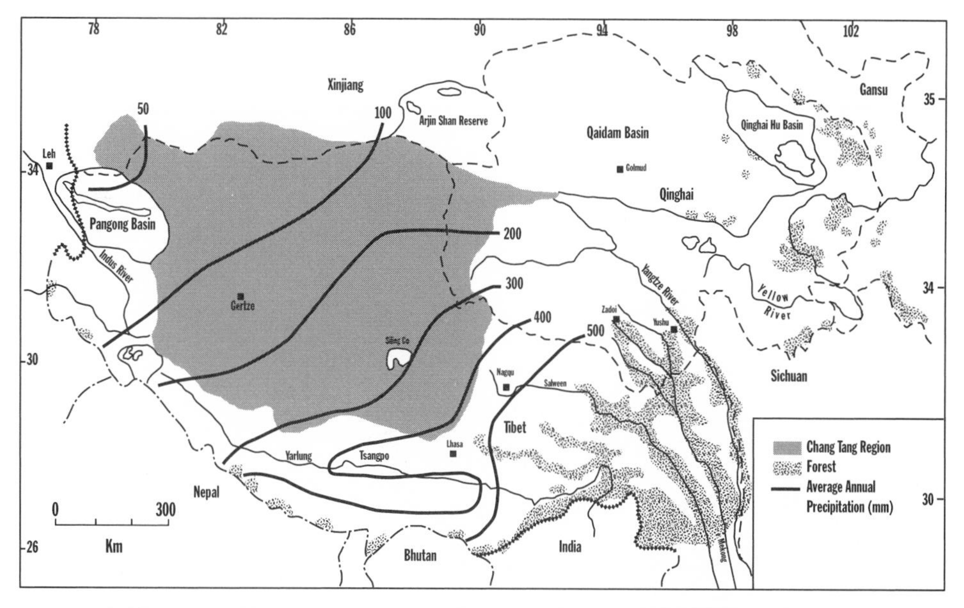

2.2. The principal lake basins and the distribution of forests on the Tibetan Plateau. The extent of the Chang Tang, as defined in this report, is also shown, as is the average annual precipitation there.

in the northwest Chang Tang the difference was of similar magnitude.

The Tibetan Plateau consists of several distinct topographic regions determined by drainage patterns and the parallel mountain chains that dissect it. Only the eastern and southern parts have outlets to the ocean. In the east the headwaters of the Yellow, Yangtze, Mekong, and Salween Rivers lie in the uplands of Qinghai and Tibet, and in the south the Yarlung Tsangpo (which downstream becomes the Brahmaputra), Indus, and Sutlej also flow to the sea (figs. 1.2 and 2.2). In addition, the watersheds of several south-flowing rivers such as the Arun lie north of the Himalaya, a striking indication that they are older than these mountains, having maintained their ancient channels during the uplift and in the process created gigantic canyons. Similarly, along the northern rim of the plateau the Ulug He slices north through the Kunlun Shan, but it, like all rivers there, vanishes in the sands of the Taklimakan.

Much of the plateau comprises lake basins that vary in size, have no outlets, and are fringed by mountains and hills (fig. 2.2). The Qaidam in northern Qinghai is the largest of these basins. About 650 km long and up to 350 km wide and at an elevation of 2600–3000 m, the Qaidam was once a huge lake, of which only several small lakes and a few marshes survive. Adjoining the Qaidam to the east lies the basin of Qinghai Hu; the lake is 4200 km^2 in size and lies at an elevation of 3200 m. And to the

Table 2.1 Main vegetation types in the Chang Tang Reserve (334,000 km²)

Vegetation type	%
Stipa spp.	29.6
Kobresia spp.	3.8
Stipa spp. and *Carex moorcroftii*	18.0
Carex moorcroftii	8.9
Ceratoides spp. and *Carex moorcroftii*	11.4
Sparse mountain vegetation	16.7
Glaciers and bare mountaintops	9.8
Lakes	1.8

Note: Calculated by D. Miller based on map in *Rangelands of Tibet* 1992.

west of the Qaidam is the Arjin Shan Reserve, which comprises a series of shallow basins encircled by mountains. The two largest lakes in the reserve are the Ayakkum Hu at 4120 m and the Aqqikkol Hu at 4250 m. Along the western edge of Tibet are the large Panggong and the small Mapam Yumco Basins. The remaining area, a block of terrain about 900 km long and 700 km wide and lying mostly in Tibet, is for purposes of this report defined as the Chang Tang (fig. 2.2).

Bordered in the north by the Kunlun and in the south by the Gangdise and Nyainqentanglha Ranges, the Chang Tang has many lakes, from mere ponds to Nam Co and Siling Co, which are over 1000 km² in size. Since drainage of these basins is internal, salts and minerals have over the millennia concentrated in the lakes, making most of them brackish or saline. Fan (1981) classified these lakes as carbonate and sulfate types, with the latter often having high concentrations of Na_2SO_4 or $MgSO_4$. The terrain varies from valleys hemmed in by rugged ranges, and rolling hills separated by broad, shallow valleys, to enormous flats, the landscape becoming more spacious and higher from south to north. For instance, in the south the lakes generally lie at an elevation of 4300–4500 m, whereas in the north they lie at around 4800–5000 m. The more than 575 lakes in the Chang Tang Reserve make up at least 1.8% of its total area. About 50 peaks are over 6000 m and the highest is 7167 m. Glaciers and mountains above the limit of vegetation make up about 9.8% of the reserve (table 2.1, fig. 2.7). The remaining 88% or so represents potential wildlife habitat.

As already noted, in early postglacial times the basins were filled, and lake waters lapped against hills and peaks, which like peninsulas and islands rose among these inland seas. Siling Co today is about 1350 km² in size, but judging by beach lines it was once around 6200 km². Fed each summer by glacial runoff, most lakes were then probably fresh. Habitat for plains ungulates was at that time relatively scarce. As lake levels receded, ever more habitat became available, creating conditions that no doubt had a major impact on the numbers, distribution, and movement patterns of animals.

Climatic changes have long had a marked influence on the vegetation of the Tibetan Plateau. Analyses of pollen from various parts of Tibet and Qinghai have revealed several cycles between warm and humid climates and cold and dry climates since the Pliocene. During the early Pliocene the northern slopes of the Himalaya and the valley of the Yarlung Tsangpo had semievergreen forests with *Quercus*, *Castanopsis*, *Podocarpus*, and other species interspersed with steppe (Wang, Li, and Zhang 1981), a forest vegetation typical today of southeastern Tibet at elevations of 1000–2000 m (Li 1985). Farther north around the Tanggula Shan, in areas that are now wholly treeless, there was also a forest-steppe, with stands of *Pinus*, *Ulmus*, *Abies*, *Quercus*, and *Cedrus*, as well as *Ephedra*, a shrub characteristic of dry steppe (S. Xu 1981). The equid *Hipparion* was a prominent member of the fauna. By the late Pliocene the climate had become cooler and drier and the steppe more widespread. Forests no doubt changed composition and advanced and retreated several times during the climatic changes of the Pleistocene. During the early Pleistocene, for example, *Quercus*, *Betula*, *Corylus*, *Picea*, and other shrubs and trees, together with such steppe shrubs as *Artemisia* and *Ephedra*, were found near Tuotuohe in the eastern Chang Tang (S. Xu 1981). Using oxygen isotope records from glacier ice in Qinghai, Thompson et al. (1989) found that mean summer temperatures have varied by up to 4°C during the past 40,000 years, a period that covers the late Pleistocene and Holocene, with the climate warmest 40,000–32,000 years ago. Later, the Pleistocene was colder and more arid (Gasse et al. 1996), and conifers replaced deciduous, broad-leaved trees (R. Xu 1981).

In postglacial times, during the Holocene, there have been three wet periods and four dry ones in the past 10,000 years according to pollen profiles from Chagcam Caka in the northwestern Chang Tang (about 33°20′ N, 83° E). Wet periods occurred 10,000, 8000–7000, and 6000–5000 years ago (Wang, Li, and Zhang 1981). The pollen profile from Huang and Liang (1981) in figure 2.3 shows that the climate was on the average more humid from about 13,000 to 4000 B.P., judging by the presence of *Pinus* as well as *Tamarix* and *Salix* shrubs in an area where no tall woody plants occur today; *Ajania* and *Ephedra* were common. Gasse et al. (1996) found similar changes in arid and humid conditions at Panggong Co during the terminal part of the Pleistocene and the Holocene. The evidence indicates that for a part of the Holocene there were stands of forest throughout the plateau, particularly along lakes and in protected localities with adequate soil moisture. A pollen profile from Nara Co southeast of Lhasa showed that southern Tibet was then warmer and more humid than the Chang Tang. *Betula*, *Tsuga*, and *Quercus* persisted until 3000 B.P., in contrast to the Chang Tang, where a pollen sample from Coqen showed no evidence of trees at that time (Huang and Liang 1981). In general, temperatures then averaged perhaps 3–4°C higher than today (Wang, Li, and Zhang 1981). The past 3000 years have become both drier and colder. Fang and Yoda (1991) showed that low temperatures and annual precipitation of less than 1100 mm limit the distribution of evergreen broad-leaved forest in Tibet, and the same two factors affect all tree growth. Trees and tall shrubs vanished from the Chang Tang in the late Holocene (Huang and Liang 1981).

This brief overview reveals that much of the plateau has had a forest-steppe or steppe vegetation since the Pliocene. During the Pleistocene the plains ungulates were at times constricted by the advance of glaciers or the rise of lakes. But of importance is the fact that for several million years the animals have had open habitats available to them. Frequent heavy snows, as were necessary to initiate major glacial advances, could have caused periodic and widespread starvation (see Schaller and Ren 1988). However, since the last glacial peak, about 18,000 years ago, conditions have probably favored the spread and survival of plains ungulates.

Today most of the plateau has a severe continental climate. But precipitation and temperature are strongly related to longitude, latitude, and elevation. Based on temperature and humidity, the plateau's climate can be divided into four categories (Liu and Wu 1981). The plateau inclines toward the southeast, and moisture in the form of the southwest monsoon comes up the gorges from the east and south. Consequently precipitation

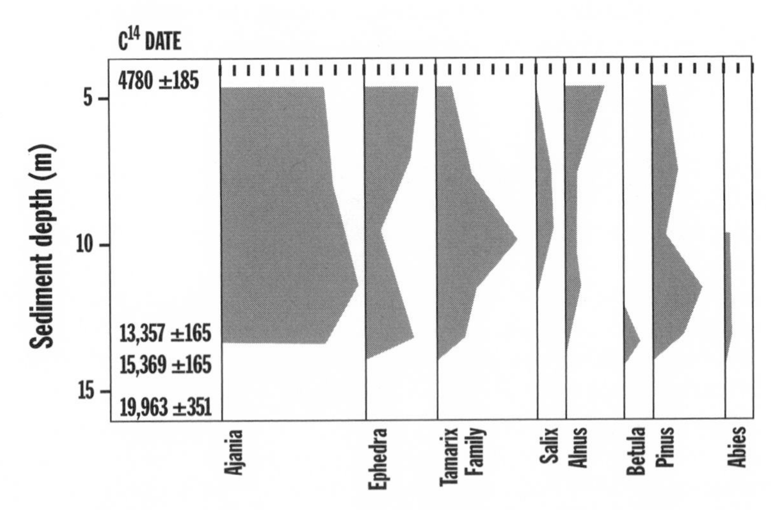

2.3. A pollen diagram of sediments from Chagcam Caka (4400 m) in the northwestern Chang Tang, dated in years B.P. (Adapted from Huang and Liang 1981.)

Table 2.2 Weather data from selected sites in Tibet

	Lhasa	Xigaze	Tingri	Dam-xung	Nagqu	Bain-goin	Gerze	Shi-quanhe
Elevation (m)	3658	3836	4300	4200	4507	4700	4415	4278
Av. °C in coldest mo. (Jan.)	−2.2	−3.8	−11.3	−9.9	−13.9	−11.2	−11.9	−12.5
Av. °C in warmest mo. (July)	15.5	14.6	10.9	10.8	8.9	8.6	12.1	13.4
Absolute min. temp.	−16.5	−25.1	−46.4	−35.9	−41.2	−35.8	−33.8	−33.9
Absolute max. temp.	29.4	28.2	22.9	25.4	22.6	21.7	24.4	25.7
Frostless days	133	127	104	67	20	9	50	85
Av. precip. (mm)	444	434	236	483	400	301.2	166	76
Av. no. days with wind velocity >17 m/s	32	59	112	66	97	71	200	—
Hours sunshine/ yr. (%)	69	74	77	64	66	61	73	78
Years of record	1951–75	1955–75	1957–69	1962–75	1954–75	1965–75	1972–75	1966–75

Source: Feng, Cai, and Zheng 1986.

decreases in a gradient from east to west and from south to north. The southeast is a humid tropical and subtropical montane area. The relatively low-lying fringe of the plateau is temperate, except in the north, with the eastern part in Sichuan humid, the southern part in Tibet semiarid, and the Qaidam Basin and far western Tibet arid. The central part of the plateau in a broad band from Qinghai west through the southern Chang Tang is subfrigid, humid in the east and semiarid in the west; the Qilian Shan in northeast Qinghai also belongs in this category. The northern Chang Tang is frigid and arid.

In southeastern Tibet the average temperature during the coldest month (January) is about 8°C; during the warmest month (July), 22°C. Average annual temperatures there are above 6°C, whereas in the Chang Tang they vary from 0 to −6°C and below (Feng, Cai, and Zheng 1986). The average annual temperature in Tian Wendian, a post at 5500 m in the Aksai Chin area, is −10.9°C (Chang and Gauch 1986). As shown in table 2.2, localities such as Lhasa and Xigaze in southern Tibet have a temperate climate with about 130 frost-free days. By contrast, Nagqu, Baingoin, and Gerze in the Chang Tang are cold, windy, and seldom frost-

Table 2.3 Weather data from two Chang Tang towns

	Bangoin (4700 m)				Gerze (4415 m)			
	Av. temp. (°C)		Precip. (mm)		Av. temp. (°C)	Precip.		
	1991	1992	1991	1992	1971–85	1971–85	1991	1992
Jan.	−14.4	−11.0	6.1	0.5	−12.3	0.6	0.0	2.5
Feb.	− 6.5	−11.5	0.3	3.4	− 9.6	1.1	0.3	0.0
Mar.	− 4.2	− 3.0	1.4	0.8	− 5.5	0.7	0.6	0.3
Apr.	− 0.9	0.0	0.8	1.2	− 0.2	2.2	0.0	0.0
May	4.5	2.5	0.5	17.6	3.8	8.1	0.5	1.6
June	7.7	6.6	104.4	34.1	9.3	19.3	17.6	26.8
July	9.5	8.1	90.8	42.7	11.7	59.8	55.6	78.1
Aug.	8.1	8.4	80.7	87.3	11.0	62.0	62.8	58.5
Sept.	5.6	7.0	57.1	56.0	7.4	31.6	44.7	21.6
Oct.	0.6	− 0.1	0.0	10.3	0.0	3.3	0.0	0.0
Nov.	− 6.7	− 6.6	0.5	0.6	− 7.8	0.7	0.0	0.0
Dec.	−10.2	−11.8	2.7	3.1	−12.0	0.4	0.0	0.1
Total			345.3	257.6		189.8	182.1	189.5

free (tables 2.2 and 2.3). Only the southeast part of the plateau has an average annual precipitation of 500 mm or more. Most of the rest receives 400 mm or less precipitation, with the far northwest and parts of the Qaidam Basin receiving not even 50 mm (fig. 2.2). Most precipitation falls from July to September, in the northern Chang Tang usually as snow, sleet, or hail.

During the spring of 1992 we were in the northwestern Chang Tang north of Aru Co at 5000 m. Between 12 June and 10 July, the average daily minimum was −5°C (−3 to −9°C) and the maximum was usually between 5 and 10°C. It snowed or hailed on 15 of the 29 days, and winds at times reached 30–35 km/hr. Later that year, between 19 July and 3 August, the average daily minimum at Aru Co was 1.4°C (−3 to +5°C) and the maximum was 13.3°C (7 to 18°C). It snowed on four days, and a mixture of sleet and rain fell on two. Southwest Qinghai had an exceptional cold spell between 23 October and 2 November 1985, with an average daily minimum of about −29°C (−23 to at least −40°C). More typical winter temperatures prevailed during our fieldwork near Shuanghu between 1 and 25 December 1991, when the average daily minimum was −25.3°C (−15 to −34°C). Bonvalot (1892) recorded −44.5°C on 18 January 1890. Winter winds are often strong. In February 1908, Hedin (quoted in Vaurie 1972) recorded speeds of 32–39 km/hr daily at 1300 hours.

Permafrost may occur at elevations above 4100 m in areas with an average annual temperature of −2.5 to −3.5°C (Tong 1981). Much of the northern Chang Tang lies within this permafrost zone. The maximum depth of the seasonal thawed sediment layer was 1–6 m in the eastern

Chang Tang along the Lhasa-Golmud highway, and permafrost thickness reached 155 m (Tong 1981). In mid-June 1992 we dug three holes at 5000 m in the plains near Toze Kangri north of the Aru Basin. Permafrost was reached at 59–63 cm, indicating only a shallow thawed layer beyond which plant roots had not penetrated.

Precipitation, temperature, and wind have a critical influence on vegetation growth. I can do here little more than make general comments about the distribution of the major plant formations. Forests on the plateau are largely found in southeast Qinghai, east Tibet, and west Sichuan, east of 94°30′–95° E, but a few patches extend west along the northern slopes of the Himalaya (fig. 2.2). The lushest forests occur around the southern flank of Namjagbarwa, where the Yarlung Tsangpo turns south into India. There Li (1985), Chang (1981), and Zhang et al. (1981) recognized five to six vertical zones (fig. 2.4). Below 1000 m in elevation is a tropical, evergreen dipterocarp forest with abundant lianas and epiphytes that grades upward into a semievergreen broad-leaved forest of *Castanopsis*, *Celtis*, and other genera. Above it, to about 2500 m, is an evergreen broad-leaved forest in which *Cyclobalanopsis*, *Lithocarpus*, and oak *(Quercus)* are conspicuous at the lower level and also pine *(Pinus)* higher up. Fir *(Abies)*, hemlock *(Tsuga)*, and spruce *(Picea)* mixed with some broad-leaved species, such as *Acer* and *Magnolia*, occur above 2500 m up to about 4000 m, although at the higher elevations most broad-leaved species cease, leaving an understory primarily of *Rhododendron* and, at the upper forest line, of gnarled birch *(Betula)*. An alpine shrubland of *Rhododendron*, *Rosa*, *Salix*, *Lonicera*, and others together with luxuriant meadows extends to around 4800 m. This zone gives way to subnival vegetation of scattered cushion plants and forbs. The forests are the home of takin, serow, goral, Asiatic black bear, leopard, and other large mammalian species that are outside the purview of this study.

In eastern Tibet and western Sichuan, among mountains within reach of the southwest monsoon, forests grow up to an elevation of 4100–4500 m. A mixed coniferous and broad-leaved forest composed predominantly of spruce, fir, and oak with an understory of *Acer*, *Lindera*, *Litsea*, *Rhododendron*, and other trees predominates. The forests in eastern Qinghai are small and patchy, composing a mere 0.3% of the province (table 2.4), and they consist predominantly of junipers *(Sabina)*, which in the Yushu area and eastern Burhan Budai Shan grow to an elevation of 4400 m. *Picea crassifolia* and *P. asperata*, with trees up to a height of 30 m, as well as *Betula*, occur locally below an elevation of 3500 m in the Anyemaqen Shan and eastern Qilian Shan. On the slopes above these conifer forests or below the tree canopy are shrub thickets often no more than 1–1.5 m tall. For example, on moist slopes near Yushu in southeast Qinghai such thickets

Table 2.4 Land use on Tibetan Plateau (in percentages)

	Cultivated land	Forest	Pasture	Other
Tibet	0.2	5.1	69.1	25.6
Qinghai	0.9*	0.3	56.1	42.8
Western Sichuan	0.7	10.6	44.4	44.3

Source: After Cheng et al. 1981.
*Cultivation increased greatly during the 1980s.

Table 2.5 Demographic profile of Tibet and Qinghai in 1990

	Tibet	Qinghai
Total population	2,196,010	4,456,946
Population density per km^2	1.8	6.2
Birth rate	27.60	22.65
Death rate	9.20	6.84
Rate of natural increase	18.40	15.81
Population doubling time (years)	38	44
Number of children per couple (1987)	4.26	2.72
Urban population as percentage of total	12.59	27.35
Illiteracy rate as percentage of population (aged 15 years and over)	44.43	27.70

Source: Based on the 1990 population census.

consist of *Rhododendron capitatum*, *Potentilla glabra*, *Spiraea alpina*, *Sibiraea angustata*, and *Salix* sp., sometimes in almost monotypic stands, at other times of mixed composition. *Caragana jubata* and *Dasiphora fruticosa* are both common on damp slopes in the Anyemaqen and eastern Qilian Ranges.

Average population density on the plateau is low (table 2.5), but over the centuries human settlements have had an impact on the shrub and forest cover. Along the Yarlung Tsangpo gnarled and procumbent junipers still cling to high slopes, no doubt the remnants of forests that were cut for firewood and building materials. The remaining shrubs such as *Caragana spinifera* are now often dug out, along with the roots, to serve as fuel. In and around the Qaidam Basin the forested area was reduced from about 255 km^2 in 1954 to 38 km^2 in 1977 (Cai, Liu, and O'Gara 1989).

Less than 1% of the plateau is under cultivation (table 2.4), although in recent years agriculture has expanded considerably in the Qaidam Basin as the population increased from about 10,000 in 1946 to 270,000 in 1986 (Cai, Liu, and O'Gara 1989). Peas, potatoes, rape, and wheat are grown mainly at low elevation and barley at the highest. In the eastern part of the plateau, fields are concentrated in the lower valleys; in western Tibet, along the drainage of the Yarlung Tsangpo and, locally, along the Indus and in the Panggong Basin around Rutog. Small patches of cultivation also occur in the Chang Tang, as at the northern end of Tangra Yumco. The upper limit of cultivation is as low as 3300 m in some parts of the eastern plateau, whereas it reaches 4400 m and in favorable localities even 4500 m along the Yarlung Tsangpo and other areas of western Tibet.

Most of the plateau consists of rangelands, one of the largest such ecosystems in the world and one that is floristically distinctive. In this high, cold, and dry environment, where the sun shines for 2500–3600 hours per year and its radiant value is an intense 140–195 kW/cm^2 (Huang 1987), the vegetation tends to be low; most plants are hardy perennials and cushion plants. Over 300 species of seed plants have been collected from the area, mostly Graminae, Compositae, and Leguminosae (Wang 1981).

Chang (1981) divided the rangelands into five major regions, to which I have added one, the Qaidam Basin: (1) the Qaidam Desert; (2) the high-cold, or alpine, meadow of eastern Qinghai and Tibet; (3) a xeric shrubland and steppe along the valleys of the Yarlung Tsangpo and Indus in southern Tibet; (4) the high-cold, or alpine, steppe in the southern Chang Tang; (5) the high-cold desert, or desert steppe, in the northern Chang Tang; and (6) the desert mountains in the northwest. Within each region are many distinct plant formations and communities, each determined by elevation, drainage, precipitation, sunlight, and other factors. Chang and Gauch (1986), for example, identified 26 plant communities in the various regions of western Tibet, based mainly on the dominance of certain graminoids and dwarf shrubs. Achuff and Petocz (1988) listed 18 communities between 3300 and 5100 m in the western half of the Arjin Shan Reserve (table 2.6).

The vegetation on mountain slopes has a relatively narrow vertical zonation and the vegetation on the plains has a broad horizontal zonation, both based on precipitation and elevation. Taking mountain slopes first, I provide three brief examples of such zonation from ranges with different amounts of precipitation (fig. 2.4). Figure 1.2 shows the location of these ranges.

1. The arid western Qilian Shan has at low elevations (3000–3900 m) a desert shrub formation with a cover of about 30% (Huang 1987). Dominant shrubs include *Dasiphora fruticosa*, *Salsola abrostranoides*, *Ceratoides latens*, and *Reaumuria kaschgarica*. Above this zone is a desert steppe with a thin (<10%) cover of low *Ceratoides compacta* and *Artemisia frigida*, tufts of *Stipa* grass, and various forbs such as *Saussurea arenaria* and *Oxytropis falcata*. Between 4200 and 4600 m is a belt sparsely covered with cushion plants, prominent among them *Arenaria* sp., *Rhodiola* sp., *Androsace tapete*, and *Thylacospermum caespitosum*. And extending upward to the snow line

Table 2.6 Vegetation formations and communities in the Arjin Shan Reserve

Formation	Community	Elevation (m)
Mountain desert	*Ceratoides compacta*	3900–4700
	Salsola abrotanoides and *Ceratoides latens*	3300–3900
Desert grassland	*Carex moorcroftii* and *Ceratoides compacta*	4300–4600
	Stipa spp. and *Ceratoides compacta*	4250–4400
	Stipa glareosa and *Ceratoides compacta*	3800–4300
Grassland	*Carex moorcroftii*	4250–5000
	Carex moorcroftii and *Stipa* spp.	4450–4600
	Stipa spp.	4250–4600
	Stipa spp. and *Carex* sp.	4250–4600
	Carex moorcroftii and cushion plant	4500–5000
	Stipa glareosa	4000–4400
	Leymus secalinus	3850–4000
Wet meadow	*Myricaria prostrata*	4250
	Leymus secalinus and *Puccinellia* spp.	3800–4000
	Kobresia capillifolia	3850
	Eleocharis quinquefolia	3850
Herb mat	*Kobresia* spp.	4600–5100
	Leontopodium nanum and *Arenaria* sp.	4400–5100

Source: After Achuff and Petocz 1988.

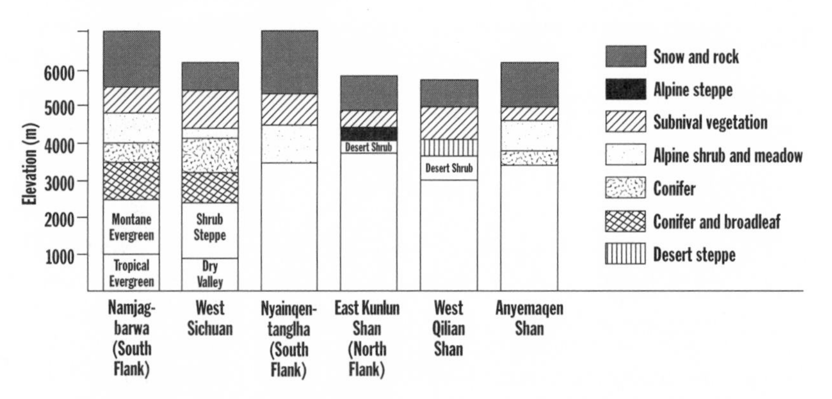

2.4. The altitudinal zonation of major vegetation formations on several mountain ranges of the Tibetan Plateau. (Adapted from Zhang et al. 1981 and Huang 1987.)

at about 5100 m is the subnival zone. Scattered in the sand and gravel and tucked into the lee of rocks are some of the same species of cushion plants found at lower altitudes: clumps of *Elymus*, *Festuca*, *Kobresia*, and other graminoids, and a variety of forbs such as *Saussurea*, *Saxifraga*, and *Draba* (Li et al. 1985; Huang 1987).

2. The north-facing slopes of the Kunlun Shan rise abruptly from the desert of the Qaidam Basin. Along the base is a zone of desert shrub: *Sal-*

sola, *Ceratoides*, *Ephedra przewalskii*, and *Sympegma regalii*. But, in contrast to the Qilian Shan, the higher elevations are relatively moist and support an alpine steppe of graminoids, particularly *Stipa purpurea*, *Littledalea racemosa*, *Festuca* sp., *Poa* sp., and others, together with cushion plants and forbs. From about 4400–4500 m to the permanent snow line is the subnival zone (Huang 1987).

3. Above the forests, above 4000 m, the Anyemaqen Shan has a lush zone of alpine shrub and meadow with a turf layer of organic material that may reach a thickness of 20–30 cm. The turf consists of a dense mat of *Kobresia*, as well as other graminoids, and many forbs—*Gentiana*, *Meconopsis*, *Anemone*, *Potentilla*, *Aster*, *Anaphalis*—whose flowers speckle the hillsides with blue, yellow, and white in summer. A typical subnival zone occurs higher up, extending to the snow line at about 5200 m.

Qaidam Basin is distinctive in its vegetation, being transitional both geographically and climatically between the Mongolian desert to the north and the high uplands of the plateau. Indeed the vegetation of the basin has a strong affinity with the deserts of central Asia as expressed by shrubs of the genera *Calligonum*, *Haloxylon*, and *Trachomitum* (Walter and Box 1983a). In general, large salt flats in the center of the basin are virtually unvegetated, and marshes and the edges of brackish ponds support *Phragmites* reeds. Low *Artemisia* and *Salsola* shrubs cover large tracts, as do *Tamarix ramosissima*, *Ephedra przewalskii*, and *Nitraria schoberi*, the last-named 2 m tall, especially on the sand-and-gravel plains near the foothills.

Lying in the rain shadow of the Himalaya between about 3500 and 4300 m, the drainages of the upper Yarlung Tsangpo and Indus in southern Tibet, as well as the desert mountains in northwest Tibet, are arid and covered with a mixture of steppe and xerophilic shrubs. *Stipa*, *Aristida*, *Pennisetum*, and *Orinus* are common grass genera, and *Artemisia*, *Sophora*, *Lonicera*, *Berberis*, *Caragana*, *Cotoneaster*, and *Hippophae* are conspicuous shrubs (Walter and Box 1983a; Chang 1981).

The rest of the rangelands—the broad valleys, basins, and rolling to hilly plains—extend from the moist east in Qinghai, at elevations as low as 3000–3500 m, to the arid west and northwest in Tibet and Xinjiang, at 5100 m and above. Three main vegetation formations extend over these vast uplands: alpine meadow, alpine steppe, and desert steppe.

1. The alpine meadow is prominent in the eastern half of this area, where annual precipitation exceeds 400 mm, particularly in southeastern and southern Qinghai, as around the twin lakes Gyaring and Ngoring (fig. 2.2). The formation extends west into Tibet, generally below an elevation of 4500 m, to about the longitude of Lhasa and Nam Co. To the west, with its increasing aridity, the alpine meadow becomes largely riparian along streams, seepages, swamps, and lakes where soils remain saturated

with meltwater from snow and glaciers (Cincotta et al. 1991). This vegetation formation characteristically has a sod layer 10–40 cm thick densely covered with short, perennial sedges (*Kobresia* spp.) and various forbs, among them *Leontopodium*, *Anaphalis*, *Polygonum*, *Primula*, and *Potentilla* (fig. 2.5). A hummock vegetation of *Kobresia* (often the 30 cm tall *K. royleana*) may develop in swampy depressions. Since riparian meadows have water readily available, they have a longer growing season and hence remain green and more palatable to ungulates than other plant communities. Livestock and certain wild ungulates such as kiang and wild yak seasonally tend to concentrate on this restricted habitat. On the extensive alpine meadow pastures of eastern and southeastern Qinghai, such as in the counties of Maqen, Madoi, Zhidoi, Zadoi, and Yushu, average livestock densities are so high—28–70 animals/km^2—that continual grazing and trampling of hillsides by the animals, coupled with solifluction, have broken the sod layer, causing extensive sheet erosion. Similar habitat deterioration is evident in Amdo County and some other parts of Tibet.

2. Lying generally between 4000 and 5000 m, the alpine steppe is cold and windy, the sun's radiation is intense, and the daily changes in temperature are great. Plant coverage is sparse, seldom more than 30%, and the soil is poor (gravel, sand, silt, and clay, alluvial or eolian in origin) and

2.5. Alpine meadow habitat borders a stream in the Aru Range. Wild yaks favor the thick turf of *Kobresia* for grazing. (August 1988)

2.6. Alpine steppe is covered predominantly with *Stipa* grass, an important feed for most ungulates, especially in winter. A solitary wild yak bull is on the plain. (December 1991)

without a sod layer. A bunch grass of the genus *Stipa (S. purpurea, glareosa, subsessilifolia*, and others) is dominant, and it is much favored by most ungulates in winter; other graminoids such as *Festuca* spp., *Poa* spp., and *Carex moorcroftii* are also abundant locally (fig. 2.6). Dwarf shrubs, usually less than 30 cm tall, include *Ajania*, *Ceratoides*, and *Potentilla*. Herbaceous plants, most under 10 cm in height, include many legumes *(Astragulus, Oxytropis, Thermopsis)* and several species of *Saussurea*, *Ajuga*, and *Corydalis*, among others, as well as such cushion plants as *Androsace*, *Leontopodium*, *Arenaria*, and *Rhodiola*. The alpine steppe extends up the hillsides to 5200–5300 m, an elevation above which only scattered subnival species persist, but there is a shift in plant dominance with altitude. *Kobresia* replaces *Stipa* as the main graminoid, and cushion plants become more abundant (see table 12.3). Although these mountains are high, often 5500 m or more, they may extend only a few hundred meters above the level of the plain.

The alpine steppe can be divided into two broad vegetation zones based on the dominance of two graminoid genera, as shown in figure 2.7 for the Chang Tang Reserve. The southern part is covered predominantly with *Stipa* spp., a vegetation type which occurs on about 30% of the reserve (table 2.1), and the central part has two prominent genera, *Stipa* spp and *Carex moorcroftii*. Since wild and domestic ungulates tend to prefer *Stipa*

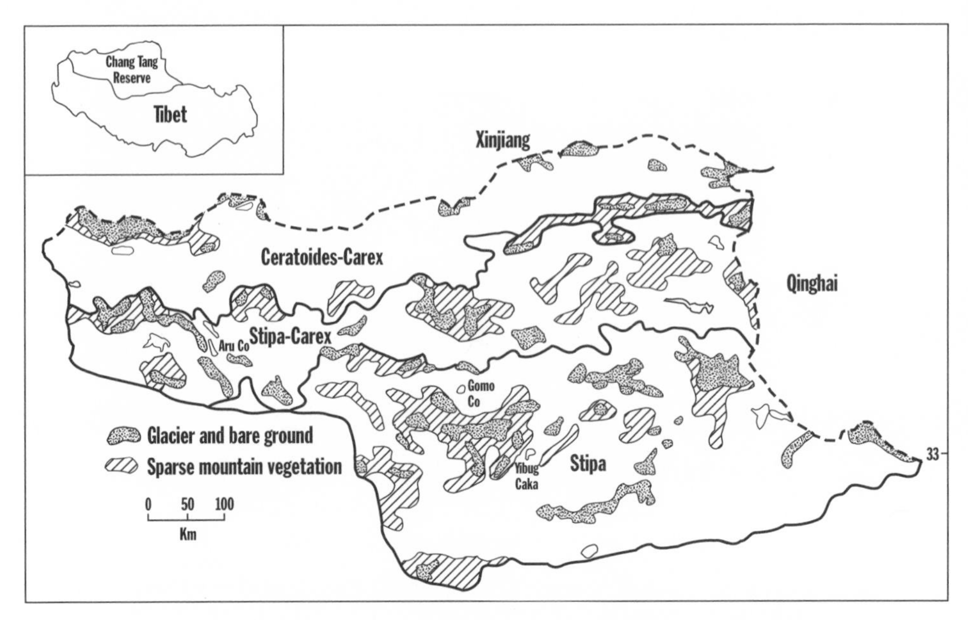

2.7. The three main steppe zones of the Chang Tang Reserve and the approximate distribution of high-elevation terrain. (Adapted by D. Miller from *Rangelands of Tibet* 1992 and from G. Joshi, pers. comm.)

2.8. Desert steppe habitat at 5000 m along the northern rim of the Aru Basin. Luotuo Hu is visible on the left and the glaciers of Toze Kangri in the distance. Chi Doa scans the terrain for chiru. (August 1992)

over *Carex* as forage, the relative abundance of these graminoids has an impact on ungulate distribution.

3. North of the alpine steppe, extending out of Tibet into Xinjiang, is a desert steppe, a bleak and arid environment with large tracts virtually devoid of vegetation. Many of the species in this formation are the same as in the alpine steppe but coverage is more sparse (fig. 2.8). The dwarf shrub *Ceratoides compacta* is dominant. *Carex moorcroftii*, coarse, sharp-pointed, and endemic to the plateau, grows in extensive swards on sandy soils. *Stipa*, *Kobresia*, *Oxytropis*, various cushion plants, and a few others occur thinly throughout this region. With forage meager, the density of wild ungulates is low, and pastoralists and their livestock leave much of the area unoccupied. Along the northwestern edge of Tibet are desert mountains, somewhat lower in elevation but largely similar in vegetation except that among the dwarf shrubs *Ceratoides latens* replaces *C. compacta*, *Ajania fruticolosa* is locally common, and *Caragana vessicolar* grows like dark, low hillocks in some valleys. Chang (1981) classified this area as a distinct vegetation formation.

Many plants in the harsh environment of the alpine and desert steppe

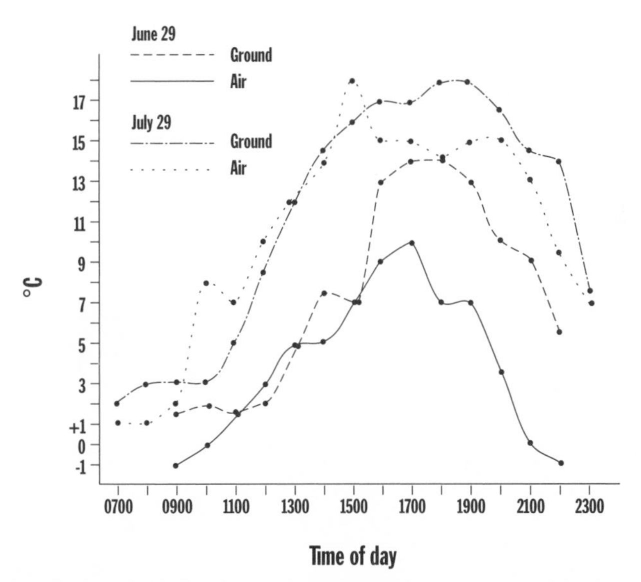

2.9. Air and subsurface (10 cm) soil temperatures during two sample days, showing retention of warmth by the soil. The 29 June temperatures were taken at 5000 m on a clear-to-cloudy day with a minimum temperature of −4°C; the 29 July temperatures were taken at 4850 m on a clear-to-partly-cloudy day with a minimum of +1°C. Dawn was at about 0700 and dusk at 2200 hours.

have distinctive adaptations to the harsh environment. By remaining close to the ground, sometimes in protected hollows, they avoid wind. The soil surface temperature rises rapidly in the intense sunlight, reaching levels higher than the air, and plants close to the ground can absorb the warmth for their physiological processes. After heating in the morning, the subsurface soil also retains warmth longer than does the air (fig. 2.9). Some plants have shiny hairs, possibly to retain humidity (Huang 1987) and to reflect heat into the interior (Larsen 1987). Various species have large taproots for nutrient storage. Cushion plants create their own microclimate by accumulating windblown soil and snow. Their crowded stems trap air, which is warmed by the sun. Huang (1987) found that the temperature inside an *Arenaria* cushion before sunrise was 1.2–3.5°C higher than the temperature at the surface.

3

Chiru (Tibetan Antelope)

Almost from my feet away to the north and east, as far as the eye could reach, were thousands upon thousands of doe antelope with their young. The mothers were mostly feeding, while the young ones were either lying down and resting, or being urged on by their mothers. All had their heads turned towards the west, and were travelling slowly in that direction, presumably in search of the fresh young grass springing up in the higher western tablelands.

Everyone in camp turned out to see this beautiful sight, and tried, with varying results, to estimate the number of animals in view. This was found very difficult however, more particularly as we could see in the extreme distance a continuous stream of fresh herds steadily approaching; there could not have been less than 15,000 or 20,000 visible at one time.

C. G. Rawling (1905)

As a boy in Germany, I had read one of Sven Hedin's books on Tibetan exploration. His descriptions of Tibetan antelopes traveling in large herds over desolate uplands where no humans live had particularly intrigued me. He referred to the animals as *orongos*, a Mongolian name, which added an even greater aura of mystery and remoteness. When I finally had the opportunity to study wildlife in the Chang Tang, the Tibetan antelope (or chiru, as it is called in parts of Tibet and in some of the early Western literature) became a main focus of the project. For a time the animals would be abundant in an area, by far the most abundant of the wild ungulates, but then they would vanish, migrating over routes and to destinations that remained unknown. To study the species was a challenge, not only a scientific one but also a logistic one, because, lacking the use of an airplane, we had to find the animals in vast terrain, much of it uninhabited. Yet research on the species was urgent. Chirus, though fully protected in China and by international convention (Convention on International Trade in Endangered Species), were being heavily hunted for their wool, which was smuggled to India for weaving and then sold primarily to developed nations as high-priced scarves and shawls (see chapter 15). Without knowledge of the chiru's basic ecological requirements, particularly the extent of its seasonal movements, it would be difficult to provide it with adequate protection.

Taxonomy

The chiru (*Pantholops hodgsoni*) is the only genus of large mammal endemic to the Tibetan Plateau. The animal looks like an antelope that has somehow strayed from the African plains. It has often been placed into the subfamily Antilopinae, and indeed Brian Hodgson in his original description of the species in 1830 refers to it as an "antelope" (Sterndale 1884). However, on the basis of various skull characters, Pilgrim (1939) and others designated it as a caprid, subfamily Caprinae, combining it with the saiga in the tribe Saigini. Recent morphological and molecular work has shown that the chiru is most closely allied to the caprids and the saiga to the gazelles (Gentry 1992; Gatesy et al. 1997), as discussed in chapter 13. For this reason, I call the animal chiru rather than Tibetan antelope. No subspecies of chiru have been designated.

Two late Miocene(?) fossils, genus *Qurliqnoria*, found in the Qaidam Basin were considered by Gentry (1968) to be related to chirus, and a Pleistocene fossil, *Pantholops hundesiensis*, a chiru slightly smaller than the living one, was described from western Tibet (Pilgrim 1939).

Description

With their rather chunky bodies and slender legs, chirus are reminiscent of antelopes. Adult males stand about 83 cm high at the shoulder (measured from the hoof tip) and weigh up to about 40 kg (table 3.1). Engelmann (1938) gave measurements of two males with a mean shoulder height of 94 cm and body mass of 36 kg. The measurements of three males given in *Research on Flora and Fauna* (1979) were 74–105 cm for shoulder height and 24–27 kg for body mass. The male's most conspicuous antelope-like feature is the long, slender, black horns, which rise almost vertically from

Table 3.1 Shoulder heights and weights of chirus

	Shoulder height (cm)		Body mass (kg)	
	Sample size	Mean ± S.D.	Sample size	Mean ± S.D.
Young (4 mo.)	7	63.4 ± 2.4	3	15.5 ± 0.9
Yearling male (16 mo.)	3	76.8 ± 1.6	2	25.5 ± 1.9
Yearling female (16 mo.)	2	67.0 ± 2.8	1	20.9
Adult male	2	83.0 ± 1.4	2	38.8 ± 2.5
Adult female	8	74.4 ± 2.2	9	25.9 ± 2.9

Note: All except 1 male and 2 females were measured in winter; most died after the blizzard of Oct. 1985, 14 of them killed by dogs.

the head, curve slightly back in the distal half, and then terminate with smooth rapier-like tips pointing forward. The horns are laterally compressed and have about 15–20 ridges along the front for two-thirds of their length. They diverge moderately, with a mean tip-to-tip measurement of 29.3 ± 4.4 cm (range 19.2–41.2 cm). Chirus have their full adult dentition at about 2.5 years of age (see table 3.5), an age when their horns reach a length of around 50 cm. Of 162 horns 50 cm long or longer, mean length was 57.3 ± 3.0 cm, with a maximum of 65.1 cm. Horn circumference near the base was about 12 cm. There was no indication that horns continue to grow annually after a male is fully adult, in contrast to other caprids whose horns become longer throughout their lives.

The coat of adult males is dense and woolly, with hairs 4–6 cm long on neck and body. The summer pelage is reddish fawn with light gray and brown tones grading to white on the underside. A diffuse white rump patch is almost hidden beneath the 13–14 cm long tail. The ventral side of the tail has long white hairs. The face and the front of the legs are dark gray. In general, the animal looks drab, without conspicuous adornments. However, by late October, prior to the rut, the males have changed into a striking winter pelage. The face is now black, as is the front of the legs, the black markings on the forelegs curving toward the shoulder and the black on the hindlegs extending back to trace the margin of the rump patch. The body color is predominantly light gray and tan with conspicuously white underside from chin to belly; some males are so light-colored that they look white at a distance.

At the age of 16 months, yearling males weigh about as much as adult females (table 3.1). At that age their horns are 23–29 cm long, but they grow rapidly, reaching 30–33 cm at 18 months and over 40 cm at 24 months. Subadult males lack the marked change between summer and winter pelage. At the age of 2.5 years, they have some dark gray on face and legs in winter, and this color deepens the following year, but not until the age of 4.5 years do they exhibit the full nuptial pelage.

Females are about 74 cm tall at the shoulder and average about 26 kg (table 3.1). They are hornless, unlike other female caprids. Their coat is fawn-colored, almost pinkish, often with rust brown on the nape, blending into the whitish underside. A pale white area encircles the tip of the muzzle and the eyes. The top of the muzzle and front of the legs are grayish. There are two nipples. The young are colored like the females.

Chirus also have two small but distinctive morphological features. The muzzle is concave and blunt and the tip looks enlarged because of the presence of a walnut-sized bulge on each side of the nostril. This bulge marks the site of an air sac, representing an enlargement of the nasal passage. In Tibet and Qinghai, the hair on the bulge is grayish brown, contrasting only somewhat with the black face of males in winter, whereas in Xinjiang the bulge is white, a highly conspicuous feature.

When examining chirus, I found no evidence of preorbital, pedal, or other glands except large inguinal glands in the groin of both sexes. The glands in adult males have an opening 5 cm long and pouch 6 cm deep, "large enough to contain a man's fist" (Rawling 1905). A waxy, yellow exudate smelling like peanut butter lines the cavity.

Status and distribution

Chirus are confined to the Tibetan Plateau, where they prefer flat to rolling terrain, although they readily ascend high rounded hills and penetrate mountains and cross passes by following valleys. Alpine steppe or similar semiarid habitats are favored, the species being rare or absent from those parts of the alpine meadow region that have an average annual precipitation of 400 mm or more. Desert steppe and other such arid areas have also been occupied, at least seasonally, but chirus occur neither in the Qaidam Basin, except at its western rim, nor in northeastern Qinghai. In Xinjiang's Tula Valley, chirus frequent elevations as low as 3250 m, but most of their range lies above 4000 m, and on the Depsand Plain in northern Ladakh they can be found as high as 5500 m (Roosevelt and Roosevelt 1926).

The only chirus outside China are in the Ladakh part of India. Approaching Karakoram Pass from the south in 1868, Shaw found that "the plain, barren as it seems, is frequented by Tibetan antelopes" ([1871] 1984). Today, chirus occur in only two small areas of eastern Ladakh, where about 200 animals, mostly males, cross the border seasonally from Tibet and the Aksai Chin area of Xinjiang (Fox, Nurbu, and Chundawat 1991a). A few chirus once strayed into western Nepal (Schaller 1977b), but there is no evidence that they still do so.

From Ladakh the species extends 1600 km eastward across Tibet and southern Xinjiang to the vicinity of Ngoring Hu in Qinghai. Its current range is divided into two areas: a northern one of about 490,000 km^2 and a central one of about 115,000 km^2 (fig. 3.1). Distribution between the two areas was continuous until recent decades, and there may still be rare contact near the western end. To further delineate distribution, figure 3.1 notes where I and others observed chirus outside the Chang Tang Reserve. The range of chirus west to east appears to be much as it was a century ago when Bower wrote: "Their habitat may be said to commence in the west at the Karakorum Pass, and extends to the Lhasa-Sining [Xining] road in the east, and probably they are found in occasional patches of suitable country still farther east" (1894). However, the range has contracted in central Tibet and eastern Qinghai.

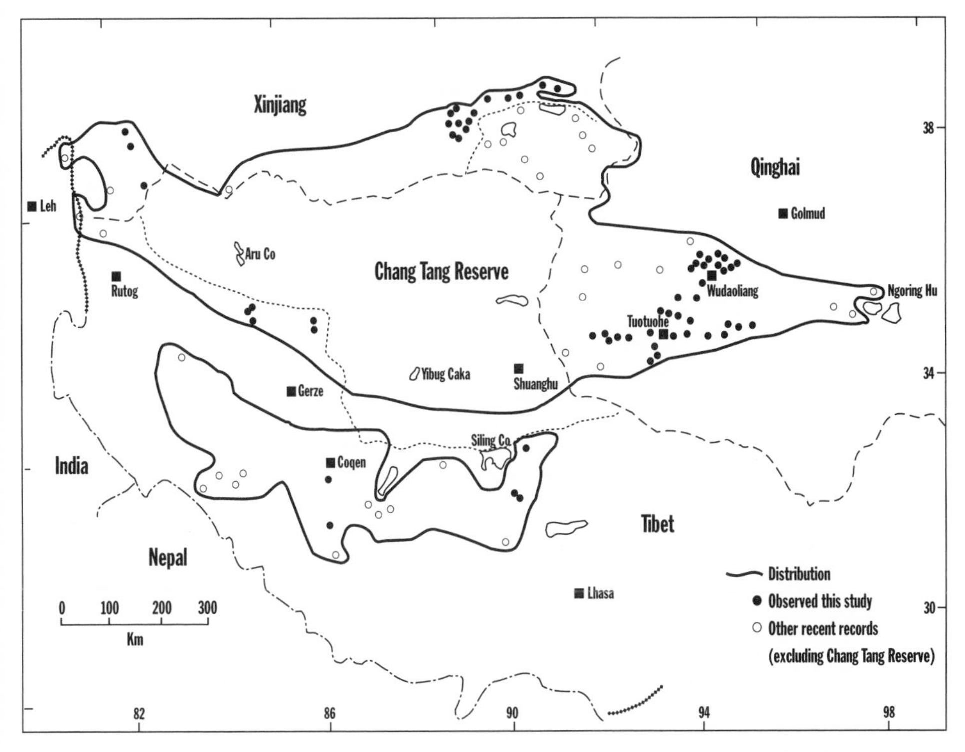

3.1. The current distribution of chiru (Tibetan antelope). (Chiru sightings within the Chang Tang Reserve are not indicated.)

In comparing chiru numbers in the past with those of today, I have only the accounts of a few Western travelers to draw upon. The most useful of these accounts are those of Sven Hedin, who often noted the relative abundance of chirus and other species. To enable future investigators to compare their data with mine, I also provide occasional details on location and density. However, the abundance of chirus in an area depends on the movement patterns of a particular population. Movements of chirus are of two types: (1) resident animals, which make only local movements, and (2) migratory animals, especially females, which may travel up to 250–300 km between summer and winter ranges. An area may be devoid of chirus at one season yet swarming with them another.

Qinghai

Chirus reach their eastern limit near Ngoring Hu, where they are rare (Kaji 1985). During a winter survey east of the Lhasa-Golmud road in 1985, we also found chirus scarce, whereas they were moderately abundant west of the road. In the Yeniugou (Wild Yak Valley), a valley of the Burhan Budai Shan west of Kunlun Pass, about 2000 chirus were found in summer and the females gave birth there (Harris 1993; Harris and Miller 1995). We transected an area of 2100 km^2 on the plains just south of the Yeniugou in November 1986 and calculated 3087 chirus, a density of 1.47/km^2 (Schaller and Ren 1988). I surmise that these animals represent a resident population which shifts in part between summer in the Yeniugou and winter on the plains.

A large segment of the remaining chirus in southwestern Qinghai appear to be migratory and part of one main population. In addition, there seems to be part of another population which enters seasonally from Tibet. Based on a survey that went from the Lhasa-Golmud road west to Ulan Ul Hu, then north to the foothills of the Burhan Budai Shan and back to the road, Feng (1991b) observed 4843 chirus, including a herd of 2000 females and young in late July, and he estimated a total population of 6000–7000 in 75,000 km^2 of southwestern Qinghai, or 0.08–0.09/km^2. This density figure and others represent crude, not ecological, density, including habitat unsuitable for chirus such as lakes and high mountains. In November 1986 we counted chirus within 50 km along each side of the Lhasa-Golmud road between Wudaoliang and Tuotuohe, covering an area of about 17,900 km^2 (excluding the census block of 2100 km^2 mentioned earlier), and observed 281 chirus, a density of 0.015/km^2, most of them in the Golo Valley north of Tuotuohe (Schaller, Ren, and Qiu 1991). However, few chirus wintered in that area, in contrast to the previous year when after a heavy snow many animals moved there from the west (see below). One wintering ground of these animals is along the upper Tuotuohe Valley, where Ma (1991) saw many chirus in November 1986 and we tallied 550 in November 1993.

Xinjiang

The Tula Valley extends for about 300 km along the southern base of the Arjin Shan and one of its subranges, the Yüsüpalik Tag. We surveyed the western half in 1987 and the eastern half in 1988. A total of 305 chirus were tallied, most of them in one concentration of males. Few chirus penetrated the mountains north of the valley. In a week's work in the Yüsüpalik Tag during July 1988 we counted 37 animals, some of them females with young. A survey of about 4000 km^2 of the Kunlun Shan and the uplands to the south of the Tula Valley, an area just west of the Arjin Shan Reserve, gave a count of 438 chirus, a minimum density of 0.11/km^2, most of them in one concentration of females (Schaller, Ren, and Qiu 1991).

Almost all travelers who crossed what is now the Arjin Shan Reserve commented on the presence of chirus (Bonvalot 1892; Hedin 1903). The western 18,000 km² of the 45,000 km² reserve were censused by Achuff and Petocz (1988). They reported 2946 chirus, most of them in several concentrations, or 0.16/km², a figure similar to ours from west of the reserve. However, their count did not include the Ayak Kum Basin, where females had gathered to give birth, nor did it cover the eastern half of the reserve, where the density may be greater, judging by the sightings of Butler, Achuff, and Johnston (1986). Feng (1991a) gave a density of 7.2/km² for one area, no doubt a local concentration. A reliable estimate for the whole reserve is unavailable. Butler, Achuff, and Johnston (1986) quoted a figure of 70,000–100,000 (1.5–2.2/km²) based on Chinese sources, but this figure is much too high, and at any rate, heavy poaching during the early 1990s greatly reduced the population (Wong 1993). The chirus in the region appear to be resident, without much contact with those in the Chang Tang of Tibet and Qinghai, judging in part by the distinctive white color of the nasal air sac.

Near the western end of the chirus' range, the species was reported at the Keriya Pass (35°40′ N, 82°05′ E) on the Xinjiang-Tibet border (Deasy 1901). It also occurs in the Aksai Chin, a desolate plateau above 5000 m in elevation, where we counted 13 animals when crossing the area in August 1988.

Tibet

Chiru remnants persist in the central Chang Tang (fig. 3.1), most of them in small, scattered herds. For example, I encountered 74 animals in the basin of Buka Co, southeast of Siling Co, in November 1991, apparently a temporary aggregation for the rut. Yet chirus were once abundant in this region. West of Dogze Co, at about 32°00′ N, 86°30′ E, Hedin observed in December 1906 "great flocks of kyangs and Pantholops antelopes" ([1922] 1991). Today the nearest chiru population winters over 75 km to the north. Near Dong Co (32°10′ N, 84°45′ E), he found that "kyangs and antelopes were numerous" during March 1908; today chirus do not frequent that valley. Just north of Gerze, in an area without chirus now, Hedin encountered them "by the hundreds, and in larger flocks than ever" in February 1908. And traveling near holy Mount Kailas in September 1907, he noted that "antelopes were seen grazing and not shy at all, as nobody kills them there." Chirus are rare in that region today.

Much of the northern Chang Tang lies within the Chang Tang Reserve. The southern margin of the reserve, south of about 32°30′ N, is almost devoid of chirus, although occasional animals are encountered southeast of Shuanghu. The animals are similarly scarce along the western border

Table 3.2 Number of wild ungulates observed during various months in the 300 km² Kongkong census area, Chang Tang Reserve

	July 1991	Sept. 1990	Oct. 1993	Dec. 1991
Tibetan gazelle	7	35	17	2
Tibetan argali	7	18	—	—
Chiru	73	199	323	393
Kiang	27	49	67	10
Wild yak	1	66	—	—

of the reserve in an area where Hedin found tracks "very numerous" and where he saw "an antelope that, perhaps only a few minutes ago, had been caught in a trap" ([1922] 1991). The rest of the reserve is occupied by chirus, but since most animals are either migratory or shift ranges seasonally, accurate counts are difficult to make. For example, a count in the 1800 km^2 Aru Basin during early August 1990, in which an attempt was made to tally all animals, revealed 635 chirus, a density of $0.36/km^2$. Shortly after this count was made, a herd of about 2000 migrating females with young arrived. Four counts were conducted in a valley and surrounding hills just north of Kongkong Co near Garco between July and December. Chiru numbers in this 300 km^2 area fluctuated from 73 to 393 (table 3.2). Our most extensive censuses were conducted on alpine steppe roughly between Shuanghu and Yibug Caka in December 1991 and October 1993 by driving through valleys and basins and by scanning hills and plains in an effort to count most animals. This area was selected because chirus concentrated there in winter and it also had large populations of kiangs and gazelles (but not yaks). In 1991, a total of about 3900 chirus were counted in 17,500 km^2 (a minimum of $0.22/km^2$), and in 1993, a total of 3066 chirus were counted in 10,500 km^2 ($0.29/km^2$). These densities are probably not exceeded elsewhere over such a large area.

Migration

Two behavior patterns are basic to understanding the migration of chirus. One is that the sexes segregate almost completely during summer, a fact noted by Rawling when he wrote that males and females "live apart in different parts of the country" (1905). Sometime in late April or May most 10- to 11-month-old males separate from their mothers and either join their male peers or the adult males, which also part from the females at that time. The adult females and their female offspring migrate north in May and June to certain calving grounds. A few male young also accompany their mothers. In June 1992, 15% of the young, now almost yearlings, migrating past Toze Kangri were males and 85% were females. Young females no doubt learn the traditional travel routes to the calving grounds from their mothers. By contrast, most males travel only a relatively short distance from their wintering areas. Seasonal segregation of the sexes is also apparent in nonmigratory populations, as in the Arjin Shan Reserve (Achuff and Petocz 1988). Thus the movement of the sexes must be discussed separately.

The second pattern is that the chirus in the northern Chang Tang of Tibet and Qinghai are divided into more or less distinct populations, each with certain wintering and calving areas. To delineate these populations and trace their travel routes represented a major and only partially successful effort of the project. Evidence for four migratory populations was found, based largely on the movements of females. In addition, several small populations or subpopulations with local movements exist. It is unknown what proportion of the total chiru population is sedentary or migratory, and to what extent, if any, animals shift from one mode to another. All data remain fragmentary and are based on populations whose numbers and travels have in recent decades been disrupted by human activity.

Migration of females

The migratory routes of the four populations could only be partially traced, as shown in figure 3.2. The precise location of calving grounds remains obscure, whereas the major wintering areas are fairly well known.

West Chang Tang population. On 21 July 1896, Deasy was camped on a plain north of the glacial massif Toze Kangri and saw that "for many miles in every direction except west, from Camp 19, in fact as far as the human eye aided by powerful binoculars could see, there were thousands of antelopes in large herds scattered about irregularly wherever there was plenty of grass" (1901). He named the place Antelope Plain. Later, Rawling (1905) reported a similar sight (see chapter epigraph). I presumed that chirus gave birth there, and we visited the plain in June and July 1992. We approached from the west past Gozha Co, an area with only a few chirus, most of them males. Some females were seen at the western end of Antelope Plain by a small lake, Yue Ya Hu, but a search of the region between 7 and 15 June east beyond Bairob Co to about 83°25′ E revealed no chiru concentration. Nomads told us that females now seldom venture west of about 81°20′ E, west of Gozha Co and Lumajangdong Co. However, in June 1996 a Chinese expedition reported that many females were giving birth near Gozha Co (W. Liu, pers. comm.), suggesting a shift in migratory movement.

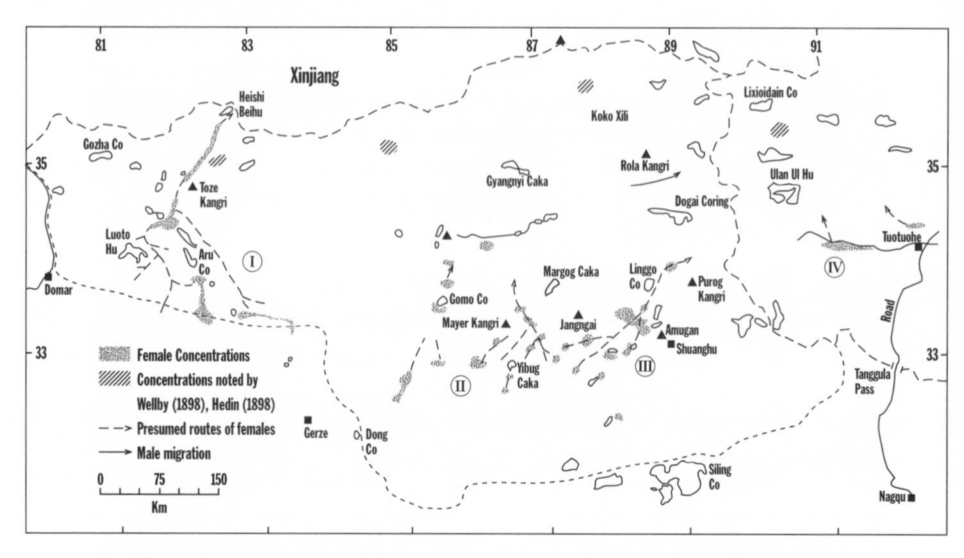

3.2. Observed concentrations of female chirus and the direction of travel by female chirus toward the calving grounds, showing the division into distinct populations: I, West Chang Tang; II, Central Chang Tang; III, East Chang Tang; and IV, Qinghai population. The one observed male migration is also indicated.

On 16 June the migratory route was found near the eastern margin of Yue Ya Hu. Moving along a narrow, specific route, herds headed northeast along a valley at the base of Toze Kangri, crossed a plain, continued northeast through the high foothills of an unnamed glacial massif, and traveled across yet another plain into hills bordering the basin of Heishi Beihu, Blackrock Northlake, so named for its volcanic rocks. At least 2200–2400 chirus passed Yue Ya Hu between 16 and 22 June. These animals were at the rear of the migration, which ceased abruptly on 23 June, when only 10 animals passed. (In addition, small female herds and a few mixed herds, totaling 396 animals, were tallied in the region.) We followed the chirus into the Heishi Beihu Basin, where tracks showed that the animals had continued northeast through the basin and apparently over low hills into Xinjiang (fig. 3.3). Almost daily snowstorms and shortage of gasoline prevented us from following. A few lagging females failed to reach the calving ground, giving birth en route. The first newborn was noted on 26 June, but parturition on the calving grounds no doubt began earlier. The concentrations observed by Deasy (1901) and Rawling (1905) were of animals returning south from the calving grounds.

In July and early August 1990 the Aru Basin contained many males and just a few females, all without young, but on 11 August a herd of about

3.3. Dawa views the Heishi Beihu Basin during our attempt to trace the migration of chirus. (June 1992)

2000 females and young arrived from the north and briefly remained around the northern edge of the basin (figs. 3.4 and 3.5). In 1992, after failing to reach the calving grounds, we returned to the basin to await this return migration. On 28 July, about 100 females and young ventured into the basin, but, as in 1990, the animals did not move south through it, avoiding this easy route. Observing from the basin's northern rim, we watched as the chirus moved southwest from Toze Kangri, passed through the hills just south of Luotuo Hu (Camel Lake), and, according to local pastoralists, crossed the Aru Range by a low pass (see fig. 1.5) into the Lumajangdong Basin and from there traveled south. During intermittent observations between 20 July and 3 August 7350–7750 females and young passed Luotuo Hu, and the total migration was no doubt over 10,000. We were also told that chirus migrate south along a route east of the Aru Basin.

On 21 August 1988, we observed many females with young from the southern rim of the Aru Basin south for about 70 km. Two years later, on 18–19 August, they were in the same area and also sporadically eastward past Kehu Co for 75 km (fig. 3.2). By contrast, when on 4 August 1992 we drove south of the Aru Basin, the female migration had not yet reached the area.

This overview indicates that the females migrate north in May and June,

3.4. Chiru females and young migrate southward in the hills north of Memar Co.

3.5. A 1-month-old chiru suckles when a migratory herd halts briefly.

calve somewhere just across the border in Xinjiang, and return south almost immediately, reaching the Aru Basin by late July or early August. Bypassing the basin, they continue south and also spread east and west over the alpine steppes, most reaching their autumn and winter range by the end of August or early September. The one-way travel distance is about 300 km.

A small percentage of the females either do not migrate or move only a short distance. Some of these females are barren, at least in the year that they were observed not to migrate, and some avoid the long trek and calve near their wintering grounds. A few females are possibly resident in the north. Hedin observed "flocks of antelopes" ([1922] 1991) at Bairab Co (35° N) on 6 October 1906.

CENTRAL CHANG TANG POPULATION. We were unable to delineate the eastern limit of the large West Chang Tang population, and figure 3.2 shows a gap between it and the Central Chang Tang population. In late August 1990, we drove north of Dong Co and then east along about 33° N latitude. About 50 km from the lake we noted a number of females and young and farther on there were at least 150 in a basin. Wildlife was scarce along our route eastward, but on a plain before reaching Yibug Caka were about 125 chirus. Females and young were also spread around Yibug Caka, east and northeast of the glaciated Mayer Kangri, and along the western end of the Jangngai Range. (A detailed survey in December 1991 showed a similar distribution.) In late September, over 200 chirus, most of them females and young, were seen around Gomo Co, and a few were to the north across a hill range as far as the slopes of Zangser Kangri. All these animals were presumed to belong to one population. Where is its calving ground?

We surveyed the central area in July 1991 and June 1994, checking the Tian Shui Valley and traveling as far north as Gyangnyi Caka (35° N, 87° E). On 11 July 1991, just after the birth season, we found a concentration of at least 800 females and young in the hills south of the Tian Shui Valley. Of these we sampled 756 (718 females, 38 young) and noted that the young-to-female ratio was an exceptionally low 5:100. There was no evidence of widespread infant mortality in that area. Perhaps many females had not been pregnant and failed to travel north or they had young elsewhere and lost them. The population probably contained more females than the ones in this concentration. The following winter the ratio of young to females was also low around Yibug Caka, a mere 12:100 (n = 1381), suggesting that many of the Tian Shui females were there. By contrast, the adjoining East Chang Tang population had a ratio of 31:100 (n = 2146), as shown in table 3.4.

Probably calving grounds exist or existed north of the area we surveyed. Traveling northwest of the Tian Shui Valley at about 35°25′ N, 85°00′ E, Wellby (1898) encountered many chirus of unspecified sex in mid-July 1896. He also commented on the presence of kiang and chiru just north of Gyangnyi Caka. These may have been mostly males, as we later observed just south and west of that lake.

East Chang Tang population. The main wintering ground of the East Chang Tang population extends in a line between the northern end of the Amu Range and the southern slopes of the Jangngai Range, a distance of about 110 km. Chirus once were common around Shuanghu, according to old residents, but few now venture east of the Amu Range. The Central and East Chang Tang populations come into contact along the Jangngai Range. When moving north to its summer grounds, the central population apparently moves around the western end of the range and the eastern population crosses over or angles around the eastern end. Most of these chirus winter in a strip of alpine steppe no more than 50 km wide.

We made an unsuccessful attempt to find the calving grounds in late May and June 1994. On 30 May, 300–350 females and their offspring were moving north across the hills toward the basin of Linggo Co, seemingly the last of the migrants, and the following day farther north we observed 650–700 animals on a pass just west of Purog Kangri. All animals were traveling northeast. A month's search over a vast area north past Rola Kangri and Dogaicoring Qangco as far as the Koko Xili Range and west past Rola Kangri to Gyangnyi Caka revealed only males. I hypothesize that the females moved east along the northern slopes of Purog Kangri and then north, possibly within Qinghai, to calve somewhere near the Tibet-Qinghai border. Hedin reported an "abundance of game, especially kulans and antelopes" (1903) west of Rola Kangri at about 35°15′ N, 87°50′ E, but we noted only males in that region. Farther north, at about 36° N just south of the Muztag massif, he noted "abundant droppings of khulans (wild asses) and antelopes."

The chirus between Shuanghu and Yibug Caka represented the main winter concentration of this population. Small concentrations also occur farther north in uninhabited terrain where we did no winter work. According to a truck driver who accompanied Shuanghu officials on hunts, both male and female chirus were found at the following localities between October and December: in the hills just south of Linggo Co, near the southeastern corner of Dogai Coring, at the base of the hills southeast of Rola Kangri, and, sparsely, near Dogaicoring Qangco. Females in these areas need to travel only a relatively short distance to the calving grounds.

As in every population, a few females do not travel north, and these are, with rare exception, not pregnant. But on 23 June 1991, one heavily pregnant female, accompanied by a 1-year-old male, was observed 50 km southeast of Shuanghu.

Qinghai population. The main wintering grounds of this population appear to be in the southwestern corner of Qinghai, where we encountered concentrations in the upper Tuotuohe and Golo Valleys. A calving ground may lie about 200 km to the northwest of these areas near the Tibetan border where Wellby (1898) saw many chirus in late July 1896 near Lixioidain Co and Feng (1991b) observed about 2500. However, it remains unclear whether this concentration is part of the East Chang Tang or the Qinghai population. (The Epilogue provides further information.)

Some chirus fail to move north in summer. In July, Hedin noted that in the southwestern corner of Qinghai "game was extraordinarily plentiful thereabouts. We saw fully half-a-dozen herds of orongo antelopes *(Pantholops Hodgsoni)*, each consisting of a score of individuals" (1903).

Migration of males

Adult and most yearling males associate little with females during summer. For example, in July 1991 and June 1994 we traveled widely within an area of about 40,000 km^2 extending from the Jangngai Range and Purog Kangri north 200 km to Gyangnyi Caka and beyond Dogaicoring Qangco. Excluding one concentration of females near the Tian Shui Valley, we tallied 1962 males, 12 females, and 1 young. In late June and July 1991, we checked the winter range of chirus between Shuanghu and Yibug Caka and tallied 146 males and 11 females which had not migrated north. A count in the Aru Basin during July 1992 gave a tally of 1050–1100 males and 52 females.

The males were widely dispersed singly and in small herds even in areas through which females did not migrate. However, certain localities had concentrations, as in the Aru Basin. Such localities were noted in the western and eastern parts of the reserve, not the central part.

In June 1994 we observed many tracks of males heading east past Yupan Hu, a small lake northwest of Dogai Coring. A week later, on 12 June, many males were encountered about 60 km east of Yupan Hu among high hills at the eastern end of Rola Kangri. A total of 507 males was counted in 17 km and the tally for the day was 718, most animals traveling slowly east and northeast. These males were about 250 km north of the main winter area of the East Chang Tang population. They appeared to have a distinct migration route different from that of the females (fig. 3.2). After several days this concentration had dispersed, some males possibly to Qinghai and others locally. For instance on 16 June, 101 males in several herds were in a shallow bay of Dogaicoring Qangco. No comparable mass movement of males was observed elsewhere.

Males exhibited at least three types of movement patterns. Some remained on their wintering grounds all summer. For example, the census block near Garco (table 3.2) had 73 chirus, all males, in July, the same area which in December would be a principal rutting site. Many males traveled at least a short distance to a summer range. Those in the Aru Basin probably belonged in that category. Nomads told us that chirus congregate for the rut in the valleys west of the Aru Basin that few males spend the summer in those valleys, and that only some remain in the basin during rutting season. Males also moved far from their winter areas, usually northward, dispersing widely, and then in autumn probably returned to places where chirus traditionally rut. Consequently males, in contrast to females, tended to be scattered throughout the range of the species.

Adaptive value of migration

Studies of such migratory ungulates of the African savannahs as white-eared kob (Fryxell 1987) and wildebeest (McNaughton 1985; Fryxell 1995; Murray 1995) have shown that movements are linked to seasonal changes in forage quality. Migrants are thought to have greater access to nutritious food than residents and to use this food more efficiently. Saiga antelope populations in Kazakhstan migrate erratically, their movements determined by a need for nutritious forage, especially in drought years, and by avoidance of areas with deep snow that may cause mass starvation (Heptner, Nasimovic, and Bannikov 1966). On the eastern steppes of Mongolia at least one large population of Mongolian gazelles travels south in October and November and remains until early March along both sides of the Mongolia-China border before traveling north again. Since heavy snows on the steppes are most likely between December and March, the migration may in part have evolved to escape them. Springbok in southern Africa once made mass movements, apparently a means of dispersing when populations are high (Child and Le Riche 1969). Migration also benefits ungulates in that they avoid the full impact of predation because a predator may not be able to follow the herds, being tied to a certain locality by small young or territorial imperatives (Fryxell, Greever, and Sinclair 1988). Why do chirus migrate? The fact that males and females have different movement patterns suggests that the behavior differs in its adaptive value for each sex. Four facts might have an influence on movements.

Food quality. Most chirus of both sexes leave their winter range just as the first spring growth appears and move north into vegetation which is still dormant or has barely begun to green. The females travel into such conditions in their last month of pregnancy, a time of nutritional stress. This tendency is especially marked in the West Chang Tang population, in whose summer range green forage is almost unavailable until after the animals give birth. Plants on the calving grounds are possibly high in phosphorus levels, an element needed for milk yields and proper development of young (Murray 1995). Certain plants may also be highly nutritious. *Ceratoides*, a common dwarf shrub, may reach protein levels of 30% in the Pamirs (Walter and Box 1983b). However, the females turn southward with their newborns soon after parturition, just at a time when new plant growth at the calving site is at its most vigorous, to end their three-month travel circuit back on the alpine steppes after the vegetation there has passed its peak nutritional state. Such a stressful sojourn during late pregnancy and lactation must have or once have had a powerful adaptive advantage, but I was unable to relate it to forage quality.

Male chirus, by contrast, migrate or move more leisurely than the females, taking advantage of the spring growth and at times concentrating at particularly favorable sites. As the summer progresses, they tend to move higher into the hills where the variety of plants is greater and growth stage younger and hence more nutritious. Few animals remain on low-lying steppe at that season.

Insect harassment. Mosquitoes, blackflies (Simuliidae), horseflies (Tabanidae), and two species of parasitic dipterous flies (Oestridae) may affect movements of caribou and survival of their young (Helle and Tarvainen 1984; Walsh et al. 1992). Only a few horseflies and the oestrid flies occur in the Chang Tang. One of these oestrid flies, the warble fly, parasitizes virtually all chirus (see below). Since the larvae often infest the animals in large numbers, they could affect health. However, the flies are active from June into September in both alpine and desert steppe, and chirus cannot escape by migrating.

Predation. Synchronous births among large ungulate aggregations reduce predation rates on newborns by swamping predators with a surfeit of vulnerable prey (Estes 1976). Animals reduce the predation risk even more by migrating. Chirus move into desolate terrain where prey densities and therefore predator densities are low for much of the year. Wolves, the chiru's principal predator, tend to be sedentary, with young in dens, at the time of the northward migration. Female chirus no doubt dilute the impact of predation by moving north en masse. But measured against this potential benefit is the stress of a long migration and probably an increased risk of death to newborns from inclement weather at high elevations.

AVOIDANCE OF DEEP SNOW. Occasional heavy snows may make it so energetically costly for chirus to obtain forage that many die of malnutrition (see below). At such times the animals make aberrant movements in search of forage quite unlike their regular annual migrations. The northward migrations in spring enter a region of lower precipitation but also colder weather, where a snowstorm on the calving grounds is more likely than a rainstorm. The migrations appear unrelated to snow cover, at least under present conditions.

VESTIGIAL BEHAVIOR. The climate of the Chang Tang fluctuated during the Pleistocene and Holocene, cold and dry periods alternating with relatively warm and damp periods. These changes were marked by the advance and retreat of glaciers, lowering and raising of lake levels, and shifts in vegetation zones (see chapter 2). Patterns of chiru migration were possibly established when it was nutritionally adaptive for females to move north, and the behavior was then retained, there being no strong selection against it. The current migratory routes sometimes suggest tradition rather than logistic pragmatism. Chirus bypass the easy route through the Aru Basin perhaps because glaciers were once so low and lake levels so high that the way was impractical. In the eastern Chang Tang, chirus possibly detour east high along the foothills of the Purog Kangri because Dogai Coring, 70 km long today, was an even larger lake in the past and barred direct access to the north.

A variety of factors may be additive (Fryxell, Greever, and Sinclair 1988), each contributing a modest advantage to the productivity of chirus, with the result that migrations persist.

Total number of chirus

Because chirus are sexually segregated by season, migratory, and widespread at varying densities, numbers can be estimated with accuracy only on the basis of an aerial census throughout the reserve at a certain time of year. It was not possible to conduct such a census. We counted animals of the East Chang Tang population on their southern winter grounds but do not know how many remained farther north; we observed the West Chang Tang population on migration but have little idea what percentage of the population passed us. Hunting chirus for their wool (see chapter 15) has been so heavy that counts made in 1990 no doubt fail to reflect numbers in 1995. The fragmentary data on numbers from inside and outside the reserve were presented earlier. My estimates are based on these data and on general impressions obtained during my travels through the chirus' range; they represent little more than guesses and should be viewed as such.

In 1992, we observed about 7500 females and young migrate past the Aru Basin, and there may have been double that number. If the dispersed males and others are added, the whole West Chang Tang population contained perhaps 25,000 animals. A part of the Central Chang Tang population winters in the Yibug Caka region, where, in 1991, we tallied over 1775 chirus; an unknown number was to the west and north. The whole population probably numbered at least 5000–7500. Many animals of the East Chang Tang population were in the Tsasang and Garco areas during winter. A census there in 1991 produced 2125 animals. These chirus and the ones outside the survey area to the north probably indicated a population of 5000 or more animals. Thus, about 35,000–40,000 chirus may have been present in the reserve during the early 1990s. In addition, a few thousand animals occurred scattered in small herds outside the reserve, raising the total for Tibet to roughly around 45,000, an estimate that is similar to the 43,000–58,000 proposed by Liu and Yin (1993).

In Qinghai, 6000–7000 chirus were estimated to be present in the southwest (Feng 1991b), and there were also several thousand animals in and around the Yeniugou and a few east of the Lhasa-Golmud highway. Possibly 10,000–12,500 chiru survived in the province. Except for a few hundred chirus in the Aksai Chin area, most animals in Xinjiang were in or near the Arjin Shan Reserve. Perhaps no more than 10,000–15,000 remained in Xinjiang.

If these figures are of the correct magnitude, then the total for the whole plateau would be 65,000–72,500. Whatever the precise number, fewer than 75,000 chirus may now remain in existence.

Population and herd dynamics

Population composition is difficult to ascertain except when animals have congregated for the winter rut. Not only are the sexes segregated and widely dispersed for much of the year but each sex also may form certain associations. Females with young and without young may each congregate, and adult males occasionally form large herds, something that is not seen among yearling males. Chirus give birth in the second half of June and early July to a single young, and these young are here termed yearlings after they reach about 1 year of age. As with many medium-sized ungulates, females probably first conceive at the age of 1.5 or 2.5 years and give birth at the age of 2 or 3 years. (Two young females, 11 months old, which were

Table 3.3 Composition of two migratory chiru populations in late August and September 1990

	Central Chang Tang	East Chang Tang
Sample size	1250	621
Adult male	278 (22.2%)	115 (18.5%)
Yearling male	78 (6.2)	29 (4.7)
Adult female	580 (46.4)	293 (47.2)
Yearling female	76 (6.1)	30 (4.8)
Young	238 (19.0)	154 (24.8)
Males: 100 females	54.3	44.6
Yearlings: 100 females	23.5	18.3
Young: 100 females	36.3	47.7

killed by poachers, were not pregnant.) Subadults grow rapidly, so much so that yearling females are often difficult to distinguish from adults after the age of 15 months, and I then place the two into one category. Each migratory population is analyzed separately because differences in habitat and vagaries of weather may have an influence on composition, especially on the percentage of young. The following age and sex classes are here recognized: young, yearling male and female, and adult male and female.

Population composition

The ratio of males to females (yearling and adult combined) varied considerably even when a population was on its winter range, apparently because males continued to arrive at these areas throughout the early winter. For example, the ratio in the East Chang Tang population increased from 45: 100 in September to 75:100 in December (tables 3.3 and 3.4). Most males are probably near the females during the December rut. But during that month in 1991 the ratio in the Central Chang Tang population was 52: 100 and in the East Chang Tang population 75 : 100 (table 3.4), a difference for which I have no satisfactory explanation. Between 1990 and 1993, the combined Central and East Chang Tang populations averaged 29% males (of which 6% were yearlings), 53% females, and 18% young. Migratory populations in some other ungulate species also tend to have a skewed sex ratio favoring females (Kelsall 1968; Fryxell 1987). The causes contributing to a higher mortality of adult chiru males remain unclear. Yearling males and females, 14–15 months old, were present in about equal number in the Central Chang Tang and in the East Chang Tang populations in 1990 (table 3.3).

The ratio of young to females varied considerably from year to year,

Table 3.4 Composition of three migratory Chang Tang chiru populations during winter

	Central		East		Qinghai
	10/1993	12/1991	10/1993	12/1991	11/1993
Sample size	770	1381	2126	2146	471
Adult male	105 (13.6%)	352 (25.5%)	444 (20.9%)	637 (29.7%)	95 (20.2%)
Yearling male	46 (6.0)	88 (6.4)	124 (5.8)	147 (6.8)	48 (10.2)
Female	428 (55.6)	839 (60.7)	1073 (50.5)	1041 (48.5)	225 (47.8)
Young	191 (24.8)	102 (7.4)	485 (22.8)	321 (15.0)	103 (21.7)
Males: 100 females	35.3	52.4	52.9	75.3	63.6
Young: 100 females	44.6	12.2	45.2	30.8	45.8

probably determined by snowstorms or other inclement weather at calving time, as has been documented for caribou (Kelsall 1968). In 1993, a benign year, the ratios in three populations were similar, about 45:100, during winter when young were at most 6 months old (table 3.4). In 1990 only a modest difference in ratios existed between the Central and East Chang Tang populations (36:100 and 48:100), whereas in 1991 the difference was considerable (12:100 and 31:100). The reason for the virtual reproductive failure in the Central Chang Tang population in 1991 is unknown, but a paucity of young was already noted in July, suggesting some disaster on the calving grounds. In August 1990, the ratio of young to females in the West Chang Tang population around the Aru Basin was 49:100 (n = 681), whereas two years later, after heavy snows during the birth season, it was somewhat lower, at 40:100 (n = 2812). Thus, under normal circumstances the ratio of young to females ranged from about 30:100 to 50:100. These low figures indicate that up to about half the young die within a month or two after birth, a situation similar to other migratory species, among them wildebeest (Sinclair 1979), caribou (Kelsall 1968), white-eared kob (Fryxell 1987), and Mongolian gazelles (this study.)

Mortality in young during their first winter continues to be high. In October 1993, the ratio of young to females was 45:100 in the East Chang Tang population (table 3.4). In late May the following year the ratio of female young to females was 15:100. Assuming that young males, most of which had by then left their mothers, were equally abundant, the ratio would be 30:100. About a third of the young had died that year in the seven months between early winter and spring.

Yearling males averaged 6% in the populations (tables 3.3 and 3.4), and if the percentage of females was similar, then annual recruitment was about 12%. Given an average of 47% adult females in the populations, this year-

Table 3.5 Tooth eruption in the lower jaw of chirus

	Incisor			Premolar	Molar		
Age (mo.)	1	2	3	4	1	2	3
4	D	D	D	D	(P)		
11–12	D–(P)	D	D	D	P	P	(P)
16–18	(P)–P	D	D	D–(P)	P	P	(P)
23–26	P	(P)–P	D–P	(P)–P	P	P	(P)–P
28	P	P	P	P	P	P	(P)–P
36	P	P	P	P	P	P	P

Note: D = deciduous tooth; P = permanent tooth; ( ) = tooth erupting.

ling figure indicates that at least two-thirds of the chirus died between birth and the age of 2 years.

Mortality

Chirus were killed by predators and hunters and they died of disease, starvation, and inclement weather. In most instances only a set of old horns or some stripped bones attested to a chiru's demise, making it impossible to deduce cause of death. Predator kills and disease deaths were seldom noted, but on one occasion, after a heavy snowstorm, starved animals provided a large sample of bodies. Trapped and shot chirus were also examined in three hunting camps.

When cause of death was known, the lower jaw was usually collected for later aging by tooth eruption and wear and by counting the annuli on the first incisor, a technique well developed for ungulates of temperate climates (Schemnitz 1980). Chirus begin to replace the deciduous with the permanent first incisor at the age of about 12 months, an age when the third molar also appears. The animal has its full permanent dentition at about 28–30 months of age (table 3.5). When older than 2.5 years, chirus could only be assigned a relative age, based on the amount of wear on the infundibulum in the center of each molar cusp. These infundibula wear off in sequence, starting with the anterior one of the first molar, and it is assumed that wear is correlated with relative age (see table 3.7). In addition, the annuli of 41 sectioned incisors were counted by Matson's Laboratory (Milltown, Montana). The annuli were in most cases so irregular or indistinct that the results could not be used. Only 16% of the teeth were aged with what was considered a good reliability. In many instances, tooth wear and annuli counts showed little relation. The oldest animal, based on annuli, died at about 8 years of age.

Table 3.6 Parasites in feces of Chang Tang ungulates (number of occurrences)

		Protozoans		Cestodes and nematodes			
	Sample size	Ento-amoeba	Coccidia*	Anopo-cephala†	*Entero-bius*	Strongyle	Tricho-strongyle
Chiru	4	1	4	—	3	—	—
Argali	1	1	1	—	—	1	—
Blue sheep	1	1	—	—	—	—	—
Gazelle	1	1	1	—	—	—	—
Wild yak	2	1	—	—	—	—	—
Kiang	2	—	—	2	—	2	2
Domestic sheep/goat	5	—	5	—	—	1	—
Domestic yak	1	—	—	1	—	—	—

Note: Analyzed by Judy Kramer, Wildlife Conservation Society.
*Possibly *Eimeria.*
†Possibly *Moniezia.*

DISEASE AND PARASITES. According to pastoralists, a disease of unknown etiology may cause widespread death among chirus. A local veterinarian said that a "bleeding disease" (possibly hemorrhagic septicemia) kills them, one that also affects domestic yaks. "A disease has destroyed nearly every antelope," wrote Rockhill (1894).

On 3 August 1992, we found two females near each other at the northern rim of the Aru Basin, one dead and the other struggling on its side unable to rise. The animals were returning from the calving grounds, and both were lactating. The fat of the bone marrow of one was partially depleted and of the other, a two-year-old, wholly depleted. The lungs of both females had large yellow splotches, which in a laboratory were diagnosed as vascular congestion with acute bacterial infection causing severe pneumonia.

A few fecal samples of chirus and other ungulates were collected in July 1991 for a preliminary check on intestinal parasites (table 3.6). Definite identification of cysts and ova is at times difficult, but the results are indicative. Chirus harbored two kinds of protozoans and a nematode *(Enterobius)*, sharing the former with several other ungulate species, wild and domestic. The parasite burden was light to moderate. However, these parasites can become pathologically significant if the infestation is heavy or the animal's nutrition is marginal.

Chirus are infested with large grubs of a warble fly (family Oestridae) which "live below the skin of the rump, thriving and fattening on the flesh underneath" (Rawling 1905). The adult warble flies (probably *Hypoderma*) are large, black, and hairy, and they have a reduced proboscis; they mainly parasitize chirus, seldom affecting livestock or other wild ungulates, according to local people. In caribou (Kelsall 1968), as probably in chirus, the fly lays eggs on the hairs at the back of the legs. The larvae bury into the skin and migrate to the lower back, where they cut breathing holes. They develop there, encased in a fibrous sac, until the following spring, when they escape through the breathing holes and pupate on the ground. According to our informants, larvae leave the chirus in June when the grass greens. Several chiru hides had over 50 perforations or scars each. Three chirus, examined between June and early August, had one or more larvae—blackish, corrugated creatures 2.5 cm long and up to 1.4 cm wide. Heavy warble fly infestations weaken caribou and in extreme cases cause death (Helle 1980).

Another oestrid fly, the nostril fly, also affects ungulates on the Tibetan Plateau. The fly deposits larvae into nostrils, and these then attach themselves in the frontal sinuses and around the throat entrance. "A species of gad-fly called *ila* tormented the horses, and made them nervous and restless by getting into their nostrils. The kulans [kiang] protect themselves against these insects by keeping their nostrils close to the ground when they graze, the orongo antelopes by spending the hot part of the day deep in the belt of sand" (Hedin 1903).

The animals modify their behavior to such an extent when an oestrid fly appears to be near them that the evolutionary impact of the parasitism must be considerable. An animal may stand with muzzle close to the ground, twitching its skin or kicking a leg, and it may run with head low and buck. Sometimes 50 or more chirus congregate and crowd on a mudflat or stand knee-deep in the shallows of a lake, remaining there for hours until in late afternoon they disperse to forage. They also cluster on patches of overflow ice *(Aufeis)* or high on snow-covered ridges. An unusually large number of males and females were solitary in September (see below). Such behavior seemed designed to avoid oestrid flies, but at the same time it affected an animal's daily activity cycle. Similar responses to insect harassment were noted in caribou (Walsh et al. 1992).

Chirus have long been known to paw bowl-shaped hollows in sandy and silty soil, roughly circular hollows 110–120 cm in diameter and 15–30 cm deep. These hollows are distinctive, quite unlike the shallow scrapes made by blue sheep and argalis on a hillside for resting, the trampled and churned depressions created by wallowing yaks, the fan-shaped pawed sites at gazelle latrines, or the irregular holes and trenches with their litter of feces dug by kiang. In addition to being regular in shape, chiru hollows lack feces. Chirus of all ages and both sexes paw hollows, and the animals lie in them as if to conceal themselves. A plain sometimes seemed empty until at our approach various animals suddenly arose from their hollows. These hollows are mostly in use during summer, especially on calm, warm days. They are on rare occasions also excavated in winter, but I did not see them in use at that season. Sometimes two hollows are within 2–3 m of each other or a young constructs one next to its mother's, but typically hollows are dispersed.

Rawling noted that "when chased or alarmed, bucks and does will often throw themselves into these holes, where their bodies remain hidden" (1905). I did not observe such behavior. From their hollows, chirus "can detect any threatening danger at a great distance" (Kinloch 1892), a reasonable assumption. After noting three fresh hollows in November 1986, we speculated that they may "have some function during the rut, animals leaving scent from the inguinal glands in them" (Schaller, Ren, and Qiu 1991). Further work did not verify this. The seasonal use of hollows, their dispersion, and the lack of feces appear related and suggest that the hollows function primarily to conceal chirus from oestrid flies, which probably find hosts by sight and smell.

Predation and hunting. Wolves, snow leopards, lynx, and probably brown bears prey on chirus (see chapter 11), but few fresh kills were found. Almost all age classes were represented among the seven wolf kills. Of note is that among the 137 chirus dead from various causes (table 3.7) only 5 were old, with infundibula worn off the first and second molars, and 2 of these were wolf kills. The kills also included a heavily pregnant female and a partially blind yearling, suggesting that wolves tended to select vulnerable prey.

In October 1985, after a heavy snowstorm, many chirus ventured near the Lhasa-Golmud road in Qinghai, where the dogs of pastoralists and road workers often killed them. Deep snow impeded escape, and malnutrition reduced the flight distance of animals to potential danger, so much so that chirus permitted my approach on foot to within 50 m or less. Fourteen of 19 dog kills were young or yearlings (table 3.7). When a dog approached one herd, the chirus bunched and fled, but after 200 m a young lagged and the dog grabbed and shook it. On another occasion, a lone female faced a dog at 6 m. The dog circled, as if trying to attack the chiru from the rear, but she turned too, always facing it, until after five minutes the dog departed. The bone marrow in the femur of 15 kills was checked and the fat in all was depleted, the marrow consisting of a reddish, gelatinous mass.

I examined 27 chirus which had been killed by pastoralists, 17 of them males. Seven (26%) of the animals were yearlings, and the remainder were

Table 3.7 Chiru mortality from various causes by age and sex

	Predator kills		Dog kills*		Poacher kills		Disease		Malnutrition†	
Age (mo.) or toothwear	♂	♀	♂	♀	♂	♀	♂	♀	♂	♀
4	—	—	7‡		—	—	—	—	81‡	
11–12	—	—	—	—	—	2	—	—	—	—
14–18	—	1	5	2	2	1	—	—	20	11
23–25	—	—	—	—	1	1	—	1	—	—
28	—	—	—	—	—	—	—	—	—	13
Permanent dentition; little or no wear	—	1	—	—	3	—	1	1	—	28
Infundibulum worn half off M1	—	1	—	1	3	1	—	1	—	10
Both infundibula off M1	1	1§	—	2	4	—	—	1	—	10
Infundibula off M1 and half off M2	—	1	—	—	—	1	—	—	—	13
Both infundibula off M1 and M2	—	1	—	—	—	—	—	—	—	3
One or more infundibula off M3	—	1	—	—	—	—	—	—	—	—
Adult molar wear not checked	—	1	—	2	4	4	1	—	3	1

*Oct. 1985.
†Oct.–Nov. 1985. See table 3.8.
‡The young were not sexed.
§Lynx kill; all others are wolf kills.

adults in their prime. Illegal hunting of chirus is discussed in chapter 15.

INJURIES. One female with a broken foreleg was seen, the only major injury noted. During the rut, males may fight until "blood flows freely" (Rawling 1905). "The combat is fierce, and the long sharp horns inflict terrible wounds, often causing the death of both antagonists" (Prejevalsky 1876). However, we did not observe such vigorous fighting.

DEEP SNOW. On 17 October 1985, a blizzard deposited about 30 cm of snow over southwestern Qinghai, the heaviest such snowfall in at least 30 years. We visited the area from 22 October to 10 November 1985 and again from 18 to 29 November 1986 to assess the impact of that snowfall on the wildlife. Most work was done along about 250 km of the Lhasa-Golmud road from near Kunlun Pass south past Tuotuohe. The northern part of this area near Wudaoliang was less affected by the snowfall than

3.6. An adult male chiru in winter pelage paws a crater in the snow to expose forage. (October 1985)

the vicinity of Tuotuohe. Snow confined us near the road in 1985, except for a 9-day cross-country survey to the east on a tractor, but in 1986 off-road travel presented few problems.

Sun and wind usually clear snow away from exposed sites within days. However, unseasonal calm and cold followed this blizzard; temperatures at night dipped below −40°C. Unable to obtain forage except by digging craters in the snow with sweeps of a foreleg, the animals expended much energy to obtain little nourishment (figs. 3.6 and 3.7). Many chirus left their traditional winter range and moved east to the vicinity of Tuotuohe. Travel in knee-deep snow further depleted their energy reserves. Chirus are in some respects not well adapted to cope with snow. The live weight of an animal divided by its total track area provides a measure of relative adaptation to life on top of snow. Asiatic ibex have an average supporting area of 848 g/cm^2; mouflon sheep, 662 g/cm^2; and a chamois, 200 g/cm^2 (see Schaller 1977b). The lower the figure, the easier travel on snow. Chirus, which seldom have to contend with deep snow, have a supporting area of 411 g/cm^2, based on hoof measurements of two females. Caribou, which often encounter snow and are well adapted to it, have a supporting area of 140–180 g/cm^2 (Kelsall 1969). Given the unusual snow conditions, many chirus were physically weak by late October and began to die.

We examined 12 probable malnutrition deaths around Tuotuohe in 1985: 1 adult male, 1 yearling male, 4 females, and 6 young. The marrow of 7 of these animals was checked and in every instance it was devoid of fat. This small sample suggested that the blizzard particularly affected the young. In 1986, we discovered the skeletons and mummified bodies of 193 chirus within about 15 km in the upper Tuotuohe Valley. This sample, as the one in 1985, contained fewer adult males and more young than would be expected from the population composition (table 3.8). From these remains we collected a sample of 88 female jaws: 11 yearlings, 76 adults, and 1 unaged animal (table 3.7). All age classes were represented, but only 3 of the animals were old, suggesting that few chirus reach an advanced age.

Since a large animal has a lower metabolic rate and lower food requirements per unit weight than a small animal (Moen 1973), an adult male chiru would require proportionately less energy than a young chiru to travel in snow and dig for food. Furthermore, adult males probably had ample fat deposits just before the rut (one male had much kidney and mesenteric fat on 22 November 1986), whereas females and young had rela-

3.7. A 16-month-old male chiru forages on a patch of open ground. (October 1985)

Table 3.8 Actual and expected number of malnutrition deaths among 193 chirus examined at Tuotuohe, Qinghai

Age and sex class	Actual number	Expected number
Adult male	3	32
Yearling male	20	8
Female	89	102
Young	81	51

Source: After Schaller and Ren 1988.
Note: The expected number is based on a sample of 1726 chirus classified in October 1985. Chi-square test showed significant difference from the expected (χ^2 = 300.54, d.f. = 3, $P < 0.001$).

tively low reserves, the former after migration, pregnancy, and lactation, and the latter because of rapid body growth. Had the blizzard occurred after the rut, after this period of great energy drain, many chiru males might also have died. For example, a severe winter killed disproportionately many pronghorn males after the rigors of the rut had depleted their fat reserves (Martinka 1967). Dead chirus had grass in their rumen, indicating that malnutrition, rather than lack of food, caused death. Much of this grass consisted of coarse stems, a plant part normally not eaten by chirus in winter because of its negligible nutritional value. (Other ungulates at Tuotuohe were also affected by the blizzard. Many gazelles and a few kiangs died, as did nearly two-thirds of the 86,000 sheep and goats in the area and half of the 8200 domestic yaks.)

At least 5 of 9 females examined in 1985 were lactating at death, indicating that starving young were able to supplement their diet with milk. Analysis of milk from a wolf-killed female in October 1993 showed 52% fat (on a dry-matter basis) and 30% proteins, a composition typical of, for example, pronghorns and deer (E. Dierenfeld, pers. comm.).

The composition of the chiru populations around Tuotuohe and Wudaoliang differed in 1986 and 1985 (table 3.9). The proportion of adult males was higher in both populations in 1986. Small sample size may be partly responsible for this difference, but low male mortality in 1985 probably contributed to it. The blizzard had, however, a significant impact on number of young. The ratio of young to females was 50–58:100 in 1985, whereas a year later no young were seen around Tuotuohe and only a few (17:100) near Wudaoliang. Because females entering the 1985 rut were in poor condition, many may have failed to come into estrus or to conceive, and fetal survival rates may have been low. Whatever the reason, the bliz-

Table 3.9 Composition of two Qinghai chiru populations in the winters of 1985 and 1986, showing impact of the October 1985 blizzard

	Tuotuohe		Wudaoliang	
	10/1985	11/1986	10/1985	11/1986
Sample size	1528	274	198	968
Adult male	253 (16.6%)	146 (53.3%)	37 (18.7%)	295 (30.5%)
Yearling male	65 (4.3)	3 (1.1)	17 (8.6)	105 (10.8)
Female	809 (52.9)	125 (45.6)	91 (46.0)	485 (50.1)
Young	401 (26.2)	—	53 (26.7)	83 (8.6)
Males:100 females	39.3	119.2	59.3	82.5
Young:100 females	49.6	—	58.2	17.1

Note: The Tuotuohe population is migratory; the Wudaoliang population probably is resident.

zard not only killed many young in 1985 but also caused virtual reproductive failure in 1986.

One catastrophe decimated chirus, altered the age and sex composition of populations, and reduced reproductive success for two years. Such events, though rare, may have a long-lasting impact on the dynamics of populations. Przewalski reported a severe blizzard in this same region in 1870 (Prejevalsky 1876). Hedin (1903) found an area littered with skeletons, probably the consequence of a heavy snowfall, as did Bonvalot when he came to "a valley strewn with the bones of animals such as arkars [argalis], koulans, yaks, orongos" (1892).

East African migratory ungulates such as wildebeest and white-eared kob are regulated by food abundance, specifically by the amount of nutritious grass in the dry season, with predation affecting primarily the nonmigratory species (Sinclair 1975; Fryxell 1987). Chiru populations are most drastically affected, though only partially regulated by, inclement weather. Spring snows probably cause heavy mortality of newborns, and occasional winter blizzards deprive animals of access to food. Wolves may once have had a considerable impact on chirus, but predators have been decimated in recent years. During this century, hunting has reduced and continues to affect chiru numbers to such an extent that the long-term impact of natural regulatory mechanisms is now of secondary importance.

Herd size and composition

Przewalski noted that chirus are "found in small herds from five to twenty, or forty head, rarely collecting in large troops of several hundred" (Prejevalsky 1876). While this is true, herd dynamics greatly varies by sex and season as herds change in size and composition and animals shift associations. As with many ungulate species, the only long-lasting association is between a female and her young, a bond which may persist well into the offspring's yearling stage. All other associations appear to be unstable, lasting from minutes or hours to perhaps days. Three types of herds were evident: males (yearling and adult) were often either alone or in all-male herds; female herds consisted of only females (yearling and adult) with their young; and mixed herds contained one or more animals of both sexes. I attempted to count each herd and to classify every member, but this was not always possible with shy, large herds.

Male herds. Except near and during the rut, adult males forage and travel separately from the females for much of the year or associate with them only briefly and casually. Most male young leave their mothers when 10–11 months old and join others of their age or associate with adult males until early winter, when, like the adult males, many join mixed herds.

The size of male herds varied considerably in different parts of the Chang Tang Reserve during summer (table 3.10). Males concentrated in the Aru Basin, with over 1000 there in July and August 1992. Sixteen percent of the males were solitary, and of these 77% were adult and 23% yearling. Over half of the males were in herds of 2–20, the largest herd numbered 82 animals, and the mean was 6.6 (Schaller and Gu 1994). However, large herds in which it was not possible to classify every individual or which contained a female or two were not included in these computations. For example, 149 chirus stood clumped on a patch of *Aufeis* and of these 134 were adult males, 13 were yearling males, and 2 were females. Other large herds which may have included several females numbered 103, 140, 181, and 364; one herd contained an estimated 450. If added to table 3.10, such herds would raise mean and median size greatly. Male herds with over 100 animals were not encountered in other parts of the reserve. In Xinjiang's Arjin Shan Reserve, Achuff and Petocz found that "the largest herd observed . . . comprised 356 male chiru dispersed over the South Aqik Lake Plain in a number of smaller bands" (1988).

North of the Aru Basin in the Toze Kangri region, the largest male herd contained 12 individuals (table 3.10). That June thousands of females and their offspring migrated through the area. Over 2000 animals passed Yue Ya Hu, but no adult males and only a small number of yearling males were with them. Of 230 adult males tallied in the region, only 15 (6.5%) were in mixed herds, indicating little association between the sexes even when females were abundant. In the Central Chang Tang population, male herd size did not exceed 13, and more animals were alone (40%) than in the other samples. Over 800 females and young were concentrated at one site, with only 5 adult males and 1 yearling male among them. Herd sizes

Table 3.10 Size distribution of male chiru herds on the summer range

	Population			
	West (Aru Basin)	West (Toze Kangri)	Central	East
Month/year	7–8/1988, 1990, 1992	6–7/1992*	6–7/1991, 1994	6/1994*
Total animals sampled	1211	253	423	1679
Number of solitary animals	193	46	169	167
Number of herds	155	52	87	206
Mean herd size	6.6	4.0	2.9	7.3
Median herd size	10.5	4.5	2.5	11.4
Herd sizes				
1	15.9%	18.2%	40.0%	9.9%
2	9.2	16.6	22.7	6.2
3	7.4	9.5	13.5	6.6
4	4.6	11.1	9.5	4.5
5	4.5	7.9	5.9	7.4
6	3.5	11.9	—	4.3
7	3.5	5.5	1.6	4.6
8	3.3	6.3	1.9	3.8
9	2.2	3.6	—	2.1
10	2.5	—	4.7	3.6
11–20	12.8	9.5	0.2	17.9
21–30	10.5	—	—	8.4
31–40	2.8	—	—	4.3
41–50	—	—	—	5.4
51–100	16.9	—	—	10.9

*Young males, 11–12 mo. old, were considered yearlings.

in the East Chang Tang population were similar to those in the Aru Basin; the largest herd numbered 98 animals.

After leaving their mothers, yearling males typically associated with adult males, as an example from a June survey of the East Chang Tang population showed. Of 1679 males alone or in male herds classified, 18.9% were yearlings. Most yearlings (80%) were in herds with at least one adult male. Few of the others were solitary and the rest were with 2–4 peers. When both adults and yearlings were in a herd, the latter composed about 10–50% of the total number of males (fig. 3.8), a variation without obvious pattern except that yearlings seemed to prefer the company of adult males. Yearlings probably learned travel routes and favorable feeding sites by following adults.

Males tended to remain separate from females even after the sexes began to share the winter range. In late August and September many males were alone (30.7%), mean herd size was a low 3.7, and 74.7% of all males (*n*

= 479) remained solitary or in male herds. Changes occurred in October, the onset of winter, when most solitary males joined male herds, raising mean size to 9.2 (table 3.11). However, 70.2% of all males (n = 705) remained separate from the females, a figure similar to the previous month.

Data during November, with the rut imminent, were collected only in Qinghai. In late October and early November 1985, a time of deep snow, chirus atypically migrated in search of forage, often in mixed herds. Only 30.4% of the males (n = 395) were alone or in male herds at that time. Of these, few (2.8%) were alone, and male herds were of about average size with a mean of 6.4 (table 3.12). Conditions in November 1986 were more normal. Fifty-two percent of the males (n = 547) were not associated with the females, and of these 7.5% were alone and the rest in herds with a mean size of 5.0. That same month in 1993, we tallied 148 males along the upper Tuotuohe Valley, and of these 80% were in male herds (none were solitary) with a mean size of 9.9 and a range of 4–27. There was thus considerable variation between areas in the percentage of males that associated with females in mixed herds during November.

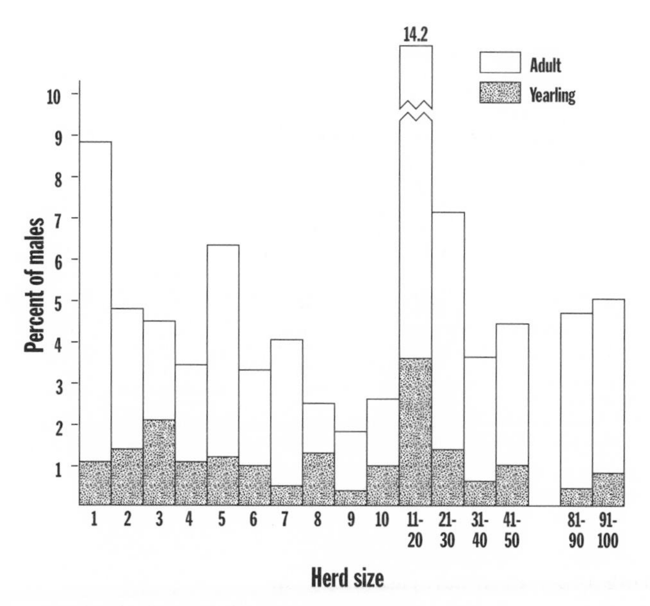

3.8. Percentage of male chirus of the East Chang Tang population in various herd sizes, based on a sample of 1362 adult and 317 yearling males.

Table 3.11 Size distribution of male chiru herds in autumn and early winter; Central and East Chang Tang populations

	Month/year	
	Late 8–9/1990	10/1993
Total animals sampled	358	495
Number of solitary animals	110	18
Number of herds	67	52
Mean herd size	3.7	9.2
Median herd size	5.3	15.9
Herd sizes		
1	30.7%	3.6%
2	9.4	3.2
3	14.2	4.9
4	6.7	3.2
5	2.8	8.1
6	5.0	4.9
7	7.8	2.8
8	4.5	3.2
9	2.5	3.6
10	—	—
11–20	16.3	33.4
21–30	—	5.5
31–40	—	13.1
41–50	—	—
51–60	—	10.3

Table 3.12 Mean chiru herd sizes (excluding solitary individuals) in Qinghai and Xinjiang, 1985–88

	Qinghai*				Xinjiang (late 5–6/1987)	
	10–11/1985		11/1986			
	No. herds	Size (range)	No. herds	Size (range)	No. herds	Size (range)
Male herd	28	6.4 (2–22)	49	5.0 (2–14)	40	7.4 (2–38)
Female herd	55	5.7 (2–20)	23	5.2 (2–11)	26	7.2 (2–18)
Mixed herd	31	36.4 (3–279)	68	11.8 (2–44)	15	11.5 (3–48)

Source: Schaller, Ren, and Qiu 1991.
*Data from the Tuotuohe and Wudaoliang areas have been combined.

During the December rut, most chirus congregated at certain sites where males chased each other and courted females, creating constantly shifting social contacts (fig. 3.9). The number of chirus in major concentra-

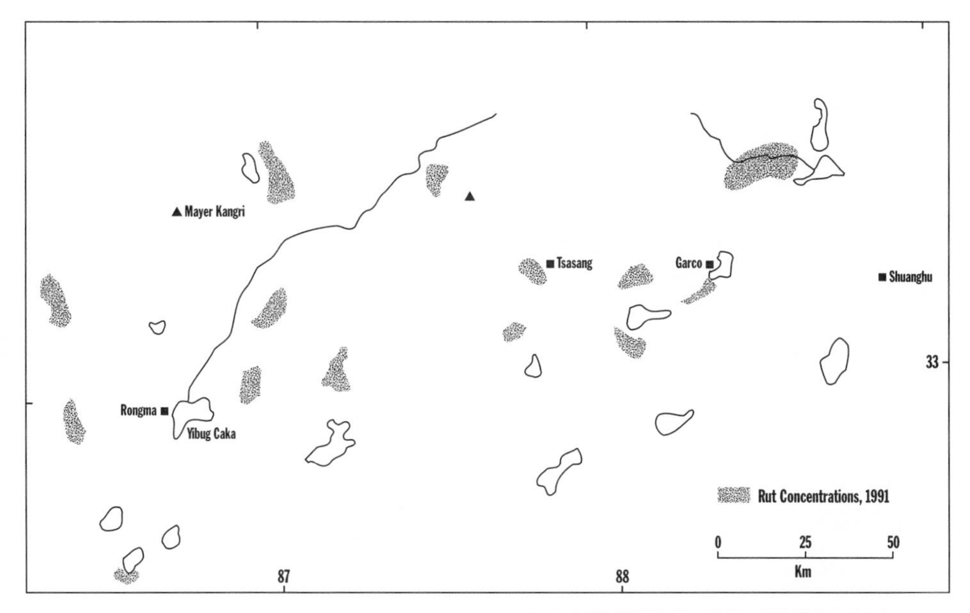

3.9. The location of 13 rut concentration areas of chirus, December 1991. The animals east of 87°30′ E are part of the East Chang Tang population, and those to the west are part of the Central Chang Tang population.

tions ranged from about 100 to over 1000, spread at times over a kilometer or more. Each concentration included a variable number of adult males, from about 2 to 8 males per 10 females, and an even more variable number of yearling males (table 3.13). Having spent the preceding months mainly with adult males, many yearlings now rejoined females, and 69.9% of yearlings ended up in mixed herds. Large male herds disbanded, and many males (32.6%) were again alone or in small male herds (21.5%), though all were near females on the courtship site. The most striking aspect of herd composition now was the many small mixed herds (mean size 4.7), each with just 1 adult male. These herds were harems, but often fleeting in composition (see chapter 14). Of the mixed herds, 79.2% had only 1 adult male associated with them and 13.7% had 2–10; only yearling males were in 7.1% of the mixed herds (table 3.14). All mixed herds with 4 or more adult males were relatively large, containing at least 9 members.

Chirus were also scattered over the range away from the rutting concentrations. The herd composition of these animals resembled that at court-

Table 3.13 Ratio of chiru males to females in rutting concentrations, in the Central and East Chang Tang populations, December 1991

Approx. size of concentration	No. of animals sampled	Adult ♂♂: 100 ♀♀	Yearling ♂♂: 100 ♀♀
Central			
150	138	49	22
300	214	55	12
275	227	35	5
150	62	58	24
300	157	47	13
450	272	31	7
200	117	38	11
100	74	36	4
East			
1250	598	81	12
140	129	26	0
400	190	84	23
150	101	72	77
350	294	21	6
125	107	49	38

ship sites, with mean herd size small and only 1 adult male in most (90.6%) mixed herds (table 3.14). Over a half of the adult males were not in a herd with females even during the rut.

Female herds. Females may change their herd size even more drastically than the males, ranging from a solitary life to a crowded migrating herd of a thousand or more. The tail end of the northward migration of the East Chang Tang population in 1994 consisted of a few small herds, totaling about 75 animals, and four herds with 150–300 animals each for a total of 900–1000. Five adult males were among them. When the females of the West Chang Tang population passed Toze Kangri in June 1992, 65.2% of the animals were in herds with more than 50, and the three largest herds contained 189, 237, and 262 individuals. Over 7000 chirus of the same population, but now including newborns, traveled south past the Aru Basin in late July and early August. Except for 6 solitary animals, 3.2% were in herds of 2–50, 8% in herds of 51–100, 26.6% in herds of 101–500, and 62.2% in herds exceeding 500, with the largest at least 1000.

Such large herds were evident only during rapid migratory movements. As soon as the animals returned to their autumn and winter ranges on alpine steppe, they dispersed. By late August and in September, female herds seldom numbered more than 20 individuals (mean 4.0). Solitary animals were unusually common (15–20%) during this season, most of them

Table 3.14 Herd size and composition in chiru rutting concentrations and outside such concentrations in the Central and East Chang Tang populations, December 1991

	In concentrations	Outside concentrations
Number of chirus sampled	2885	439
Number of male herds	68	8
Number of female herds	127	27
Number of mixed herds	294	53
Mean size (range) of male herds	3.0 (2–11)	3.3 (2–8)
Mean size (range) of female herds	4.0 (2–15)	4.1 (2–15)
Mean size (range) of mixed herds	4.7 (2–29)	4.7 (2–9)
Number of yearling males	173	20
% solitary yearling males	5.8	20.0
% yearling males in male herds	24.3	25.0
% yearling males in mixed herds	69.9	55.0
Number of adults males	766	118
% solitary adult males	32.6	38.1
% adult males in male herds	21.5	22.0
% adult males in mixed herds	45.9	39.9
% mixed herds with 1 adult male	79.2	90.6
% mixed herds with 2 adult males	8.2	7.5
% mixed herds with 3–4 adult males	4.1	—
% mixed herds with 5–10 adult males	1.4	—
% mixed herds with only yearling males	7.1	1.9
% females and young in mixed herds	70.4	60.1

females but also a few young which may have become separated from their mothers (table 3.15). During this same period a high proportion of males were also solitary. Of 1434 females and young sampled in late August and September 1990, 91.8% were alone or in female herds. Solitary females, like the solitary males, joined herds in October, and mean herd size doubled. Of 2164 females and young sampled, 78% were in female herds, indicating a shift toward mixed herds since the previous month. The hectic activities of the rut disrupted herds, and mean female herd size declined from 7.6 in October to 4.0 in December. Only about 30–40% of the females and young were now in female herds, compared to 78% in October. Still, about a third of the females were sexually segregated on the courtship grounds.

Summary

Endemic to the Tibetan Plateau, the chiru is found primarily on the alpine steppe of northwest Tibet. This project plotted the current distribution

Table 3.15 Size distribution of female chiru herds on the autumn and winter range, late August to October

	Population		
	West	Central	East
Month/year	Late 8/1988, 1990	Late 8–9/1990	10/1993
Total animals sampled	282	1317	1687
Number of solitary animals	55	191	20
Number of herds	55	293	218
Mean herd size	4.1	3.8	7.6
Median herd size	4.0	3.8	11.4
Herd sizes			
1	19.5%	14.5%	1.2%
2	13.5	21.3	5.5
3	13.8	13.9	5.5
4	14.2	9.4	7.8
5	5.3	8.0	5.3
6	2.1	4.1	5.7
7	2.5	4.8	3.3
8	—	3.0	3.8
9	—	2.7	2.7
10	3.5	1.5	4.7
11–20	11.3	10.0	23.9
21–30	—	—	17.0
31–40	14.2	2.6	4.0
41–50	—	—	5.6
51–60	—	4.2	—
61–70	—	—	4.0

of the species, it discovered that there are at least four migratory populations and a number of resident ones, and it found that during spring the migratory females separate from the males and travel up to 300 km north into desolate and uninhabited terrain to calve, whereas most males move only relatively short distances, the two sexes remaining essentially segregated except during the midwinter rut. I was able to document herd sizes and composition, such as females migrating in compact herds with several hundred or even over a thousand animals together, and a male attempting to maintain a harem of several females during the rut. Mortality of young was high, up to half dying within a month or two after birth, though death rates varied between populations and from year to year. A heavy snowfall that covers forage may cause mass death in chirus from malnutrition, as we observed in Qinghai. Our most important discovery was that the chirus were being slaughtered illegally by the thousands for their fine wool, which, after being smuggled to India, is woven into high-priced scarves and shawls (see chapter 15). Although the chiru remained numerically the dominant wild ungulate in our principal study area, the Chang Tang Reserve, it was killed wholly unsustainably there and elsewhere. The project created worldwide awareness of the urgent need to protect the species.

In spite of our efforts to study the chiru, the data remain fragmentary and preliminary, which is also the case for other ungulate species in the following chapters. No adequate chiru census has been done, and I can do little more than guess at the total number, perhaps fewer than 75,000. The great herds of 15,000 animals or more that Western explorers reported a century ago are now gone, decimated in the past few decades, and with them we have lost a magnificent opportunity to study, conserve, and manage the species while it retained its traditional numbers and patterns of behavior. I am reminded of the North American bison, whose millions had vanished before the ecology and behavior of the animals were investigated and efforts were made to save the species. The chiru now offers a similar last chance. The populations should be censused accurately and monitored, and fecundity and mortality should be assessed on an annual basis. Why do females migrate to the barren north, give birth, and immediately return south to better pastures, making a long, stressful trek when pregnant and lactating? Nutrition, inclement weather, and avoidance of wolves and noxious oestrid flies which parasitize chirus can all be invoked as an explanation but offer no satisfactory answer. So far even the location of the calving grounds of the migratory populations remains unknown. This chapter offers a baseline of information about the chiru, but it also represents a call for action, for the pressing need to study the species while it still roams the Chang Tang in moderate numbers.

4

Tibetan Argali

On the wild bleak uplands of Thibet, where for hundreds of miles not a tree is to be met with; where in every direction, as far as the eye can reach, there is nothing but a vast expanse of barren soil, rock, and snow; where there is no shelter from the glare of a cloudless noon, nor from the freezing winds that sweep the naked hills with relentless force towards the close of day; here, in the midst of solitude and desolation, where animal life has apparently to struggle for existence under every disadvantage, is the home of this great wild sheep.

Alexander Kinloch (1892)

THE ARGALIS *(Ovis ammon)* of central Asia occur in the Pamirs, along the Tian Shan and Altay Mountains, on the Tibetan Plateau, and on various other ranges from Russia, Kazakhstan, Tajikistan and Afghanistan east into Mongolia and China. They are the largest of the sheep, some rams exceeding a shoulder height of 110 cm and body mass of 100 kg. With long, thin legs and a compact but lithe body, argalis are adapted to open terrain, to escape danger through fleetness. They are usually found on high rolling hills and plateaus and on relatively gentle mountain slopes. Adult rams have massive, curled horns, with the tips pointing forward and often flaring outward (fig. 4.1). Argali subspecies vary somewhat in size and appearance, from the rather small Karatau argali *(nigrimontana)* to the almost mythical Marco Polo sheep *(polii)*.

Having surveyed Marco Polo sheep along the eastern margin of its range in Pakistan and China, I was interested in studying Tibetan argalis *(hodgsoni)*. To my regret these animals were uncommon, indeed so rare that weeks of travel often provided no more than a weather-worn skull or two. Of the six subspecies recognized by Geist (1991), the Tibetan argali is so far the only one listed as endangered in Appendix I of the Convention on International Trade in Endangered Species (CITES). Given the precarious status of this subspecies, I concentrate on numbers and distribution rather than on natural history, about which the project discovered little new.

Taxonomy

In April 1988, four hunters from the United States shot four argalis in the northeast corner of the Tibetan Plateau in Gansu. They were at the west-

4.1. The skull of a Tibetan argali ram in the Yako Basin. The animal was about 7.5 years old at death.

ern end of the Yema Nanshan, a subrange of the Qilian Shan. Agents of the U.S. Fish and Wildlife Service confiscated the trophies as belonging to an endangered animal, the Tibetan argali. Four biologists, including myself, identified the hides and horns, agreeing that they were those of Tibetan argalis. But the hunters claimed that Tibetan argalis occur only along the southern margin of the plateau, south of the Tanggula Shan, and that in the north, in the Arjin Shan and adjoining ranges to the east, another subspecies, *dalai-lamae*, is found. The vast central part of the plateau was ignored in the discussions. After a two-day taxonomic review, Wang, Li, and Song (1988) provided a Chinese response to the dispute by stating that the trophies belonged to *dalai-lamae* but that another subspecies, *darwini*, also inhabited the same area. Bunch, Mitchell, and Maciulis (1990) created yet more confusion when they referred to the shot animals as *jubata* in spite of the fact that *jubata* is generally recognized as a subspecies, the Shanxi argali, found north and northeast of the Tibetan Plateau (Geist 1991). Since conservation laws are based on taxonomy, precision in identification is essential (Geist 1992). How many argali subspecies exist on the plateau?

Statistical variations in horn shape and angle, body size, and pelage color have all been used to establish subspecies. Nutritional, as well as individual, differences affect horn and body size. For instance, I noted that Tian Shan argali rams (*karelina*) and Altay argali (*ammon*) living on the lush alpine meadows in the western parts of these ranges tended to have larger and more flaring horns than those in the arid eastern parts. Pelage characteristics have also been uncritically used in taxonomic work (see Allen 1940), yet these change with age, sex, and season. The best available criterion for dividing argalis into subspecies is the pelage of mature rams during winter (Geist 1991). At that season, Tibetan argalis have a long-haired white ruff that terminates abruptly at the grayish brown shoulder. A large white rump patch surrounds the tail. The rump patch is distinctly separated from the body, which is dark grayish brown near the back and lighter on the sides. The dark pelage extends down the front of all four legs. The backs of the legs are white, as are belly and face. A dark lateral stripe divides the upper parts from the white belly. The black-tipped tail is small, less than 6 cm long without the terminal hairs, the shortest tail among argalis (Geist 1991).

I observed several mature rams in the Chang Tang Reserve in mid-September when they were in transition from a summer to winter coat. The neck was white but without a marked ruff, and the outline of the rump patch was indistinct. The upper parts of the body and the front of the legs were light grayish brown except for elongated dark brown hairs on the hump. A gray line traced the front of the shoulder. The face was grayish brown, but with a white patch on the rostrum. The belly was white. Such a slight difference in pelage could once have been the basis for designating a different subspecies.

The perception that the Tibetan argali is confined to the southern part of the plateau along the Himalaya from Ladakh to Sikkim (Blanford 1888–1891; Tsalkin 1951; Sopin 1982) is based on an inadequate search of the published information. Lydekker (1898) already suggested that the subspecies "extends to the Kuenlun" in the north, and subsequently several expeditions observed, collected, and named *hodgsoni* in various northern parts of the plateau. Allen (1939) reported *hodgsoni* from near Serxii in western Sichuan and from several localities in Qinghai, namely along the Yangtze upstream from Yushu, near the headwaters of the Yellow River, and north of Donggi Co. Farther north, Przewalski observed Tibetan argalis in the southern foothills of the Burhan Budai Shan (Prejevalsky 1876). Kozlov wrote that "the stone sheep, or argali (*Ovis hodgsoni*), is fairly common in those parts of the Nan Shan that we explored" (1899), referring to the same region where the disputed argalis were shot in 1988. Leche (1904) described two adult *hodgsoni* rams that were shot by Sven Hedin's expedition. One was killed in what is now the Arjin Shan Reserve and the other farther east in the Arjin Shan (about 94° E). These records of Tibetan argali extend the range of the subspecies into the area that is supposedly occupied by *dalai-lamae*, *jubata*, and *darwini*. Since geographic separation is often used as a criterion for dividing two variable forms into subspecies, this supposed sympatry among argalis requires an explanation.

During his fourth expedition, in the winter of 1884–1885, Przewalski (1888, quoted in Geist 1991) shot a small argali ram just north of the Arjin Shan Reserve, and this animal was later designated as the type specimen of *dalai-lamae*. A few years later, Hedin collected one of his adult *hodgsoni* rams close to the same locality (Leche 1904). As Geist (1991) has shown, the type specimen of *dalai-lamae* is actually a juvenile *hodgsoni* ram. Starting with de Pousargues (1898) and Lydekker (1898), biologists had questioned the validity of *dalai-lamae*, considering it a synonym of *hodgsoni*, but much of this century passed before Geist (1991) clarified matters by pointing out that the taxonomic confusion resulted from a failure to distinguish juveniles from adults.

The Shanxi argali (*jubata*) is or was found in an arc across northern China from Hebei and Shanxi west through Shaanxi and Inner Mongolia to Gansu. Its pelage differs from that of the Tibetan argali in, for example, having a light gray, rather than white, neck and a white rump patch that does not surround the tail (Geist 1991). Solely on the basis of horn characteristics, Nasanov (1923) reclassified Hedin's *hodgsoni* as *jubata*, and Sopin (1982) relabeled the same specimens again but as *dalai-lamae*, showing the futility of using only the horns of a few specimens to designate subspecies. There is no evidence that *jubata* occurs on the Tibetan Plateau.

The taxonomy of argalis in southern Mongolia and northern China remains unsettled. The Tian Shan with its *karelina* and the Altay with its *ammon* converge in the Gobi, where *darwini* occurs on scattered low ranges. Once I stood on a hill near this convergence in Great Gobi National Park and could easily see the mountains that each of these three subspecies occupy, only plains of modest extent separating them; in nearby China, to the south and east, was the habitat of *jubata*, a subspecies that has sometimes been lumped with *darwini* (Allen 1940). Another subspecies, *kozlovi*, is said to occupy parts of western Inner Mongolia (Heptner, Nasimovic, and Bannikov 1966; Sopin 1982), but the taxonomic status of this animal remains uncertain (Geist 1991). Further investigation is needed to determine whether hybrid zones between subspecies exist in this area, as they do for urial in Iran (see Schaller 1977b), whether there are as yet undescribed subspecies, and indeed whether the subspecies concept has validity in this region.

The narrow Heixi Corridor separated argali habitat along the southern margin of Inner Mongolia and parts of Gansu from the ramparts of the plateau. It is unknown what subspecies inhabit or inhabited these hills,

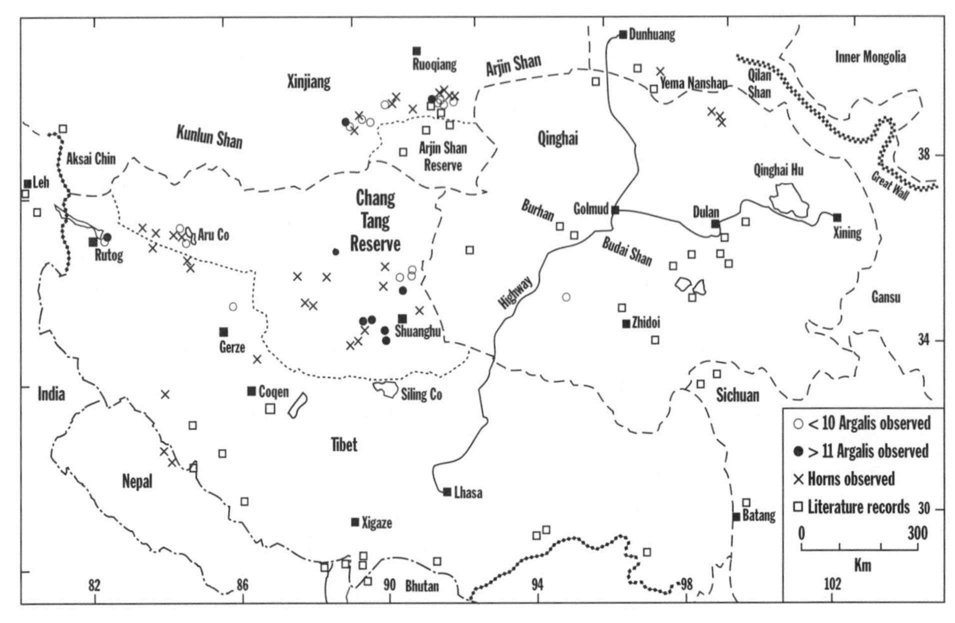

4.2. Distribution of Tibetan argalis, based on my observations of animals and horns and on selected records from the literature.

jubata, *kozlovi*, and *darwini* having all been claimed for that general region by various authorities. Argalis could in the past have crossed the corridor in either direction. But the Great Wall, which has its terminus in Jiayuguan at the western end of the corridor, was built during the Qin dynasty (221–206 B.C.) and was for centuries a barrier to argali movement (fig. 4.2).

Based on the evidence from the literature and my own observations, only one subspecies, *hodgsoni*, occurs on the Tibetan Plateau, and it occupies all of it. Argalis are troublesome to classify but careless science has created confusion for a century. And taxonomic issues will be ever more difficult to resolve as the species becomes increasingly fragmented in its distribution. Indeed, argalis in the western half of Inner Mongolia have been reduced to so few scattered herds that this vast area may soon be devoid of the species.

Description

Adult rams have a shoulder height of about 118 cm and they weigh an average of 105 kg (Schaller 1977b). Their distinctive white-ruffed winter coat has already been described. In their summer coat, rams are a light grayish brown, sometimes darker on the neck, with an indistinct margin between white rump patch and adjoining body hair, and there is a faint side stripe separating the upper parts from the white belly. The horns of rams are heavily ribbed and average 39.4 cm (S.D. 2.9, n = 4) in circumference at the base in animals 6–7 years and older; three crania with horns weighed 7.2, 9.9, and 10.4 kg, and one weight of 18 kg has been reported (Macintyre 1891). Horn length in sheep increases steadily with age, the oldest animals usually having the longest horns as measured along the outside curve. The longest horn we found measured 102 cm, that of a 9.5-year-old ram, considerably less than the 145 cm listed as a record (Dollman and Burlace 1922). Adult rams typically have their horn tips heavily broomed or broken. Annual horn growth in an animal decreases with age, as does variability in the amount of growth (fig. 4.3).

Marco Polo sheep, the Tibetan argali's nearest neighbor in Taxkorgan to the west, differ somewhat in pelage pattern—for example, the rump patch extends onto the thighs—and in horn size and shape. The horns are more everted at the tips, longer (the record is 191 cm), thinner in circumference (mean 36.7, S.D. 2.2, n = 62), seldom broken at the tips, and average annual growth is significantly greater (fig. 4.3).

Female Tibetan argalis resemble males in summer pelage and have light grayish brown upper parts, darker along the back, and white bellies and rumps. They stand 104–112 cm high at the shoulder and weigh an estimated 68 kg (Ward 1924), a third less than adult rams. One female had horns 46 cm long with a basal circumference of 19 cm.

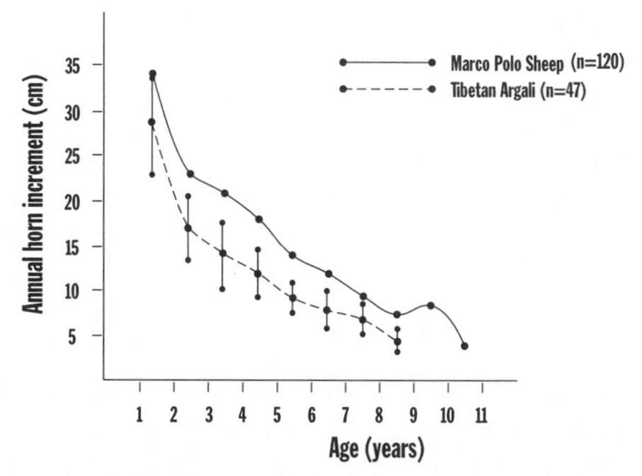

4.3. Mean length of annual horn increments of male argalis, comparing Tibetan and Marco Polo argalis. The standard deviations shown for Tibetan argalis illustrate how variability in growth decreases with age. Sample size for each age class varies; the smallest sample is that for the oldest animals. The maximum sample size for Marco Polo sheep was 120; for Tibetan argali, 47.

Status and distribution

Outside China

At its western limit the Tibetan argali occurs in Ladakh, India, and in the Aksai Chin area of Xinjiang to just east of Karakoram Pass at 35°30′ N, 77°50′ E (Stockley 1928). About 250 km of rugged terrain separate these *hodgsoni* from the nearest *polii* in the Taxkorgan Reserve. An estimated 200 argalis exist within their range of 10,000 km^2 in Ladakh (see fig. 5.4), with the densest population, about 51 animals, in the 170 km^2 Gya-Mira Reserve (Fox, Nurbu, and Chundawat 1991a). The precarious status of the animals in Ladakh is due "to wanton shooting by trigger-happy people and poachers" (Ganhar 1979), military personnel (Fox, Nurbu, and Chundawat 1991b), and other outsiders (Kinloch 1892). "When travellers were few . . . Hodgson's sheep were easy to get. Now this is all changed" (Ward 1923). Hunting was permitted even after 1975, the year in which the subspecies was placed in Appendix I of CITES. In 1975 and 1976, for example, several American hunters killed this sheep in Ladakh (Valdez 1983).

Just southwest of Leh, Tibetan argalis are barely sympatric with another species of wild sheep, the small Ladakh urial (fig. 5.4). The total Indian population of Ladakh urial numbers 1000–1500 animals, and the animals occur mainly in a narrow band along the Indus and Shyok Rivers, a range of about 1500 km^2. There they occupy dry hillsides covered with *Ephedra*, *Capparis*, and other xerophytic shrubs, ecologically separated from the high alpine meadows that are favored by argalis and also by blue sheep (Mallon 1991; Fox, Nurbu, and Chundawat 1991b).

Farther east in India, argalis once strayed out of southwestern Tibet into the Spiti and Lahul areas (Stockley 1928), and a few still occur in northern Sikkim near the Tibetan border (Shah 1994).

In Nepal, small argali populations were once found in the Dolpo district (Schaller 1977b) and about 80–100 animals still occur in northeast Mustang near the Tibetan border (D. Miller, pers. comm.). The species has also been reported from western Bhutan (Gee 1967), but at most only rare stragglers probably survive there.

Within China

The eastern limit of *hodgsoni* distribution lies in western Sichuan on tongues of grassland that project southward into the forests between such rivers as the Yalong and Yangtze (fig. 4.2). Dolan (1938) found horns there and Kaji et al. (1993) observed several small herds.

Exploring into southern Qinghai, south of the Burhan Budai and Anyemaqen Ranges, Przewalski noted: "We first saw these sheep beyond the Burhan Buddha, and afterwards as we penetrated farther into the country, but they are not common" (Prejevalsky 1876). In the same general area, around the upper Yangtze and the headwaters of the Yellow River, Schäfer (1936) reported argalis in "relatively large number," and Kaji (1985) found horns there. West of that area we saw two argalis, a subadult ram and a ewe. During extensive surveys around Zadoi and Zhidoi in southeastern Qinghai, we found no argalis, not even old horns. Local people told us that most of these sheep had disappeared 3–4 decades ago. In the eastern Chang Tang, west of the Lhasa-Golmud highway, Feng (1991b) tallied 15 argalis during extensive cross-country travel in the summer of 1990 and his estimate for the whole region of 75,000 km^2 was 200–300 argalis.

Two populations of moderate size are known to exist in the Burhan Budai Shan. In Dulan County, just southeast of the Qaidam Basin, Zhen and Zhu (1990) counted argalis in nine sample localities totaling 105 km^2. They recorded 137 animals, a density of 1.3/km^2, mostly at 4300–4700 m in elevation. Since the sample localities were selected on the recommendations of local people, they presumably represent centers of argali abundance in the region rather than average density. West of there, in the Hong Shui Chuan Valley, Cai, Liu, and O'Gara (1989) came across fresh argali skulls in 1986; according to the nomads, argalis occur there in low numbers on the south-facing slopes of the range. Still farther west, in the Yeniugou (Wild Yak Valley), Cai, Liu, and O'Gara (1989) saw 84 argalis in August 1986. Harris (1993) studied this population in 1991 and 1992, finding that about 245 individuals frequented his study area of about 1050 km^2.

In the Qilian and adjoining ranges of northeastern Qinghai, it was reported that "the argali is also frequent" ("Central Asian Expedition" 1896) and Kozlov noted that "because of the great concentration of these animals, the mountains are often called Argalinin-ula" (1899). In 1985, along the northern flanks of the Shule Nanshan, we surveyed a block of 610 km^2. Several old argali horns were seen but no animals. Just west of this area in Gansu, Zheng et al. (1989) observed 57 argalis in the 4250 km^2 Yanchiwan Reserve. The four disputed argalis were shot by the American hunters near the western border of the reserve in 1988; "over 350 argali were seen during a 7-day hunt" (R. Mitchell, quoted in *Status Review* 1991), but some animals may have been counted more than once. Argalis are also found in the mountains west of there along the Gansu-Qinghai border in unknown numbers (W. Wang, pers. comm.).

In Xinjiang, Leche (1904), Bonvalot (1892), and others had reported argalis at the northern rim of the plateau. More recently, Achuff and Petocz (1988) surveyed the western 23,000 km^2 of the Arjin Shan Reserve and found argali numbers "low," observing only 9 animals and 3 skulls, most near Aqqikkol Hu. Northwest of the reserve, near where the Arjin Shan branches from the Kunlun Shan, we tallied 27 argalis in June 1987.

In July 1988 we ascended several valleys of the Yüsüpalik Tag, a subsidiary range of the Arjin Shan just north of the reserve. A total of 64 argalis were counted on the alpine meadows at about 4300 m. Skulls at the base of the foothills indicated that the animals moved seasonally to lower elevations.

In Tibet, the eastern part with its rugged mountains has little argali habitat. No recent information on the status of the species there is available. In the southern part, argalis once existed widely along the upper Yarlung Tsangpo and Indus Valleys. Rawling, for instance, found argalis "very numerous in all the ravines and on the slopes of the mountains lying to the north of the Brahmaputra [Yarlung Tsangpo] from its source to Tradom" (1905), at about 84° E. In 1990, along the same route, we recorded only one argali skull, on a cairn. South of Gyangze, in the upper Chumbi Valley, Bailey wrote that "in May 1909, on one day's march, I saw 17 *Ovis ammon*, 25 Burhel [blue sheep] and 64 Tibetan Gazelle" along the road (1911). Jackson (1991, pers. comm.) made a 6-week survey of the Qomolangma Reserve in 1990. He found no evidence of argalis but was told of one small herd in Gyirong County. Similarly, local informants in southern Tibet told us of isolated and usually small populations, except for one population of moderate size in the hills north of the town of Lhunze at 28°40′ N, 92°20′ E, and another just south of Yamdrok Co (W. Liu, pers. comm.). The usual reply to our query about argalis was a negative shake of the head, sometimes with an added "not anymore."

Just east of Rutog among low hills, in a small area known for its argalis, we saw two herds in August 1988, one with 23 animals and the other with 6.

Within the Chang Tang Reserve, Rawling noted that "near Aru Tso large numbers of ewes and young were daily found grazing in the neighbouring ravines" (1905). During 39 days of research in the basin we saw three ewes and several skulls. North of the basin, as far north as the Xinjiang border, we surveyed one 8000 km^2 block of terrain for a month in June and July 1992 without finding any sign of argalis. Hedin ([1922] 1991) observed argalis near Laxong Co (34°20′ N, 85°10′ E). But, in general, the species then as now was scarce in the western part of the reserve, and we observed it mainly in the eastern part (fig. 4.2). In September 1988, herds with 13 and 16 individuals were seen in the foothills of the Amu Kangri Range southwest of Shuanghu. In September 1990, two small male herds and a mixed herd of 11 were observed at 5000 m among steeply rolling hills near Garco, and in the same hills, but farther to the east, we found a herd of 18 in October 1993. North of there, in and around the foothills of Purog Kangri, five herds with a total of 28 animals were observed in late May and early June 1994. In the Yako Basin, 12 argalis were seen, including a herd of 9 rams.

Tibetan argalis have a wide distribution on the plateau, but their occurrence is highly sporadic, and much seemingly good habitat is devoid of animals. As Bower noted after traversing the Chang Tang, "I doubt *Ovis ammon* abounding anywhere; that they are found scattered over a very wide stretch of country is undoubted, but, unlike yak and antelope, nowhere did we find them very common" (1894). Hedin ([1922] 1991) found them similarly sparse and sporadic. Populations are generally small, with fewer than 100 individuals. The four largest reported populations are in the Yüsüpalik Tag, in the Yema Nanshan region, and in two localities of the Burhan Budai Shan. Information from local people attested to the scarcity of argalis, as did the rarity with which we found skulls. Horns remain visible for at least a decade after an animal's death, yet we saw them so seldom that we took note of each one. When we found several skulls in one area, we usually saw argalis as well. Skulls are useful indicators of abundance, at least of abundance in the recent past. In the western part of the Taxkorgan Reserve we measured 136 horns of male *polii* in one month, more horns than we tallied on all our surveys on the plateau.

Most argalis seem to occur in high rolling or broken but not rugged hills and on the upper slopes of mountain ranges rather than on the plains and low hills that cover much of the plateau. When crossing the Chang Tang, Bower (1894) found horns as far as 88° E, noting that they prefer hills to open valleys. The rarity with which travelers like Hedin (1909) and Rawling (1905) mention argalis also suggests a discontinuous distribution in such habitat. Mountain ranges which once seemed to have had large argali populations—Gangdise, Himalaya, Arjin Shan, Burhan Budai Shan—also have human population centers along their bases, and meat hunting was no doubt a main cause in the decline of these animals. But hunting alone probably did not eliminate the sheep from the sparsely inhabited parts of the Chang Tang, although itinerant gold miners were once found in the most remote areas (Wellby 1898). Furthermore, the animals are difficult to hunt: "The senses of the *argali* are keener than those of any other animal in Tibet, and it is an exceedingly wary animal, although hardly ever hunted; the Mongols finding it useless to attempt shooting them with their matchlocks" (Prejevalsky 1876). Small, isolated sheep populations are highly vulnerable to genetic and environmental mishaps, especially if they number fewer than 50 individuals (Berger 1990). Diseases transmitted by livestock could also have had a serious impact, as they have on North American sheep (Lawson and Johnson 1982). Once eliminated from an area, argalis may be exceedingly slow at resettling it. American sheep are poor at dispersing and colonizing (Geist 1971), and argalis, though not bound to rugged terrain, also tend to be sedentary. Fox, Nurbu, and Chundawat (1991b) documented the establishment of a new argali population, and Ward believed that the animals "wander in small flocks from place to place" (1924). But the persistence of small populations only in certain localities argues against rapid dispersal. "Once having selected its ground, there it will remain; and a herd of them has been known to frequent one mountain for a succession of years" (Prejevalsky 1876). Schäfer (1936) considered rams to be wanderers and ewes to be sedentary. Small and highly fragmented, most of the surviving populations are threatened with extinction from a variety of causes but particularly from hunting.

Except for about 200 animals in Ladakh and a few in Nepal and Sikkim, all Tibetan argalis are within China, where, as this brief overview shows, most populations are so small and fragmented that many will probably vanish in the coming decades. The argali is by far the rarest member of its wild ungulate community, but as with the other species, an accurate estimate of numbers cannot be made. In Tibet, perhaps 400–500 argalis were in the Chang Tang Reserve and possibly 1500 elsewhere. Qinghai and neighboring Gansu had several local populations of moderate size, but total numbers in these provinces may not exceed 3000–4000. Argali densities in Xinjiang were generally low, and I would guess that no more than 1000 animals occur there. Including some populations in western Sichuan and those outside China, the total number of Tibetan argalis could be as low as 7000.

Population and herd dynamics

The birth season is in late May and early June. In Xinjiang we encountered small young on 6 and 9 June, and in the Burhan Budai Shan 2 newborns, still unsteady on their feet and apparently the first of the season, were observed by Zhen and Zhu (1990) on 1 June. Bailey (1911) saw newborns in June and early July. Schäfer (1936) also placed the birth season in early June. However, Kozlov (1899) found that in the Qilian Shan "lambs appear by the end of April," suggesting some regional variation in the argali's annual cycle. With a gestation period of about 150 days (Schaller 1977b), the rut would extend from the second half of December into January in most areas. Kozlov placed the rut into November, when rams "approach until a short distance apart, then rear on their hindlegs and, leaping at each other, clash with their foreheads" (1899).

Of 232 argalis observed on the plateau, 202 were classified by sex and age. The ratio of rams to ewes (adult and yearling combined) was 59:100. A greatly skewed sex ratio favoring females can also be noted in the data of Zhen and Zhu (1990) and Harris (1993). The ratio of lambs to ewes was 41:100. A single young appears to be the rule, although twins have been reported from Ladakh (Ward 1924).

Like *polii* and other argalis, the animals have a relatively short life, seldom reaching 10 years, as shown by the age rings of horns of rams found in the field (fig. 4.4). The wolf is a principal predator. In Xinjiang, three wolf droppings contained argali hair and we examined the remains of a ram killed by wolves. Stockley (1928) noted that once when snow depth in Ladakh exceeded 46 cm during the winter of 1910–1911 the wolves easily captured argalis and Tibetan gazelles.

Except during the rut, adult rams tend to form separate male herds (Macintyre 1891; Kozlov 1899). We tallied eleven male herds ranging in size from 2 to 9 (mean 4.4) separate from, but usually among the same hills as, the ewes. Schäfer (1936) reported male herds of 2–15, Zhen and Zhu (1990) counted one with 16 rams, and Cai, Liu, and O'Gara (1989) one with 52 rams. Mixed herds of females and young with an occasional subadult ram number up to 35–40 individuals (Schäfer 1936). Zhen and Zhu (1990) found that the average size of thirteen such herds was 7.2 (2–16). We observed 3 solitary females and eighteen female or mixed herds with an average size of 10.0 (2–31) animals.

Harris (1993) found that the Yeniugou population of about 245 animals seemed to be divided into six subpopulations, or bands as he called them, which tended to remain in a particular area. The composition of herds within a band changed but the band itself seemed to remain distinct over a period of weeks. One band, the largest, with at least 73 members, ranged over about 31 km². It contained 11 rams, 36 ewes, 6 yearlings (male and

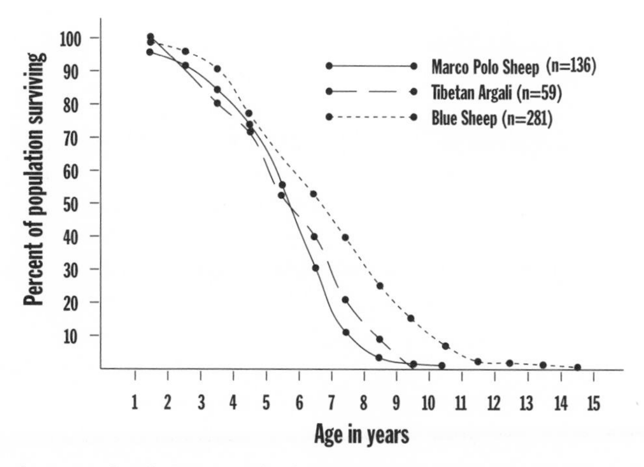

4.4. Survival curves of male Marco Polo sheep (from the Taxkorgan Reserve), Tibetan argali, and blue sheep, based on annuli counts of horns found in the field. Yearlings are probably underrepresented because their small horns are difficult to find.

female combined), and 20 lambs. The largest observed herd numbered only 43 individuals.

Conservation

The precarious status of *Ovis ammon* with its various subspecies has been internationally recognized. The World Conservation Union lists most subspecies as vulnerable or endangered, CITES places the Tibetan argali into Appendix I and all other subspecies into Appendix II, and the U.S. Fish and Wildlife Service is considering listing the species as endangered in all its native countries. Given this anxiety, it is a matter of concern that no adequate research has been done to clear up the taxonomic confusion created in the past when unreliable morphological criteria were used to designate subspecies. The validity of taxa can now be tested with molecular techniques, by using, for example, mitochondrial d-loop sequences along with microsatellite markers. Species conservation requires good systematics.

As our work on Tibetan argalis has shown, populations are small and highly fragmented even in areas where hunting pressure is low and much suitable habitat appears to be available. In the Chang Tang, factors other than just hunting seem to limit population growth and dispersal. Possibly disease introduced by livestock affects the animals, but no research has been done on this subject. Indeed even the status of Tibetan argali has been only vaguely documented, a lack that hampers a focused conservation effort.

Surveys in other parts of China also revealed that argali populations are small, scattered, and declining, as shown particularly in Inner Mongolia (Wang and Schaller 1996). But, in contrast to the Chang Tang, hunting for subsistence and commercial sale has in these areas been the main reason for the drastic decrease in argali numbers during the past few decades. The lack of official action in response to this massive decline is perplexing because argalis represent a valuable commodity, with foreign hunters paying high fees to kill trophy-sized rams. For example, in Mongolia a license to kill a Gobi argali costs at present $25,000, and one to kill an Altay argali, $30,000. Between 1967 and 1989, foreign hunters killed 1630 argalis there. Hunting companies in Mongolia characteristically established a seasonal camp in an area until large rams had been decimated and then they moved elsewhere. Lack of protection and indifference to sustainable harvesting have caused such a decline of argalis, particularly in the Altay Mountains, that most camps have been closed and only 15 hunting licenses were issued in 1994.

Argalis could readily and profitably be managed as trophy animals in selected areas. However, a substantial proportion of the money derived from hunts should benefit conservation and the local people directly. I find it deplorable that so much money is being spent on killing argalis and so little to study and conserve them. Wealthy hunting organizations, such as Safari Club International in the United States and the International Council for Game and Wildlife Conservation in France, could show a commitment to the argali's future by funding comprehensive status surveys and developing and implementing long-term management plans for the species in certain areas.

5

Blue Sheep

Although in some ways I impatiently await the rut, an aspect of bharal [blue sheep] life history about which nothing is known, I am also content just being near the animals. Toward noon they become less and less active and after pawing a bed, each animal reclines to chew cud, its eyes mere slits of satisfaction. The sun is often warm at this time of day, shade temperatures climbing above freezing, and the hills vibrate with heat waves. Beyond, to the north, are the enormous coppery cliffs of Shey canyon and the rounded dome of Purple Mountain; on the horizon a frieze of snow peaks guards the Tibetan border. . . . Resting tranquilly near the bharal, I know also that animals are wild only because man has made them so. Wolf, snow leopard, and all other creatures could be as tame as these bharal if only we would permit them to be.

George Schaller (1980)

WITH STOCKY BODY and stout legs, blue sheep, or bharal, are designed for climbing in rocky terrain. In build they resemble such cliff dwellers as American mountain sheep and Asiatic ibex rather than the slender-limbed argalis. Their mixture of sheeplike and goatlike traits once created confusion about their evolutionary relationships, and one focal point of my research in the Himalaya during the 1970s was to determine whether blue sheep are behaviorally and ecologically closer to *Ovis* or to *Capra.* My conclusion was that they are goats with sheeplike traits, and recent molecular work has further helped to elucidate the issue, as discussed in chapter 13.

During our travels on the plateau, especially when in search of snow leopards, we often encountered blue sheep. Since blue sheep are the snow leopard's principal prey there, the two species are ecologically bound to each other. We censused blue sheep in several localities and collected data on their population dynamics, but we did not observe them intensively because their behavior has been described earlier (see, e.g., Schaller 1977b; Wilson 1984). Fortunately blue sheep remain common locally, and in fact, they are the most abundant wild ungulate in the mountains.

Taxonomy

Blue sheep (*Pseudois nayaur*) extend over a vast range, from the Karakoram in the west across the plateau to Inner Mongolia in the east (fig. 5.3). Two poorly defined subspecies are sometimes recognized, *P. n. nayaur* in the west and *P. n. szechuanensis* in the east, but Groves (1978) considers these invalid. Schäfer (1937a) shot several blue sheep in the gorge of the Yangtze River near Batang which were smaller than usual. Typical *nayaur* males weigh 60–75 kg and females 35–45 kg, whereas Yantze males weigh 28–39 kg and the one female weighed was 25 kg. The horns of the Yangtze males are thinner and have less of an inward curve, and the tips turn up more than those of the usual males (Schäfer 1937a; Allen 1940). Schäfer (1937a) thought that he had discovered a new form but he did not name it. Later, Groves decided that "the distinctiveness and isolation of the Dwarf Blue Sheep . . . suggests that, as a provisional measure at least, it should be classified as a full species," *P. schaeferi* (1978), a designation also accepted by Wu et al. (1990).

The so-called *schaeferi* occurs for an unknown distance north, south, and west of Batang mainly along the steep, arid, lower slopes of the Yangtze River gorge between 2600 and 3200 m in elevation (Wu et al. 1990). Above these blue sheep is a forest zone that extends 1000 m upward to alpine meadows where *nayaur* is said to occur. The idea that a narrow forest zone can segregate a mobile large mammal to such an extent that two such closely related forms could maintain themselves as distinct species deserves a little skepticism. It is true that blue sheep tend to avoid forest, although I observed a herd enter a juniper stand in Qinghai to reach a spring. Allen (1940) speculated that the Yangtze blue sheep are stunted because of poor nutrition in their arid environment.

Does an arid environment affect size of blue sheep? We observed blue sheep in a variety of habitats, some of them extremely dry, but failed to note any marked difference in body size. Horn growth in males, however, may provide a rough indication of habitat quality in wild sheep (Geist 1971). For example, Asiatic ibex from the western Tian Shan, with its fine alpine meadows, grew significantly more horn each year than those in the spartan environment of the Taxkorgan Reserve (Schaller et al. 1987). Contrary to expectations, annual horn increments of blue sheep males from arid areas such as the Arjin and Kunlun Shan, Qilian Shan, and Taxkorgan Reserve were similar to those of animals from the luxuriant mountain meadows around Zadoi (fig. 5.1). However, basal circumference of horns of males 6 years old and older showed minor regional variation. The circumference of horns from the arid Arjin Shan averaged 22.35 ± 1.30 cm ($n = 20$), significantly ($P < 0.05$) smaller than the Zadoi horns, with 24.25

± 1.82 cm (n = 26), and Taxkorgan horns, with 25.21 ± 1.03 (n = 12). The Arjin Shan horns did not differ significantly from the Qilian Shan horns, whose circumference was 23.7 ± 2.32 cm (n = 8). Thus, in blue sheep the circumference of horns, rather than the length, may be a better indicator of habitat quality, and as noted, *schaeferi* has notably thin horns.

Groves (1978) compared the tooth row of a large *schaeferi* skull (68 mm) with the tooth rows of 3 *nayaur* skulls of comparable age (64–72 mm) and found them to be of similar length. Craniometric studies in red deer revealed that animals that have the same tooth row length have genetically the same body size, and that regional variation in body size was environmentally induced (see Geist 1971). Further work is obviously needed to elucidate the taxonomic position of the small blue sheep in the Yangtze River area.

Description

Adult males are robust and handsome, about 80–91 cm tall at the shoulder, with a sleek grayish brown to slate blue pelage. The ventral surface of the neck, the chest, and the front of the legs are dark gray to black. The neck is markedly swollen during the rut, giving the animal a bulky shape. A conspicuous black flank stripe separates the upper parts from the white belly. The rump patch, the inside and the back of the legs, and the tip of muzzle are also white, as is a spot on the knee and above the hooves; the eyes too are fringed with white. The hair is short without beard, ruff, or other hairy appendage. The smooth horns sweep up and out and then curve back before curling up at the tip. They are massive and relatively short, the record being 84 cm; the longest horn measured during this study

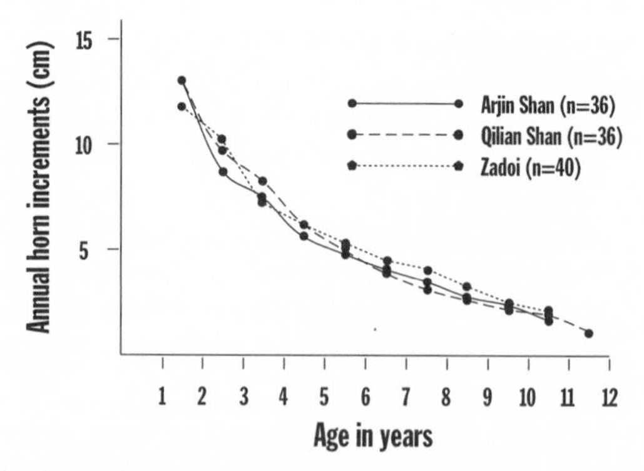

5.1. Mean length of annual horn increments of male blue sheep from three areas of the Tibetan Plateau with different habitat quality. Note the similar growth rates.

5.2. Blue sheep blend well into hillsides. A male lies among the boulders, and a female with young stand near him.

was 68.5 cm. Horn tips are sometimes broomed but seldom broken. Average distance between horn tips was 66 (57–82 cm) in males of 6 years and older (n = 37). Females resemble males except that they have gray, instead of black, markings and their short horns, 10–20 cm long, project first up and then out (fig. 5.2). Wang and Hoffmann (1987) provide a detailed description of the species. Although seemingly marked in a striking pattern, blue sheep are remarkably inconspicuous, blending so well into their environment that they are often difficult to spot.

Status and distribution

Blue sheep are highly tolerant of environmental extremes. They can be found on sun-bleached desert mountains in searing heat at elevations below 1200 m and on windy and cold slopes at 5300 m. In some of their habitats it seldom rains, whereas in others 1500 mm of precipitation may fall in a year and snow may lie 1 m deep in winter (Wilson 1981). They can be found in hills that are mere hillocks and on the grandest mountains in Asia. But their varied terrain has several features in common. Blue sheep inhabit treeless slopes or remain in the alpine meadow and shrub zone above timberline. They prefer relatively gentle hillsides (less than 40°

slope) covered with grasses and sedges (Wilson 1981) but with nearby cliffs into which to escape in times of danger (Oli 1996). Animals are seldom more than 200 m from a rocky retreat (Fox et al. 1988).

The western limit of blue sheep distribution lies in several valleys of Khunjerab National Park in Pakistan and the eastern part of the Taxkorgan Reserve in China (fig. 5.3). From there the animal's range extends east along the Kunlun Shan and the northern flanks of the Karakoram and across the eastern half of Ladakh (fig. 5.4). Groves stated that the blue sheep "is absent from the main plateau itself" (1978). Actually they are found throughout the plateau, with the Kunlun Shan and Arjin Shan forming the northern border and the Himalaya the southern (see Bartz 1935). The species has also penetrated the gorges of the Himalaya and occupied the southern slopes of that range in various parts of India, Nepal, and Bhutan (Schaller 1977b; Wilson 1985). On the plateau, blue sheep occur on all major ranges—Gangdise, Tanggula, Aru, Jangngai, Anyemaqen, and others—as well as on many massifs and low rocky ridges. We sometimes came upon a herd seemingly far from typical habitat, but there always was

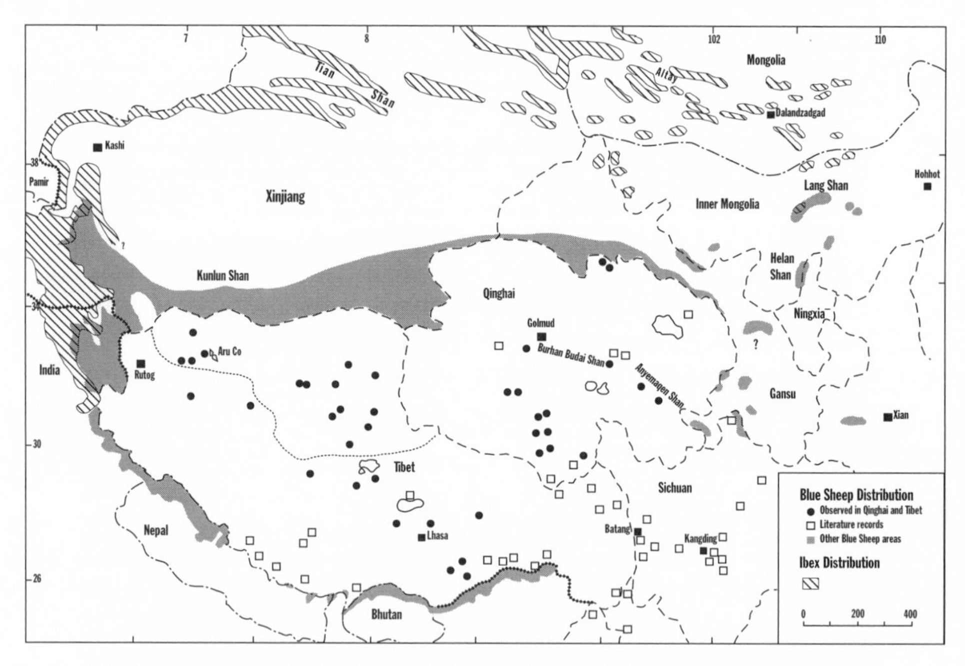

5.3. The distribution of blue sheep and ibex in China (excluding ibex in northern Xinjiang) and neighboring areas, showing range overlap in the two species. The distribution of blue sheep in Tibet, Qinghai, and western Sichuan is indicated by locality records from this project and from the literature; the distribution of the species in Gansu is probably wider than shown.

an isolated cliff in the vicinity, which the animals had reached by traversing gentle hills from rock outpost to outpost. In Sichuan and Gansu, along the eastern edge of the plateau and its marginal mountains, blue sheep occupy many rugged ridges, as on Gongga Shan near Kangding (Young 1935), on most passes between Kangding and Batang (Bailey 1945), and in southern Gansu (Wallace 1913).

Blue sheep penetrated north and northwest from the plateau across Gansu into Inner Mongolia almost to the border of Mongolia, occupying

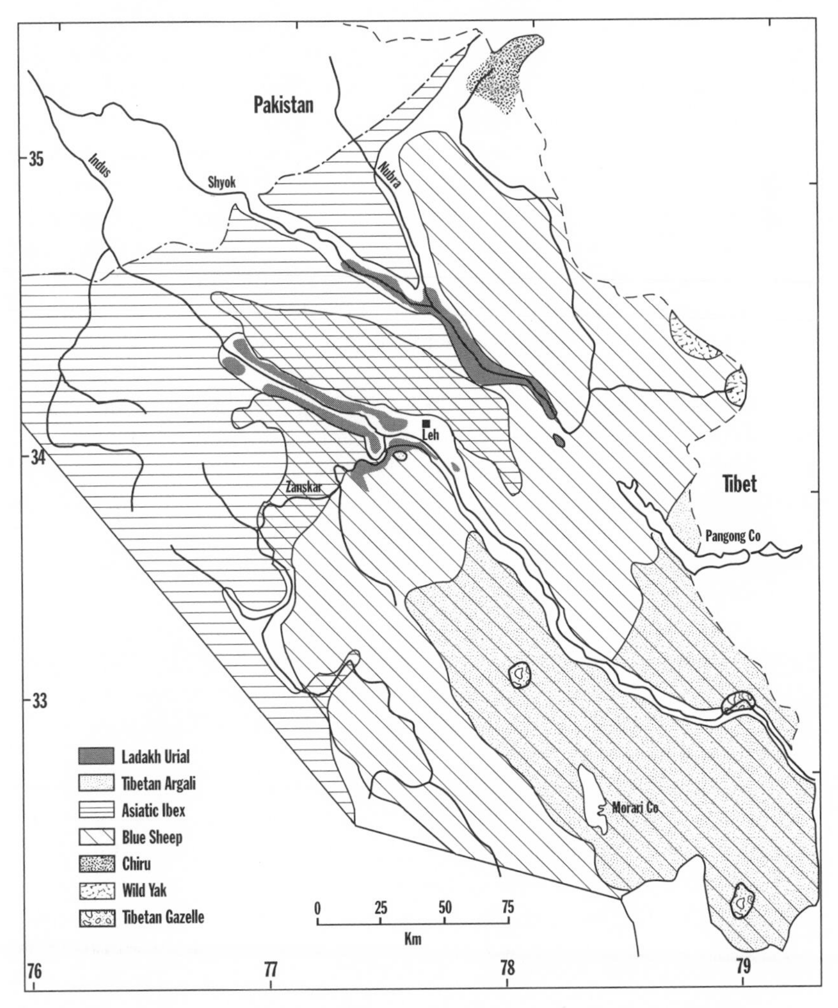

5.4. The distribution of wild ungulates in the Ladakh region of India. The kiang (not shown) has a distribution similar to the Tibetan argali. (Adapted from Fox, Nurbu, and Chundawat 1991a and Mallon 1991.)

a number of the ranges that rise above the semidesert and desert. These ranges include the Longshou Shan, Dongda Shan, Lang Shan, Zhuozi Shan, and the Helan Shan, where Przewalski (Prejevalsky 1876) first reported them. Their eastern limit in that region lies in some massifs at about 109°20′ E, just east of the Lang Shan; for unknown reasons they are not found in the nearby Ulan Shan and Daqing Shan. However, a late Pleistocene fossil was found in Hebei, 140 km northwest of Beijing, about 400 km east of the animal's current range (Wang and Hoffmann 1987).

Minimum crude densities of blue sheep in several areas of Ladakh were 0.7–1.4/km^2 (table 5.1), and the calculated total was 11,000 animals in their whole range of 38,000 km^2 (Fox, Nurbu, and Chundawat 1991a). Densities in four areas of Nepal ranged from a low of 0.7 to a high of 6.6–10.2/km^2.

We attempted to obtain total counts of animals in selected mountain blocks in Xinjiang and Qinghai by walking along ridges and valleys and scanning all slopes; we usually spent one to three days in a particular area. One block in Xinjiang's Taxkorgan Reserve had few blue sheep (0.2/km^2) and another a moderate density (2.5/km^2). Several blue sheep densities have been reported from farther east in Xinjiang. Gu (1990) estimated 10,000 blue sheep in the Arjin Shan Reserve. (However, "10,000" is also a Chinese hyperbole that means countless.) During extensive surveys in that reserve by Butler, Achuff, and Johnston (1986) and Achuff and Petocz (1988), only 44 were observed, and the latter authors noted that an "estimate of 10,000 animals is hardly possible." Luo and Gu (1991) calculated a population of 3948 blue sheep in 703 km^2—a density of 5.6/km^2—in the arid Arjin Shan northwest of the reserve. The area was subsequently promoted as a site for foreign hunters. The population figure is much too generous judging by my surveys in these mountains. In Qinghai, we selected survey areas after we had been told that many blue sheep inhabited them. Therefore, densities are representative of the best, rather than typical, conditions. Densities ranged from 1.1/km^2 in the arid Shule Nanshan to 3.6–5.3/km^2 in the Anyemaqen Shan and limestone massifs in southeastern Qinghai (table 5.1).

In the Aru Basin of northwest Tibet, blue sheep were found only in the Aru Range along the west side. We checked all suitable terrain in 1990, except the most southern part, and counted 121 blue sheep. Only 22 of these were adult males, so few that we presumed that many had moved elsewhere for the summer. Local informants affirmed that many males appear in winter. Our estimate was 200 blue sheep in summer and perhaps 300 in winter after the seasonal influx of males. Crude density of blue sheep in the 1800 km^2 basin was 0.1–0.2/km^2.

Table 5.1 Minimum crude densities and population compositions of blue sheep in several areas

Locality	Date	Area (km²)	No. of animals	Density (km²)	Ratios (/100 ♀♀) Ad. ♂♂	Yearling ♂ and ♀	Young
Ladakh							
Shang Reserve	7/1986	190	235	1.2			
Hemis Nat. Park	11/1984	1200	1236	1.0			
Math Valley	7/1989	85	60	0.7			
Stock Valley	7/1989	75	108	1.4			
Nepal							
Dhorpatan Reserve	1977	960	700–740	0.7–0.8	102	80	83
Manang District	4–5/1990	105	697–1071	6.6–10.2	93	33	51
Lapche	3/1972	35	50	1.4	116	82	88
Shey	11–12/1973	550	500–700	0.9–1.3	134	29	40
Xinjiang							
Taxkorgan Reserve (Raskam)	5–6/1985	150	31	0.2			
Taxkorgan Reserve (Mariang)	5–6/1985	120	265	2.5			
Qinghai							
Shule Nanshan	9–10/1985	610	657	1.1	71	28	65
Anyemaqen (east)	9/1986	170	906	5.3	0	37	64
Anyemaqen (west)	7/1984	50	237	4.7	15	22	53
Zadoi (north)	8/1986	220	798	3.6	90	29	65
Zadoi (south)	9/1986	185	949	5.1	134	37	75
Yushu	8/1984	140	222	1.6			
Tibet							
Aru Basin	7/1992		87		7	26	28
Noorma Co	9/1990		147		25	38	53
Jangngai Shan	7/1991		71		72	48	31*
Qomolangma	3–6/1992		543		99	40	56

Sources: Ladakh data: Fox, Nurbu, and Chundawat 1991a. Dhorpatan: Wilson 1981. Manang: Oli 1994c. Qomolangma: Lu, Jackson, and Wang 1994. Other locales: this study and Schaller 1977b.
*Some females were still pregnant.

Counts without density estimates also indicate that blue sheep are quite abundant in some places on the plateau. In Gansu's Yanchiwan Reserve, Zheng et al. (1989) observed 469 blue sheep. I counted 300 blue sheep in the Balong Valley of the Burhan Budai Shan southwest of Dulan on 25–26 November 1986, during an incomplete survey of the drainage. In southeast Qinghai, near Gar monastery, Harris (1991) observed 286 blue sheep in about 4 km² in 1990 and he stressed that such local abundance was due to the protection provided by the monastery. Similarly, Kaji et al. "counted a total of 991 animals in only two days" around a temple in that region

(1993). In just three small adjoining valleys of a range bordering Noorma Co, I counted 191 blue sheep on 2–3 September 1990, my highest such number in Tibet, and near Garco 143 were tallied on 29 October 1993.

I would guess that at least 10,000 blue sheep inhabit the Chang Tang Reserve, but beyond that I cannot make realistic estimates of numbers. Pu (1993) gave figures of 34,500–49,600 blue sheep in western Tibet and surmised that these represented 70–80% of the total Tibet population, but his total estimate appears to be too low, judging by the wide distribution of the species in eastern Tibet.

Sympatry of blue sheep and ibex

The distribution of blue sheep is probably affected by that of the Asiatic ibex.[1] Ibex are similar to blue sheep in body proportions and size, habitat choice, and ability to survive in a variety of environments. Studies of chromosomal evolution indicate that blue sheep represent an early independent lineage from ancestral caprid stock (Bunch and Nadler 1980). One can hypothesize that blue sheep had their origin on the plateau and expanded to the northeast as ibex arrived from somewhere in the northwest and west to occupy the Altay, Tian Shan, and other ranges that blue sheep had not yet reached (fig. 5.3). The two species are perhaps ecological competitors in that the sparse and patchy habitat in the mountains cannot easily support both of these generalized feeders. The presence of one may have restricted the other from expanding its range both horizontally and vertically. The demarcation between the two is often abrupt. On a small range bordering the main Taxkorgan valley, blue sheep occur primarily on the eastern slopes and ibex on the western; similarly, blue sheep inhabit the northern flanks of the Karakoram and ibex the southern. On the north-facing slopes of the Kunlun Shan, at about 37°N, 77°E, local people told us that ibex occupy the middle-to-lower slopes at 2800 m and blue sheep the highest at 4600 m, and we saw horns in support of this. It is unclear how far ibex extend eastward along the Kunlun Shan. About 1100 km east of where we found evidence of ibex is the Qiman Tag, a spur of the Arjin Shan along the northern border of the Arjin Shan Reserve. Achuff and Petocz (1988) reported that an ibex skull had supposedly been found in that range in 1982. The local people we interviewed there knew of no such animal as the ibex. In *Economic Birds and Mammals in Qinghai* (1983), *Capra ibex* is said to occur "in high mountains of Qilian Shan." The easternmost known

1. The Asiatic ibex is either considered to be a subspecies of *Capra ibex*, *C. i. sibirica*, or a distinct species, *C. sibirica*. *C. sibirica* is said to have four subspecies within the area here discussed: *alaina* in the Tian Shan and western Xinjiang, *dementievi* in the Kunlun, *hagenbecki* in the Gobi, and *sibirica* in the Altay. The validity of these subspecies needs investigation.

range of the ibex on the plateau is in the southwest corner along the Sutlej River in India soon after it leaves Tibet (fig. 5.3); whether ibex enter Tibet there is not known. The only record for Tibet is of one set of horns obtained near the Indian border north of Panggong Co (Liu and Yin 1993), an area from which ibex have not been reported previously (see fig. 5.4).

Several areas of overlap between blue sheep and ibex are known. In Ladakh, Fox, Nurbu, and Chundawat (1991a) found that blue sheep occur in the eastern half and ibex in the western half and that the two species occupy several of the same ranges in the center (fig. 5.4). In the Taxkorgan Reserve, the two species overlap along several tributaries of the Yarkant River. On two occasions we observed the species on the same slope, once foraging as a mixed herd (Schaller et al. 1987). In a small part of Pakistan's Khunjerab National Park adjoining the Taxkorgan Reserve, ibex and blue sheep overlap but with the former in more precipitous and broken terrain (Wegge 1989). Another contact area is in the Spiti Valley of India, where Rees (1995) observed mixed herds of ibex and blue sheep. Ibex penetrated from the north and west into northern Gansu and Inner Mongolia, where, for example, they are found in the Mazong Shan. But on only one mountain range, the Lang Shan, are the two species found together (Wang and Schaller 1996).

Population and herd dynamics

We attempted to classify all individuals within a herd as young, yearling (1–2 years), adult female, and adult male, but often the shyness of the animals made this impossible; yearling males and females were also difficult to distinguish at a distance. Although fragmentary, our data supplement those published earlier by Schaller (1977b), Wegge (1979), Wilson (1981), Oli and Rogers (1996), and others.

The sex ratio of adults differed greatly from area to area. Sex ratios in Nepal were about equal or showed a small preponderance of males (table 5.1), a typical situation in wild sheep and goats with low predation pressure. Several populations on the plateau also had a relatively high male-to-female ratio (70:100 or more), whereas others had extremely few males (25:100 or fewer). In a sample of 827 animals combined from several areas in Qinghai, Kaji et al. (1993) found a ratio of 13:100 (yearling males were included with the females), whereas in a sample of 632 animals from western Sichuan it was 52:100. Subsistence hunters may prefer males because these provide more meat, but this was unlikely to be the main reason for the disparate ratios. Blue sheep populations appeared to have two patterns of male behavior. Kinloch noted that "the males for the most part separate from the females during the summer months, but mixed herds may be seen at all seasons of the year" (1892). This is correct, but in some areas most males remain in the same area as the females whereas in others they seek separate summer ranges. An extreme case was in the eastern Anyemaqen Shan, where we saw only 8 adult males among 906 blue sheep in 170 km^2. Based on his research on bighorn sheep, Geist (1971) explained such spatial segregation on the basis of reducing competition for forage between males and females with their offspring. Wegge (1979) and Wilson (1984) observed that on the fine alpine meadows of Dhorpatan seasonal segregation did not occur and this was true in the Zadoi area as well. But if forage abundance and quality stimulate segregation, it remains unclear why the sexes in the lush Anyemaqen Shan separate widely but those in the arid Shule Nanshan do not.

Published accounts differ regarding the months when blue sheep are said to mate and give birth. In the Dhorpatan area of Nepal the main rut, as defined by intense courtship and mounting, extended from about 15 December to 28 January 1976 with a peak in early January. After a gestation period of 160 days, the first newborn was observed on 27 May, and peak parturition was 13–18 June but with some young born into early July (Wilson 1984). The main rut in the Shey area of Nepal began about 29 November 1973 (Schaller 1977b). In the central Himalaya, Bailey (1911) observed that the birth season is in June and July, indicating roughly a January rut. In Inner Mongolia the rut is in November and the young are born in May (Prejevalsky 1876). In western Sichuan, according to Schäfer (1937a), blue sheep rut in October and "lambs are dropped in early May" (Allen 1939), months that fail to match the known gestation period. In Gansu, the rut is in January and "lambs are dropped in May" (Wallace 1913), again a problem with duration of pregnancy.

In the Taxkorgan Reserve, we saw no newborns in May 1985; the following year we noted newborns and small young between 7 and 22 July. On the plateau, we observed both small young and pregnant females in the Jangngai Shan on 4 July 1991 and elsewhere a newborn on 11 July. Births were generally estimated to occur between about mid-June and mid-July, indicating a rut from late December to early February. Two young in the Lhasa zoo were born during the first half of June. Bailey (1911) reported a captive birth in Gyantze on 8 August, which he considered exceptionally late. High altitudes and seasonal patterns of precipitation delay the nutritious spring growth that is so important to females during late pregnancy and lactation. The differences in parturition dates, which vary by at least a month, can probably be ascribed to the availability of high-quality forage.

Weather and level of nutrition can have a marked effect on reproductive success in any one year, as many ungulate studies have shown. Reproductive success in blue sheep, as expressed by the ratio of young to 100 adult females, varied greatly from area to area. In Nepal, the ratio ranged from 40:100 at Shey to 83:100 at Dhorpatan, and on the plateau it was a low 28:100 in the Aru Basin to a high 75:100 in Zadoi (table 5.1). Females have as a rule only single young. On the rare occasion when a second young trailed a female, it may have been a temporary association. Young may also aggregate into crèches: once I observed a herd of 3 females with 10 young and another time 5 females with 21 young. In areas with good habitat, as at Dhorpatan, a small percentage of yearling females conceive and give birth at the age of 2 years (Wegge 1979; Wilson 1981). Range conditions probably explain some of the variability in the ratios of young to females. Populations on good habitat tend to be vigorous, producing many young, whereas those on poor habitat are often stagnant (see Geist 1971). The Shey blue sheep were crowded for the winter into a small overgrazed range (8.8–10.0/km^2). There, although protected by religious sentiment, they seemed to reproduce poorly but lived a long time, judging by the many old males. The Aru Basin also had few young for unknown reasons. Perhaps the small sample was biased or perhaps heavy snows that year during the birth season affected survival.

Fecundity in a blue sheep population probably remains fairly constant from year to year, as it does in ungulates whose food supply is quite predictable (see Schaller 1977b), and the difference between the number of young and yearlings provides a rough estimate of mortality. The ratio of yearlings (male and female) to adult females ranged from 26:100 to 40:100 in most instances. An exception was in Nepal, where a large count revealed a ratio of 80:100 (table 5.1). The decline between young and the yearling class in the Qinghai populations was 42–58%, at Shey 27%, and at Noorma Co 28%. Young, which need to invest much energy in body growth, no doubt enter the stressful winter with low fat reserves, making them susceptible to malnutrition, disease, and predation. At Dhorpatan in Nepal about 50% of the blue sheep died between birth and 2 years of age, most during winter (Wegge 1979).

Adults have a relatively long life as determined from the growth rings of horns of males found in the field on the plateau (fig. 4.4); most animals had been killed by hunters. Yearlings are not fully represented in the sample because their small horns (<15 cm) are easily overlooked. Over 80% of the males died between 4 and 11 years of age. The sample comprised 10% young males (1–4 years), 73% prime males (4–10 years), and 17% past-prime males (11–15 years), a longevity pattern similar to that found in Taxkorgan except that past-prime animals were more abundant there (Schaller et al. 1987). The death rate of females at Dhorpatan was estimated at 22.5% per year (Wegge 1979). In contrast to blue sheep, argalis seldom lived beyond the age of 9 years (fig. 4.4). At that age, the first molar

Table 5.2 Sample blue sheep herds of mixed composition observed during summer in Qinghai and Tibet

Location	Date	Adult ♂♂	Yearling ♂♂	Adult ♀♀	Yearling ♀♀	Young	Total
Zadoi	8–9/1986	2	—	4	1	2	9
		5	2	6	3	4	20
		4	1	9	1	8	23
		10	1	8	1	5	25
		17	1	8	1	6	33
Yushu	8/1984	2	3	21	8	5	39
Siling Co	9/1988	3	—	9	3	6	21
		2	2	10	1	6	21
Noorma Co	9/1990	1	3	40	7	34	85
		2	4	28	12	2	48

of argalis tended to be heavily worn, whereas that of blue sheep usually did not reach this stage for another year or two. The contents of predator droppings (chapter 11) revealed that snow leopard and wolves often preyed on blue sheep. In Nepal, common leopards also ranged upward to 4000 m and killed this species (Wilson 1981).

In daytime, animals are most active in the early morning before 0930 and in the late afternoon after 1600 but they often have a feeding bout around midday as well (Schaller 1977b; Wilson 1984). When foraging they are easier to observe than when reclining, and they are less likely to detect a stalking naturalist. Disturbed animals bunch up and dash out of sight or toward a cliff.

In blue sheep, as in other caprids, the basic social unit and only stable entity consists of a female and young and sometimes a yearling too. But the animals are gregarious, associating in herds of up to 200 (Stockley 1928) and even 400 (Schäfer 1937a). In the Helan Shan the animals were usually in herds of 5–50 but with up to 100 (Prejevalsky 1876). In the Dhorpatan area the largest group comprised 44 individuals (Wegge 1979); at Shey, 61; in the Taxkorgan Reserve, 54; in the Qomolangma Reserve, 35 (Lu, Jackson, and Wang 1994); in the Aru Basin, 52; near Garco, 71; at Noorma Co, 85; in the Anyemaqen Shan, 139; and in the Zadoi area, 165; and at Donggi Co in Qinghai, 175 (Kaji et al. 1993); to give a few examples. However, herd size depended on season, population size, habitat condition, hunting pressure and disturbance, and other factors. A few solitary individuals aside, herds consisted of three types: male herds, female herds with or without young and yearlings, and mixed herds with subadults and adults of both sexes. Table 5.2 presents the composition of a sample of mixed herds.

Adult males may form herds with as many as 40 members (Schäfer 1937a). Cai, Liu, and O'Gara (1989) noted a herd of 39 males in the Bur-

Table 5.3 Percentage of adult male blue sheep in male herds and mixed herds during summer in Qinghai and Tibet

	Total no. of animals	Total no. of adult ♂♂	% ♂♂	% ♂♂ in ♂ herds	% ♂♂ in mixed herds
TIBET					
Several localities	535	121	22.6	76.0	24.0
QINGHAI					
Qilian Shan	347	94	27.1	21.3	78.7
Anyemaqen Shan	942	25	2.7	0	100.0
Zadoi	947	342	36.1	19.0	81.0

han Budai Shan, and one herd in the Taxkorgan Reserve contained 52 males and 2 eleven-month-old young of unidentified sex. On the plateau during summer we tallied one solitary male and 33 male herds with a mean of 7.5 (2–26) members. Around Zadoi and the Qilian Shan—one habitat lush and the other arid—about 20% of the males were in male herds and the rest in mixed herds; the few males we saw in the Anyemaqen Shan were all in mixed herds (table 5.3). Of 143 blue sheep classified near Garco in late October 1993, there were 19 adult males, 9 yearling males, 83 adult and yearling females, and 32 young; all males were associated with the females. During winter at Dhorpatan before and during the rut, Wilson (1984) found that 90% of the males were in mixed herds, 9% in male herds, and 1% solitary, and Schaller (1977b) reported that at Shey 66% were in mixed herds before the rut and 80% during the rut.

Female herds averaged 14.5 animals (n = 40) and mixed herds 23.2 animals (n = 57). These averages do not include many large herds, because we could not classify every individual. We tallied a total of 5818 blue sheep in 198 female and mixed herds during summer and autumn on the plateau, for an average of 29.0 animals per herd. Two-thirds of the herds had 25 or fewer members, but only 27.5% of the individuals were in these groups (fig. 5.5). Another 28.0% of the individuals were in herds with 26–50 members and the rest were in large herds. Kaji et al. (1993) tallied 85 herds in Qinghai and Sichuan, with an average of 33.2 (range 2–175) members. In the Manang area of Nepal, Oli and Rogers (1996) noted a mean herd size of 15.6 but with significantly smaller herds in winter than at other seasons.

Conservation

Blue sheep populations are affected heavily by hunting, so much so that many small ones have no doubt been exterminated. Subsistence hunting

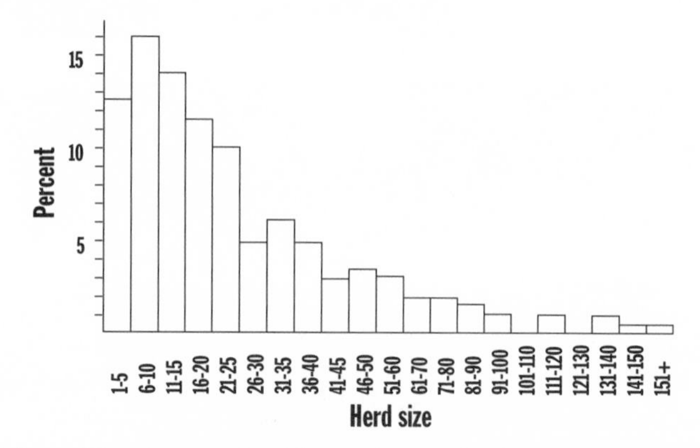

5.5. Herd sizes of blue sheep on the Tibetan Plateau. The data are based on 198 herds; they exclude solitary males and male herds.

for this species was and still is widespread. Horns are often found around tents and corrals. Near Zadoi, I came upon a tent with three men chiseling Buddhist scriptures into stones that they would later place upon prayer walls. A freshly shot female blue sheep was part of their larder. Blue sheep are sometimes surrounded on an isolated cliff by a group of men and then killed; I was shown one such site with seven fresh heads. Or, as a hunter explained to Goldstein and Beall: "as soon as I spot blue sheep on a mountain slope I turn loose my dogs. Their job is to corner one of the *na* among the crags, and bark loudly to lead me to the spot. . . . Once I get there I have plenty of time to set up my rifle and shoot" (1990). In western Nepal, local people place poisoned bamboo spears along trails to kill blue sheep and other wildlife (Jackson 1979a). In Nepal, foreign hunts for blue sheep began near Dhorpatan in the Dhaulagiri Range in 1971 (Wilson 1981), and the selection against large males has greatly affected the age structure of some populations (Wegge 1979). Similar hunts were initiated in Qinghai in 1985 (see chapter 15). Starting in 1958 and continuing for two decades, blue sheep were also killed commercially in Qinghai. About 100,000–200,000 kg of blue sheep meat were exported yearly to Europe, principally to Germany. Local records from Dulan showed that 6641 blue sheep were killed between 1981 and 1985 and 64,050 kg of meat were purchased by the government from that area (Cai, Liu, and O'Gara 1989). However, with blue sheep still widespread and locally abundant, the species is ideally suited for a well-regulated management program that could include a sustainable annual harvest in certain areas for subsistence and commercial purposes.

6
Tibetan Gazelle

> The men went out to shoot antelopes (*Pantholops Hodgsoni* and *Procapra picticauda*), which were to be seen grazing here and there in the wide valley. A herd of wild asses (*Asinus kiang*) were seen across the stream opposite our camp, and on the other bank of the lake was a large herd of wild yaks. The fabulous quantity of wild mammals to be found everywhere in north-east Tibet can be accounted for by the almost complete absence of their worst enemy—man.
>
> Pyotr Kozloff (1908)

FIVE SPECIES OF GAZELLE, including the saiga, occur in central Asia. Of these the Tibetan gazelle is endemic to the Tibetan Plateau, a small and graceful animal which is or was almost ubiquitous, the most frequently seen wild ungulate in many areas of the alpine steppes and meadows. Along the northern edge of the plateau, the range of the Tibetan gazelle overlaps with that of the goitered gazelle and, in the northeastern corner, with that of Przewalski's gazelle. This chapter focuses on the Tibetan gazelle but also provides information on goitered gazelles for comparison.

Taxonomy

Two genera in central Asia are usually given the popular designation of "gazelle." One is *Gazella*, represented by the goitered gazelle, and the other is *Procapra*, with two definite species: the Tibetan and Przewalski's gazelles. The Mongolian gazelle is often designated *Procapra* as well, but sometimes it has been placed into its own genus, *Prodorcas* (Pocock 1918; Allen 1940), as discussed in chapters 13 and 14. The skulls of *Gazella* can be distinguished from those of *Procapra* by the presence of a preorbital depression and short and broad, rather than long and narrow, nasal bones (Groves 1967). The tails of *Gazella* are relatively long and thin, whereas those of *Procapra* are stumpy, to name one conspicuous external difference. *Saiga*, though in the past often classified with the chiru in the tribe Saigini of the subfamily Caprinae, also belongs with the gazelles, subfamily Antilopinae, as shown by recent morphological and molecular studies (see chapter 13).

6.1. A female and two adult male Tibetan gazelles wade through snow following the October 1985 storm. The large white rump patch and short tail of the species are evident.

There are no definite subspecies of Tibetan gazelles, *P. picticaudata*. However, at least minor differences in the species exist within its range, with, for example, the animals in southeastern Tibet near Mount Kailas and in the Yüsüpalik Tag in Xinjiang seemingly larger than elsewhere.

Description

The Tibetan gazelle has a compact body and slender legs and stands about 60–65 cm high at the shoulder. Its coat is sandy brown to grayish brown, grayer in summer than in winter. The fronts of the legs are light gray, and the inside of the legs and the belly are white. The animal lacks conspicuous facial markings and a lateral stripe. The rump patch is white, large, and heart-shaped, and its hairs are erectile, often fanning out as the gazelle flees. A band, light rust in color, borders the rump patch, which surrounds the short (8–9 cm) black-tipped tail (fig. 6.1). Four adult females that died following a blizzard weighed 13.2–15.0 kg, and three males weighed 14.1–14.5 kg. Przewalski gave a weight of 16 kg (Prejevalsky 1876). Only males have horns. These are slender and curve up, back, and then up again, and they are ridged as far as the distal quarter; they diverge relatively little and

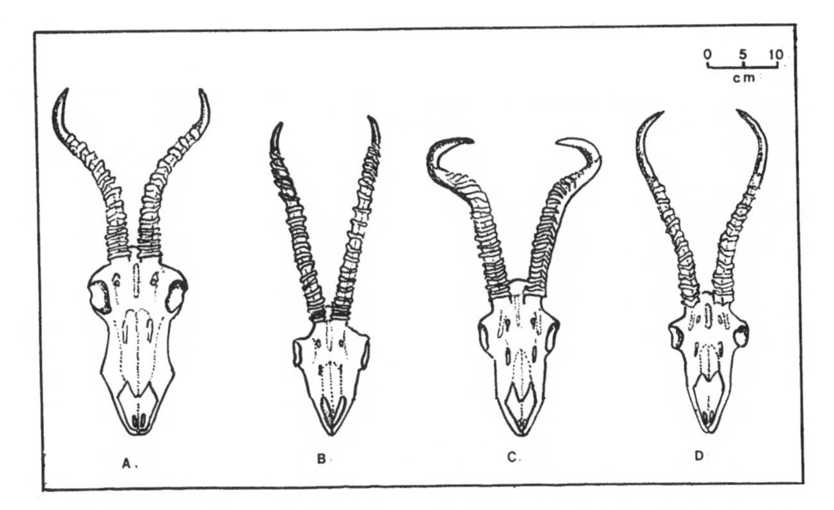

6.2. Horn shape and length of adult males of the four gazelle species in China: A. Mongolian gazelle; B. Tibetan gazelle; C. Przewalski's gazelle; D. Goitered gazelle.

the tips often bend in slightly (fig. 6.2). Horn length of adults averaged 28.9 cm (26.0–32.2 cm; n = 21), and the distance between tips averaged 13.1 cm (10.3–16.8 cm, n = 10).

Status and distribution

Tibetan gazelles are animals of open landscapes, of plains, hills, and even mountains, where they may be found both in broad valleys and high on ridges above timberline if the terrain is not precipitous. Their principal habitat is alpine meadow and alpine steppe. They avoid much of the desert steppe and other arid areas, which have few of the small forbs that are their main forage.

Outside China, small gazelle populations occur in northern Sikkim (Shah 1994) and Ladakh. Gazelles were once quite common in Ladakh, but early this century Burrard found that "they are getting more scarce every year" because of hunting (1925), and they are now close to extinction, with perhaps fewer than 50 individuals (Fox, Nurbu, and Chundawat 1991a) in a few localities (fig. 5.4).

The eastern distributional limit of the species is in western Sichuan and Gansu (fig. 6.3). "The *Gowa*, or Tibetan gazelle, is constantly met with in grass country all over Eastern Tibet," in an area that is now western Sichuan (Teichman 1922). There the species occurs "as far south as four days below Batang and an undetermined distance south of Litang, and as

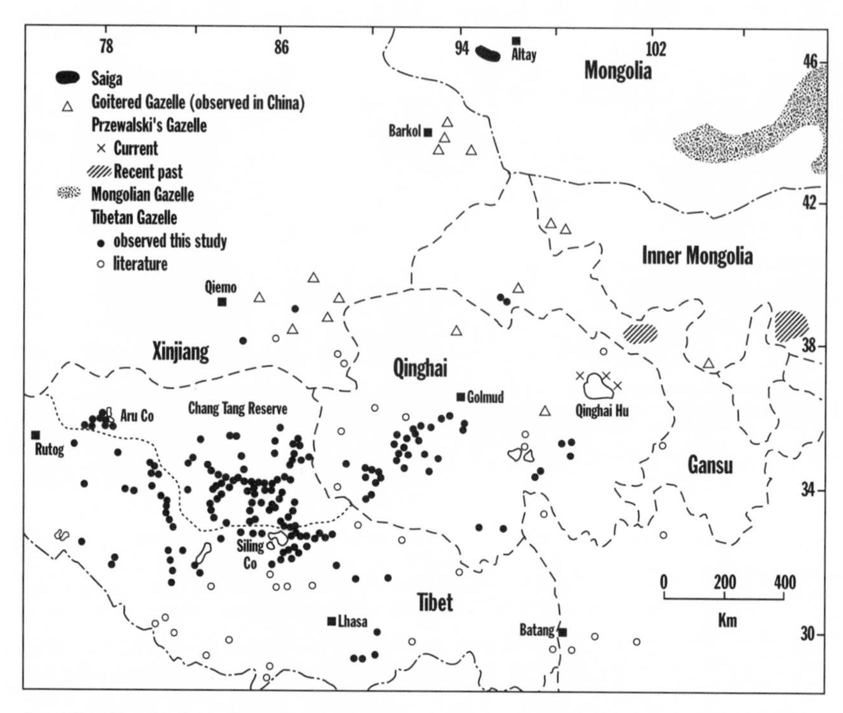

6.3. The distribution of gazelles on the Tibetan Plateau and neighboring areas, based on personal observations and some literature records.

far east as Sungpan" (Allen 1939). In the late 1980s, Kaji et al. (1993) found the animals still fairly common in northwest Sichuan. In Gansu, we observed gazelles only in the Qilian Shan, on a high plain in the Yanchiwan Reserve.

The alpine meadows around the headwaters of the Yellow River in eastern Qinghai once represented the finest wildlife area in central Asia, as conveyed by Kozloff's (1908) quote at the beginning of this chapter and by other travelers (Migot 1957; Prschewalski 1884). Gazelles were abundant there until the 1950s. Today one can traverse these grasslands for hours and seldom glimpse a gazelle. I came across a small aggregation just east of Donggai Co, and the species persists around Gyaring Co (Kaji et al. 1993). We tried to obtain a count of gazelles in two census blocks totaling 220 km^2 in the Anyemaqen Shan by walking valleys and ridges and scanning all slopes. The effort yielded 6 gazelles. Three census blocks totaling 545 km^2 near Yushu and Zadoi produced 14 gazelles. In two valleys of the Burhan Budai Shan—in the Hong Shu Chuan area and the Yeniugou—Cai, Liu, and O'Gara (1989) estimated a gazelle density of 0.78

animals/km^2. Harris (1993) calculated a density of 0.39/km^2 in the Yeniugou. Gazelles were relatively sparse in the eastern part of the Chang Tang. We transected a block of 2100 km^2 by car near Wudaoliang, counting all gazelles in a census-strip width of 600 m, and calculated 121 gazelles, for a crude density of 0.06/km^2. Similarly, Feng (1991b) surveyed gazelles in five areas totaling 1740 km^2 west of our site and derived a density of 0.08/km^2. In spite of low average densities, gazelles could readily be observed in certain localities, such as in the valleys which bisect the region. For example, on 3 November 1993 we tallied 31 gazelles in 22 km along the Tuotuohe Valley just west of Tuotuohe. Gazelles persist even along the Lhasa-Golmud highway: on 2 November 1993 we counted a total of 52 gazelles in about 30 km in a broad valley bordering the road as it winds through the Tanggula Shan.

The plateau in Xinjiang is arid, not a favored habitat of Tibetan gazelles. In a survey within an area of 6000 km^2 we tallied just 4 gazelles. In the Arjin Shan Reserve, intensive surveys in the western half (23,000 km^2) yielded 9 gazelles (Achuff and Petocz 1988); in the eastern half a brief survey reported 115 gazelles (Butler, Achuff, and Johnston 1986).

In Tibet, gazelles have a patchy distribution in the mountainous eastern part (Feng, Cai, and Zheng 1986) and along the foothills of the Himalaya (Bailey 1911; Waddell 1905; Jackson 1991; Piao and Liu 1994). During October 1995 we traveled widely between the valley of the Lhasa River south to the border of India and Bhutan. Gazelles were seldom encountered—only 23 were tallied—except on the steppes around Chigo Co (28°40′ N, 91°40′ E), where 253 were observed, including a concentration of 114 in a 1 km^2 area.

Gazelles are widespread in the Chang Tang, though now scarce on the alpine meadows of the eastern part where human and livestock densities are high. Farther west, on the *Stipa* steppes, Bower noted that "the Tibetan gazelle (*Gazella picticaudata*) is even more widely distributed than the antelope, but is nowhere so numerous as antelope are in certain places" (1894). Most travelers did little more than mention them. For example, at Lumajangdong Co, "*Pantholops* antelopes were very numerous in flocks on the floor of the latitudinal valley, and amongst them a few individuals of the small *Gazella* antelopes were grazing" (Hedin [1922] 1991).

We attempted to count all gazelles in certain areas by crisscrossing accessible parts by car and by walking along ridges and valleys in search of animals. The 1800 km^2 Aru Basin had a crude density of 0.07 gazelles/km^2 (table 6.1); the 300 km^2 Yalung Basin, 0.04/km^2; and the 300 km^2 census block near Garco, up to 0.12/km^2 (table 3.2). A minimum count in a 10,500 km^2 area between Shuanghu and Yibug Caka gave 404 gazelles, or 0.04/km^2 (table 6.1). But gazelles are difficult to spot at a distance and

Table 6.1 Minimum crude densities of plains ungulates observed in three census blocks, based on attempted total counts by vehicle

	Shuanghu–Yibug Caka*		Aru Basin	Toze Kangri
	12/1991	10/1993	(7–8/1990)	(6–7/1992)
Total area (km^2)	17,500	10,500	1,800	8,000
Vegetation zone	*Stipa*	*Stipa*	*Stipa-Carex*	*Carex-Ceratoides*
Number (number/km^2)				
Chiru	3,900 (0.22)	3,066 (0.29)	635 (0.36)	Not censused
Tibetan gazelle	352 (0.02)	404 (0.04)	125 (0.07)	0
Kiang	1,224 (0.07)	1,229 (0.12)	212 (0.12)	20 (0.003)
Wild yak	13	2	681 (0.39)	73 (0.009)

*The two censuses covered the same central part, but the 1993 census did not include some peripheral areas.

the actual density in that large area may have been at least twice that figure. Piao and Liu (1994) observed 1823 gazelles in 11,281 km of driving through various parts of Tibet and calculated an average density of 0.34/km^2, a figure which is not representative, because gazelles are often concentrated in basins or valleys. Since vehicles also favor such terrain, estimates based on one transect through an area may skew results. On crossing a basin southwest of Shuanghu in 1990, we counted 77 gazelles in 63 km within 300 m on each side of the vehicle, a density of 2.0/km^2. Driving east from Siling Co toward Amdo near the southern reserve border, we tallied 159 gazelles in 140 km, or 1.7/km^2. By contrast, we drove cross-country west from near Dogai Coring to the Yako Basin and in 162 km failed to see a gazelle. No gazelles were seen during a survey of an 8000 km^2 area of desert steppe in the Toze Kangri area north of the Aru Basin.

The extremely low density of gazelles in the northern quarter of the reserve is of significance to conservation. It emphasizes that the species can survive in appreciable numbers only on good pastures, and most of these have been occupied by pastoralists. Gazelles are scarce on these rangelands only because they have been and still are much hunted. Over 100 years ago, Przewalski wrote that the animal "is extremely wary, especially in those districts where it has learnt to fear man. . . . Its swiftness is amazing; it bounds along like an india-rubber ball, and when startled seems absolutely to fly" (Prejevalsky 1876). Such behavior has enabled gazelles to survive. But they are still killed widely and casually for food. As Rawling noted, "the flesh is delicious" (1905).

Gazelle numbers in the Chang Tang Reserve

The reserve was divided into a grid to determine gazelle numbers. For each quadrat our approximate travel distance and gazelle density (animals/

Table 6.2 Mean densities of Tibetan gazelles and kiangs in the three major vegetation zones of the Chang Tang Reserve, based on vehicle transects

	Vegetation zone		
	Stipa	*Stipa-Carex*	*Carex-Ceratoides*
No. of quadrats sampled (see fig. 6.4)	57	41	11
Tibetan gazelle			
No. of quadrats without gazelles	28	27	11
Mean density/km^2	0.26*	0.07	0
Kiang			
No. of quadrats without kiangs	26	17	7
Mean density/km^2	0.39*	0.15	0.15

*Biased result; figures presumed too high (see chapters 6 and 10).

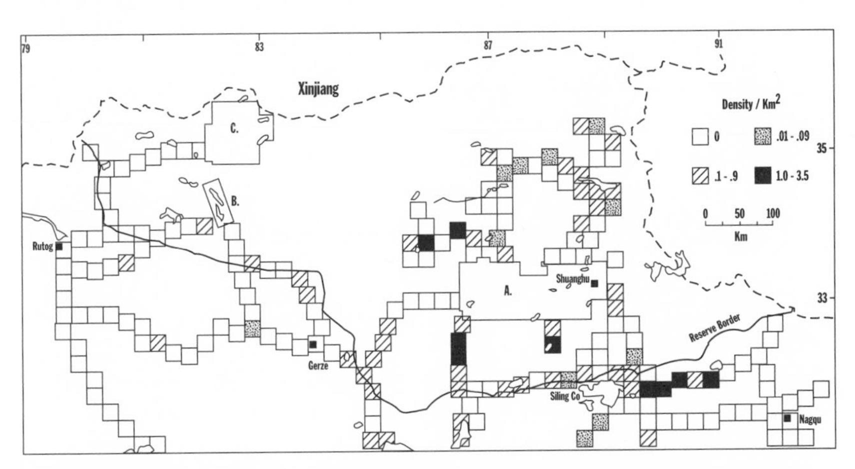

6.4. Mean density (per km^2) of Tibetan gazelles in and adjoining the Chang Tang Reserve, based on vehicle transects. A quadrat in the grid is 625 km^2, but only a fraction of each quadrat was transected. The densities reflect the patchy distribution of species. The three blocks marked A, B, and C indicate areas in which total counts were attempted.

km^2) were recorded, the latter determined by counts of animals within a total transect width of 600 m. Some routes were driven more than once but calculations were usually based on the first trip. Certain quadrats were barely entered, whereas travel in others was extensive, and this affects the calculations. The results indicate that gazelle distribution was patchy and density seldom exceeded 1.0/km^2 (table 6.2, fig. 6.4). Density calculations were based on the three vegetation zones (see fig. 2.7). The *Stipa* zone in the southern half had the highest mean density (0.26/km^2), the *Stipa-Carex* zone had an intermediate density (0.07/km^2), and the *Carex-Ceratoides* zone, or desert steppe, had the lowest (none/km^2). The results for the *Stipa-Carex* zone corresponded to a census in the similar habitat of the Aru Basin, and in the *Carex-Ceratoides* zone neither transect nor census revealed gazelles (table 6.2). The transect result of 0.26/km^2 in the *Stipa* zone is much higher than the census figures of 0.02–0.04/km^2 in the Shuanghu–Yibug Caka area. I think the former density is too high because travel routes often coincided with good gazelle habitat, and the latter too low mainly because the census was not intensive enough for these small ungulates. The actual density in this zone probably was between 0.10 and 0.13/km^2.

To convert density to gazelle number, the extent of each vegetation zone was calculated (see table 2.1); lakes, glaciers, and barren peaks (11.6% of the reserve) were excluded. With a reserve size of 334,000 km^2 and a gazelle density of 0.10–0.13/km^2, the *Stipa* zone (130,260 km^2) would have 13,026–16,934 gazelles, and at a density of 0.07/km^2, the *Stipa-Carex* zone (78,490 km^2) would have 5494 gazelles, for a total of 18,520–22,428. (At the official reserve size of 284,000 km^2 the total would be 15,748–19,071 gazelles.)

The number of Tibetan gazelles in Tibet or on the plateau as a whole cannot be estimated with any degree of accuracy, but the total may be around 100,000.

Sympatry with other gazelle species

The goitered, or black-tailed, gazelle (*Gazella subgutturosa*) has a vast range in Asia, from the Arabian peninsula to Mongolia. The subspecies *G. s. yarkandensis* is said to occupy the flanks of the mountains surrounding the Taklimakan Desert and *G. s. hillieriana*, the deserts to the north, including those in Mongolia, and to the south, including the Qaidam Basin of the Tibetan Plateau (Groves 1969). I observed goitered gazelles in several localities on or near the plateau (fig. 6.3) but am unclear which subspecies. Although the geographic ranges of Tibetan and goitered gazelles overlap, they use distinctly different habitats, the former preferring alpine meadow and alpine steppe and the latter semidesert in which shrubs are an important component of the vegetation (fig. 6.5). In two areas of the plateau where we noted overlap in the species—in the Arjin Shan and in the Qaidam Basin—the goitered gazelles were on arid flats and Tibetan gazelles in the surrounding mountains. Similarly, in Gansu's Yanchiwan Reserve, Tibetan gazelles were on an upland steppe at 3300 m and above, and the goitered gazelles were in desert and semidesert below (Zheng et al. 1989).

The least-known gazelle in central Asia is Przewalski's gazelle (*Procapra*

6.5. A goitered gazelle male browses in Mongolia's Great Gobi National Park. His long tail and small goiter are visible.

przewalskii). Allen (1940) considered it a subspecies of Tibetan gazelle, but measurements by Groves (1967, 1985) showed it to be a valid species with two subspecies. *Przewalskii* is larger than *picticaudata*, weighing about 25 kg (Jiang et al. 1994) and its stout horns are distinctive, curving first sharply outward then inward near the tips (fig. 6.2). The known range of the species in the past century has been confined to three small areas. The animals in the Ordos region of Inner Mongolia and in central Gansu, where Wallace (1913) found the animals so common in 1911 that they "kept us in meat," appear to be extinct, judging by our surveys there in 1996. In the third area, on the plateau around Qinghai Hu (fig. 6.3), three populations with a total of fewer than 200 animals persist, most of them in a strip of sand dunes and steppe near the lake (Jiang et al. 1994). These gazelles, like goitered gazelles, are mainly found in arid habitats, and this separates them ecologically from Tibetan gazelles in the Qinghai Hu region.

Population and herd dynamics

With the sexes segregated for much of the year and at times even occupying different terrain, unbiased population samples of Tibetan gazelles were difficult to obtain. Males sometimes aggregated, making them conspicuous, whereas females tended to be dispersed, especially during the birth season. For example, in late May among some hills north of Siling Co, we tallied 40 gazelles in scattered herds, of which 36 were males. Yearling males were easy to classify by their short horns, which reached a length of about 10 cm by the age of 12 months. But it was often difficult to distinguish yearlings from adult females by size after they had reached the age of 15 months, and I combined them at that age into one category. Young Tibetan gazelles, like other gazelles (Estes 1991), hide for at least two weeks after birth (fig. 6.6). In the Aru Basin, we observed four crouched young between 27 July and 9 August and noted the first young of the season following its mother on 6 August. Consequently population samples during July and August greatly underrepresented young.

Males usually outnumbered females in the population samples obtained from late May to August (table 6.3). If yearlings and adults of each sex are combined, the ratio is 121 males to 100 females, no doubt a biased result. The September to December samples had a ratio of 81 males to 100 females (table 6.4), a figure which probably reflects the composition more accurately. An October sample of 238 gazelles at Chigo Co in southeast Tibet showed a ratio of 77:100. Yearlings were evenly divided between the sexes at the age of 1 year.

6.6. A young Tibetan gazelle crouches with neck extended and ears retracted.

Table 6.3 Population composition of Tibetan gazelles during spring and summer on alpine steppes in the Chang Tang Reserve and outside the reserve in Tibet

		Number					Ratio	
Month/year	Sample size	Ad. ♂	Yrl. ♂	Ad. ♀	Yrl. ♀	Young	Ad. ♂:100 Ad. ♀	Yrl.:100 Ad. ♀
Aru Basin region								
8/1988	60	25	7	21	—	7	119	33
7–8/1990	221	81	26	83	11	20	98	45
7–8/1992	98	62	1	23	11	1	270	52
Total	379	168	34	127	22	28		
East part of reserve								
6–7/1991	119	39	14	49	17	—	80	63
5–6/1994	170*	92	10	46	16	—	200	70
Total	289	131	24	95	33	—		
Outside reserve								
7–8/1990	49	21	8	13	7	—	162	115
6–7/1991	48	23	—	20	8	—	115	40
5–8/1992	121	44	14	48	14	1	92	58
Total	218	88	22	81	29	1		

Note: Newborns are usually not observed with females before August; yearlings here are 10–13 mo. old.
*The sex of 6 yearlings was not determined and these are not shown in the column.

Table 6.4 Population composition of Tibetan gazelles during autumn and winter in the Chang Tang Reserve and other parts of the Chang Tang in Tibet and Qinghai

		Number				Ratio*	
Month/year	Sample size	Ad. ♂	Yrl. ♂	♀ (ad. & yrl.)	Young	♂:100 ♀	Young:100 ♀
East part of reserve							
9/1988	171	64	13	68	26	113	38
9–10/1990	349	81	43	148	77	84	52
12/1991	466	84	29	244	109	46	45
10/1993	699	225	34	281	159	92	57
Total	1685	454	119	741	371		
Outside reserve							
9/1988	77	18	8	38	13	68	34
12/1991	111	38	4	62	7	68	11
10/1993	160	46	17	60	37	105	62
Total	348	102	29	160	57		
Southwest Qinghai							
10–11/1985	95	63	5	16	11	425	69
11/1986	103	28	6	51	18	63	35
11/1993	127	26	6	65	30	49	46
Total	325	117	17	132	59		

*Adults and yearlings combined.

The difference in observed sex ratio between summer and winter indicated that males and females had somewhat different seasonal movements. Four censuses in the same area near Garco showed much fluctuation, numbers varying from 2 to 35 (table 3.2). However, the extent of movement is probably small; gazelles could almost predictably be encountered in certain localities such as on particular hills or in a part of a basin.

Single young were the rule; no evidence of twinning was observed. The ratio of yearlings (male and female combined) to adult females usually ranged from 40:100 to 70:100 (table 6.3). Since only about half of the adult females had a surviving year-old offspring, mortality of young was high. The ratio of young to females averaged 47:100 late in the year when the young were 1.5–5 months old. However, both yearling and adult females are included in this ratio because the two age groups were not classified separately at that season. If it is assumed that yearling females comprise the same percentage (21.7%) as in summer (table 6.3) and this number is subtracted from the female sample, then the ratio of young to adult females is 60:100. The young-to-female ratios show considerable variation by area and year, an indication either of differences in mortality or of sampling error. The steppe habitat of Tibetan gazelles is relatively constant, unlike, for example, the great variation in the quality of blue sheep habitat on the plateau. But inclement weather can have an impact on survival of young. Many gazelles died after the October 1985 blizzard in Qinghai (Schaller and Ren 1988; Kaji et al. 1993) and the females that survived no doubt entered the rut in poor condition. The following year, the ratio of young to females was only 35:100. For unknown reasons, gazelles just south of the Chang Tang Reserve showed unusually poor survival in 1991, with a ratio of 11:100.

The Tibetan gazelle "generally moves in small herds of five or seven (seldom as many as twenty), though solitary males are often seen" (Prejevalsky 1876). Herd size remains similar today (fig. 6.7). The largest herds in the Chang Tang numbered 25–26, but at Chigo Co we observed a herd of 46. Piao and Liu (1994) reported herds with over 30 animals, and Kaji et al. (1993) reported one with 48. Sometimes several small herds were close to each other—one such aggregation comprised 33 gazelles—but they were distinct at the time of the count. In such situations, herd composition may change frequently, the only stable association being between a female and her offspring: the young and sometimes also a yearling female.

Most males, both yearling and adult, remained apart from the females between May and December. With most births between mid-July and early August and an estimated gestation period of 5.5–6 months, judging by other gazelle species of similar size (Estes 1991), the main rut would begin in about mid-January. "During their breeding season, which begins towards the close of December and lasts a month, the males chase one another from the herds, but we never saw them fighting like the orongo [chiru]" (Prejevalsky 1876).

Most yearling and adult males were solitary or in male herds with up to 5 members and occasionally as many as 10–12 (figs. 6.7 and 6.8). Engelmann (1938) reported a male herd of 25–30. Fewer males were in herds with 6 or more individuals in winter (17.1%) than in summer (31.1%), but average herd size, excluding solitary males, remained about the same (3.4 and 3.6). More males were solitary in winter than in summer probably

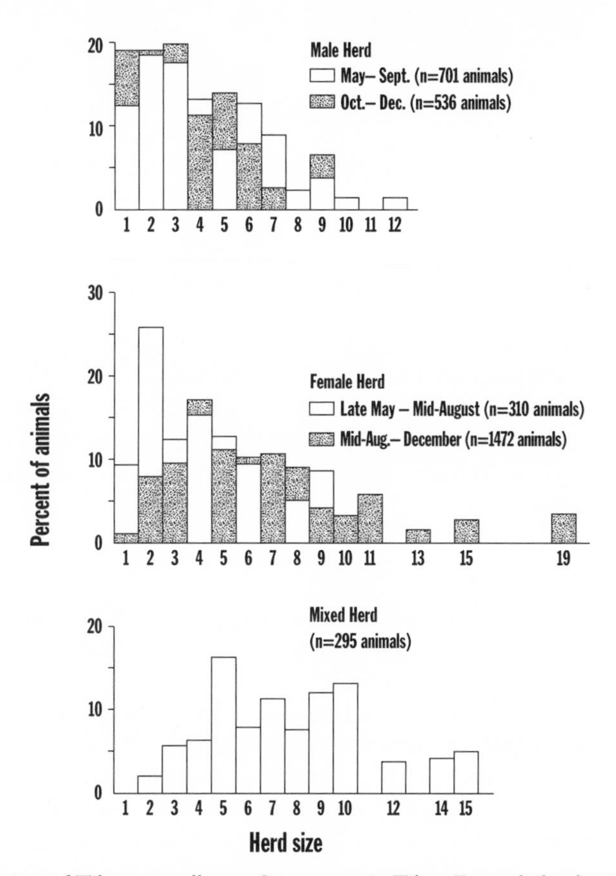

6.7. Herd sizes of Tibetan gazelles on *Stipa* steppe in Tibet. For male herds, size in winter prior to the rut is indicated separately from the spring to autumn seasons; for female herds, the size before and just after parturition is indicated separately from the rest of the year.

6.8. Tibetan gazelle males are usually solitary or in bachelor herds except during the winter rut.

because the rut was imminent and they occupied territories. Mixed herds contained only 8.3% of the total number of males sampled, an indication of the degree of sexual segregation for much of the year. Most (61%) of the males in mixed herds were yearlings. The average size of mixed herds was 6.1. Female herds, consisting solely of females and young, averaged 3.4 individuals in size during late spring and summer and 4.9 in winter. In summer, many females were alone or in twos around parturition time.

Comparisons with the goitered gazelle

Goitered gazelles were frequently encountered in the deserts and semideserts of the Gobi in southern Mongolia. Since these gazelles live in a much more arid habitat than the Tibetan gazelles, a comparison between the two is of interest. The ratio of males to females (yearlings and adults combined) in our sample during summer averaged 64:100 (n = 920) and during winter 31:100 (n = 294), a seasonal bias in observations (table 6.5) different from the one in Tibetan gazelles. Goitered gazelle males rut from mid-November until January in Kazakhstan, a period during which they occupy scattered large territories (Blank 1992). The rut in Mongolia probably began at about the same time. Most large herds divided into small ones in early November, and a male attempted to mount on 18 November. In summer many males (47.9%) were with the females in mixed herds, a pattern strikingly different from that in Tibetan gazelles. Excluding soli-

Table 6.5 Population composition of goitered gazelles in the Gobi region, Mongolia

Month/year	Sample size	Number				Ratio	
		Ad. ♂	Yrl. ♂	♀ (ad. & yrl.)	Young	♂:100 ♀	Young:100 ♀
8–9/1989	141	45	6	60	30	85	50
8/1993	100	21	2	43	34	54	79
8/1994	1047	243	43	457	304	63	67
11/1992	245	43	2	144	56	31	39
12/1990–1/1991	158	25	0	80	53	31	66
Total	1691	377	53	784	477		

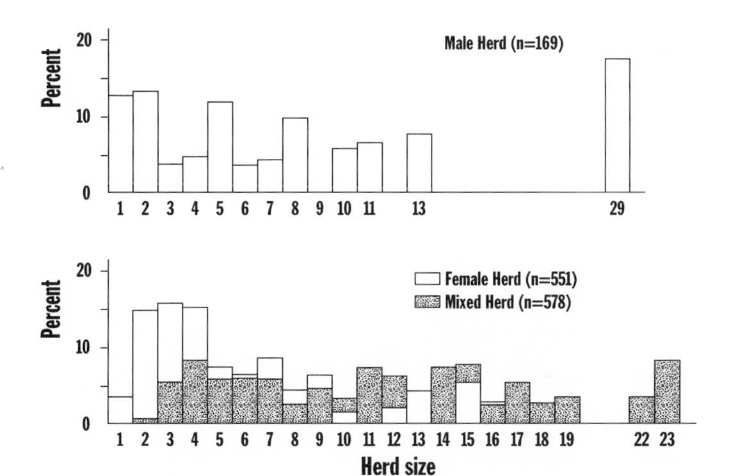

6.9. Herd sizes of goitered gazelles in the Gobi region of Mongolia during summer. The data exclude 11 large herds (25–80 animals) which fled before they could be accurately counted.

tary individuals, male herds averaged 5.5 (2–29) members; female herds, 4.2 (2–16); and mixed herds, 8.1 (2–24); figures slightly higher than those for Tibetan gazelles (fig. 6.9). These data are based only on herds whose size and composition were accurately determined. In addition, a number of large herds were observed, most of them mixed ones, but these were so shy that even accurate counts were often not possible. Of these large herds, four had 25–30 members and one had 40 members in winter; three had 30–40 and three had 70–80 in summer. If these herds were included, the average size of mixed herds would be raised to about 12. Such large herds were observed only in areas where forbs and shrubs—the main diet of goitered gazelles (see table 12.17)—were abundant. Maximum herd size in goitered gazelles thus exceeded that of any reported for Tibetan gazelles. In northern Xinjiang the largest herd in a sample of 221 goitered gazelles contained 16 individuals; the mean size of male herds was 2.0; female herds, 2.3–3.1; and mixed herds, 5.3 (Gao et al. 1996)—figures considerably lower than those from Mongolia.

7

Wild Yak

Wild cattle are of frequent occurrence in the deserts of Hither Thibet. They always live in great herds, and prefer the summits of the mountains. During the summer, indeed, they descend into the valleys in order to quench their thirst in the streams and ponds; but throughout the long winter season they remain on the heights, feeding on snow, and on a very hard rough grass they find there. These animals, which are of enormous size, with long black hair, are especially remarkable for the immense dimensions and splendid forms of their horns. It is not at all prudent to hunt them, for they are said to be extremely ferocious.

Evariste-Régis Huc and Joseph Gabet (1850)

WHILE WALKING in the Chang Tang, I occasionally met several yak bulls ponderously at rest on a hillside. They would rise and face me with their armored heads before fleeing. Their mantles of hair almost obscured their feet. Black and massive, they conveyed power and mystery; they reminded me of the rows of stone totems on Easter Island. Supremely well adapted to the harsh highlands with their thick coats, great lung capacity, and ability to clamber nimbly over rough terrain like giant goats, they are marvels of evolutionary perfection. Even their blood cells are designed for high elevations in that they are about half the size of those of cattle and are at least three times more numerous per unit volume, increasing the blood's capacity to carry oxygen (Larrick and Burck 1986). Humankind views animals with a mix of illusion and fact. To me wild yaks symbolized the boundless space of the Chang Tang and they became its icons. And they also symbolized the plight of wildlife on the Tibetan Plateau. Excessive hunting in this century has caused such a rapid decline of wild yaks that it is reminiscent of the decimation of bison in the American West during the 1800s.

Yaks have been domesticated for an unknown period (Olsen 1990). Domestic yaks were mentioned in the literature of the Western Zhou dynasty in about 850 B.C. (Cai and Wiener 1995). Yak tails were used as fly whisks by Roman ladies at the time of Emperor Domitian in the first century A.D. (Zeuner 1963), but it is not known whether they originated from wild or domesticated animals. At least 12 million domestic yaks now occupy the highlands of central Asia (Alexander 1987) from Afghanistan east to China,

Mongolia, and Russia. Their meat, milk, wool, hides, droppings for fuel, and strength in transporting loads have a crucial role in the economy of peoples in the region. The literature on domestic yaks is extensive (*Le yak* 1976; Zhang 1989; Zhang, Han, and Wu 1994; Cai and Wiener 1995), whereas little has been published about the wild progenitor of domestic yaks. Western explorers have observed them, often along the sights of a rifle (Prschewalski 1884; Rockhill 1891; Hedin 1903; Schäfer 1937c) and biologists have commented on them in the context of general wildlife surveys (Achuff and Petocz 1988; Schaller, Ren, and Qiu 1991; Feng 1991a). But detailed information about the species did not begin to appear until the mid-1990s (Schaller and Gu 1994; Miller, Harris, and Cai 1994; Schaller and Liu 1996).

Because information about the status and natural history of wild yaks was scarce when this project was initiated, I made an effort to study the species. Observing yaks was difficult, in part because populations survive mainly in remote places. And although the bulky, black forms of yaks could readily be spotted from as far as 5 km or more, the animals could also perceive a person or car at a great distance and they then usually bunched up and fled fast and far. Often we were able to do little more than count the animals and determine the number of young before a herd vanished from view.

Domestic and wild yaks may be in the same area but the latter can be distinguished by its larger size, shyness, and, in many individuals, coat color. Wild yaks are typically black, whereas domestic ones often have either patches of white, gray, or brown or an entire coat of one of these colors. In a sample of 1518 domestic yaks, 51.8% were black, 40.4% had white patches, and 7.8% had other color combinations. I considered any yak with aberrant colors to be feral or hybrid, except in one special case (see below).

Taxonomy

The yak belongs to the tribe Bovini with the bisons (*Bison*), buffaloes (*Bubalus*, *Syncerus*), and cattle (*Bos*), but its exact relationship to the others has been a matter of dispute. In 1766, Linnaeus named the domestic yak *Bos grunniens* (the "grunting ox," for its distinctive vocalization) to indicate its relationship to cattle. It was later placed into its own genus, *Pöephagus*, to emphasize its difference from other bovines. Olsen (1990) supported this generic classification on morphological grounds, noting that yaks differed as much from bison and cattle as cattle do from bison. Mitochondrial DNA sequences indicate that yaks are more similar to American bison than to domestic cattle, but that "the percent divergences among *Bison bison*, *Bos grunniens*, and *Bos taurus* . . . are comparable to those calculated among

7.1. The grayish muzzle and long mantle of hair of wild yaks are conspicuous in this lone bull.

congeners of the perissodactyl genus *Equus*" (Miyamoto, Tanhauser, and Laipis 1989). On the basis of a cladistic analysis of Bovini skulls, Groves (1981) also included the whole lineage into the genus *Bos*. Yaks occupy an intermediate position between *Bos* and *Bison* in some aspects of behavior as well (Schaller 1977a). The species designation *grunniens* is generally accepted, although Corbet (1980) used *mutus* with the explanation that *grunniens* was based on domestic animals and that the earliest valid name for a wild form should be used, in this instance *mutus* as applied by N. Przewalski in 1883. However, I retain *Bos grunniens*. No proposed subspecies have found general acceptance.

Description

The wild yak is massive with sturdy legs and a conspicuous hump that rises abruptly behind the neck and tapers down to a level back. It is black with rust-brown overtones except for gray toward the tip of the muzzle. A long fringe of hair on the lower neck, chest, sides, and thighs drapes the lower parts like a skirt. In bulls this fringe may be 70 cm long and hang almost to the ground (fig. 7.1). The tail terminates in a large, bushy tuft which covers the anal area and waves back and forth when, during aggressive encounters, the animal raises its tail vertically. One bull's tail with tuft was 107 cm long (Hedin 1898). A dense layer of wool grows beneath the coarse guard hairs. Small calves are dark brown rather than black. The thick coat and the low number of sweat glands (Larrick and Burck 1986) are efficient adaptations for conserving heat. Even on brisk days, yaks often stand knee-deep in ice-cold streams, apparently to cool off; they survive poorly in warm climates or below 3200 m on the Tibetan Plateau.

Bulls are considerably larger than cows. The shoulder height of wild adult bulls is about 175–203 cm and of cows 137–156 cm; the total length of bulls is 358–381 cm and that of the one cow measured is 305 cm (Prejevalsky 1876; Rawling 1905; Hedin 1898; Engelmann 1938). Engelmann (1938) listed the body mass of two males at 535 kg and 821 kg and of a female at 306 kg, and Schäfer (1937c) gave one female weight of 337.5 kg. Zhang (1984) encountered a newborn with a shoulder height of 67 cm and weight of 20 kg. The body mass of domestic yaks varies from area to area but averages about 400 kg for bulls and 260 kg for cows (Bonnemaire 1976). Thus bulls are about 35% heavier than cows, and a similar average size difference probably exists also in wild yaks.

Wild yak horns are gray to black. Those of bulls sweep out and forward then back and often somewhat inward, whereas those of cows curve more sharply up and farther back. The horns of 53 bulls and 12 cows were measured in the Chang Tang Reserve, most of them adults killed by hunters. Bull horns averaged 75.7 ± 10.7 (range 47.5–99.0) cm in length along the outside curve, 35.2 ± 3.2 (26.0–42.0) cm in basal circumference, and 42.6 ± 10.5 (26–83) cm from tip to tip. Cows had horns 55.1 ± 7.5 (37.0–64.5) cm in length, 19.5 ± 1.5 (17.5–23.0) cm in circumference, and 32.2 ± 13.2 (18–67) cm from tip to tip. Bull horns were longer, had a wider spread, and were more massive than cow horns, a difference already noticeable in yearlings.

In northwest Tibet, Deasy noted that "most wild yaks were black, brown specimens being very rare" (1901). This is the first mention of a rare mutation centered around the Aru Basin. These animals are not a mere brown—a color also found in domestic yaks—but a light golden brown, which makes them highly conspicuous among black herd members. Black females may have a golden offspring and golden females a black one. The mutation is evident from the Aru Basin area west toward the highway and north to 35° N near Bangdag Co, where we saw a golden bull. Of the yaks sampled around the Aru Basin 2.2% (n = 506) were golden in 1988, 1.4% (n = 700) in 1990, and 1.3% (n = 315) in 1992.

Status and distribution

Mid- to late Pleistocene yak fossils have been found in eastern Russia but no evidence exists that other than domestic yaks have occurred there in historic times (Heptner, Nasimovic, and Bannikov 1966). A fossil yak is also known from Tibet and a possible one from Nepal (Olsen 1990). At present wild yaks are confined to the Tibetan Plateau within China except that a few may seasonally venture into the Changchemno Valley near Panggong Co in Ladakh, India (Fox, Nurbu, and Chundawat 1991a). Until recent decades some also entered northwest Nepal (Miller, Harris, and Cai 1994). Yaks occupy or once occupied treeless uplands, including plains, hills, and mountains, from as low as 3200 m in elevation up to the limit of vegetation at 5300–5400 m. They reached their greatest abundance on the alpine meadows in eastern Qinghai. On alpine steppe, herds were also large on occasion but were more widely dispersed, and in desert steppe they were scarce (Prschewalski 1884; Bower 1894; Hedin 1903).

We visited most of the yak's past and current range but encountered few animals except in the Chang Tang Reserve. At first we attempted to determine average yak density in the reserve by counting all animals within 1 km on each side of our travel route. However, yak distribution was highly clumped, with most animals in a few widely scattered herds. To sample such herds it would be necessary to transect large areas systematically, something that usually was not feasible. Instead we attempted to obtain total counts by crisscrossing certain large blocks of terrain. This provided minimum figures for comparison between areas. Such figures are only approximate because herds may shift ranges seasonally or move out of an area when disturbed. For instance, 681 yaks were tallied in the Aru Basin in 1990 but only 315 in 1992, by which time pastoralists had occupied much of the basin. Yaks were concentrated in the parts with little disturbance. One herd fled over 20 km after detecting us.

In discussing the status of yaks, I draw on the literature, interviews with local informants, and our own surveys (see Schaller and Liu 1996).

Qinghai

Wild yaks were once extraordinarily abundant around the headwaters of the Yellow River near Ngoring and Gyaring Lakes. Przewalski found them there in herds of hundreds (Prschewalski 1884). "The hills around this plain, and also Karma-t'ang, were literally black with yak; they could be seen by the thousands," wrote Rockhill (1891). And according to Wellby, "On one green hill we could see hundreds upon hundreds of yak grazing; there was, I believe, more yak visible than hill" (1898). By the mid-1930s, Schäfer found the yaks scarce or absent in that part of Qinghai. He noted that "in the past decades, after Chinese, English, or Russian army rifles had found their way to the tents of the thieving highland nomads, wild yaks were decimated in a frightening manner," and he predicted that "complete eradication lies ahead" (1937c). Locals "hunt them a great deal, their flesh being the only meat a great part of these people use during winter" (Rockhill 1894). Caravans crossing the area also killed yaks for food, and foreign expeditions sometimes just killed. "We left the greater number of those we shot untouched, having no use for the meat in Tibet. The carcasses soon froze into a solid mass, the tough hide resisting the vultures and wolves. On our way back from the Blue River we saw them lying exactly as we left them" (Prejevalsky 1876). In 1949 Clark saw few animals but "hundreds if not thousands of the enormous white skulls" (1954).

The construction of a highway south from Qinghai Hu to Yushu during the 1950s opened the region to casual hunting, and this, together with widespread famine in China during the Great Leap Forward in 1958–1961, during which many animals were killed, decimated all wildlife. A few yaks are rumored to survive near the Sichuan-Qinghai border (X. Wang, pers. comm.). Some yaks persisted in the foothills of the eastern Burhan Budai Shan, where Zhang (1984) saw 65 in the late 1970s and G. Cai (pers. comm.) observed them in the early 1990s. In the open terrain near Gyaring Hu the last wild yak was noted in 1983 (Kaji et al. 1993). Schäfer's (1937c) prediction had almost come true. Except possibly for a few herds, the yak's eastern limit now lies near the Lhasa-Golmud highway. The highway roughly marks the transition zone between alpine meadow in the east and alpine steppe in the west. From here the yak's range extends westward through southwest Qinghai over a block of terrain of about 400,000 km^2, much of it within the Chang Tang Reserve (fig. 7.2).

Yaks were once common in southwest Qinghai. In one area "bunches of yaks were on every hill" (Rockhill 1894). But gold prospectors past and present killed much wildlife (Hedin [1922] 1991; Zhang 1984), and road access led to market hunting, with, for example, a geological survey team killing 250 yaks in 1980 (Feng 1991b). In 1992, poachers sold the meat of about 100 yaks in Xining, the capital of Qinghai ("Wild Yaks Slaughtered" 1992). Several wildlife surveys were conducted in that region during recent years. The upper Yeniugou in the Burhan Budai Shan had a population of about 1223 yaks in 1051 km^2 in 1991 and 841 in 1992 (Miller, Harris, and Cai 1994). In 1986, Schaller, Ren, and Qiu (1991) transected 20,000 km^2 on each side of the highway from the southern edge of the Burhan Budai Shan to the town of Tuotuohe. Only 9 yaks, all bulls, were seen. In 1993, we made a trip far up the Tuotuo River and found no yaks. In 1990, Feng (1991a) drove west from the highway to Ulan Ul Hu, north to the Burhan Budai Shan, and then east past Hoh Sai Hu. A total of 805 yaks were tallied, almost all in two concentrations. One concentration, with 363 animals, was found at Xijir Ulan Hu, and a second, with 435 animals, was at Hoh Sai Hu, but we suspect that the latter was part of the nearby Yeniugou population censused by Miller, Harris, and Cai (1994).

North of the Burhan Budai Shan is the Qaidam Basin, 2600–3000 m in elevation, whose salt flats and desert scrub are not yak habitat. Yaks

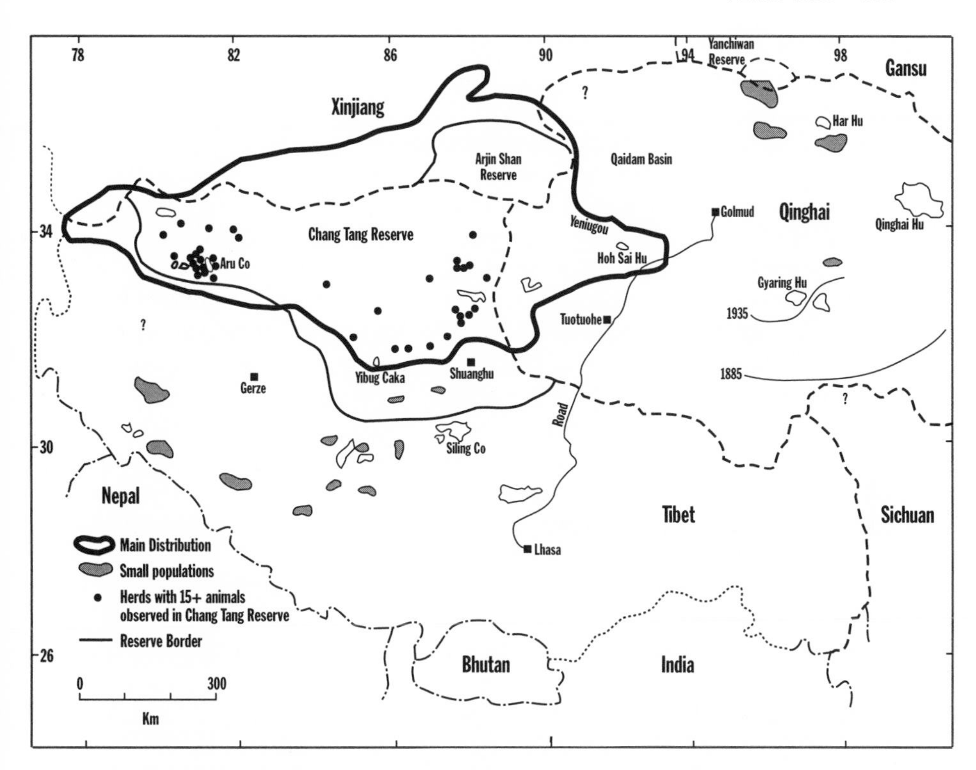

7.2. The distribution of wild yaks on the Tibetan Plateau. The easternmost limits of distribution in 1885 and 1935 are based on Schäfer 1937b. (Adapted from Schaller and Liu 1996.)

once occurred between the Qaidam Basin and Qinghai Hu, an area where Rockhill saw a "few wild asses and half a dozen yaks" (1894), but none occur there now. Toward the northern rim of the plateau, along the Arjin Shan and the Qilian Shan with its associated ranges, yaks persist in fragmented populations of unknown size. Kozloff saw "traces of wild yak" in the Arjin Shan and farther east found the animals "extremely numerous" ("Central Asian Expedition" 1896). I do not know if yaks persist in that part of the Arjin Shan, but they occur father west in the same range. Zheng et al. (1989) observed 51 yaks in the Yanchiwan Reserve of Gansu close to the Qinghai border. Local people told us that yaks enter the reserve only seasonally from the south. We censused a 610 km^2 area along the northern slopes of the Shule Nanshan just east of the Yanchiwan Reserve without finding fresh evidence of yaks (Schaller, Ren, and Qiu 1988). Twelve animals were killed for the market south of Har Hu in 1986 according to the Qinghai Forest Bureau. About three isolated populations of unknown size possibly remain in the region.

Xinjiang

We made two surveys in Xinjiang. The first traversed an area of about 6000 km^2 from the northern flanks of the Kunlun Shan across a portion of the plateau just west of the Arjin Shan Reserve. A total of 26 yaks were tallied in this barren region (Schaller, Ren, and Qiu 1991). The second survey was in the Yüsüpalik Tag, a spur of the Arjin Shan, north of the Arjin Shan Reserve. Only 23 yaks were seen, but according to local people, many animals come to that area from the reserve in winter. Sheng (1986) reported a population of about 10,000 in the 45,000 km^2 Arjin Shan Reserve, and Butler, Achuff, and Johnston (1986) quoted a similar figure. Based on transects during which 420 yaks were seen, Gu et al. (1984) calculated a total population of 5625 yaks on the mistaken assumption that the density would be similar throughout the reserve. In 1988, Achuff and Petocz (1988) conducted a survey of 18,000 km^2 covering the western half of the reserve; they saw 219 yaks and estimated fewer than 1000 in that part. No reliable estimate is available for the whole reserve, but we consider it unlikely that there were more than 3000 at that time. During the early 1990s illegal hunting by gold miners and others greatly reduced the wildlife populations in the reserve (Wong 1993).

Today's western limit of yak distribution lies between the Karakoram and Kunlun Ranges in an area known as the Aksai Chin. Once yaks were found as far as 78° E (Shaw [1871] 1984), but now they persist mainly in the eastern part of the Aksai Chin, according to our informants.

Tibet

Western Tibet has long had a large pastoral population south of about 32° N. There wild yaks have been hunted so intensively for meat, horns (used as milk pails), and other products that the animals have been exterminated or considerably reduced in number. For example, Edmund Smyth found the headwaters of the Yarlung Tsangpo "swarming with wild yaks, mostly cows" in 1864 (quoted in Allen 1982), an area in which yaks apparently persist only as remnants. The alpine meadows that stretch from the western edge of the forest past Nam Co almost to Siling Co are now devoid of wild yaks, but farther west small populations persist according to local informants (fig. 7.2). A few scattered sites were probably overlooked, and some populations may have vanished since we recorded them. At least some of the animals in this part of Tibet are either feral domestic yaks or hybrids.

Wild yaks once occurred throughout the northern Chang Tang, an area that is now within the Chang Tang Reserve. In 1891, Bower noted that they "are to be seen all over the Chang. Sometimes as many as a hundred were seen in a day, and for days together some were always in sight" (1894). But, as the observations of Hedin (1903, [1922] 1991), Deasy (1901), and

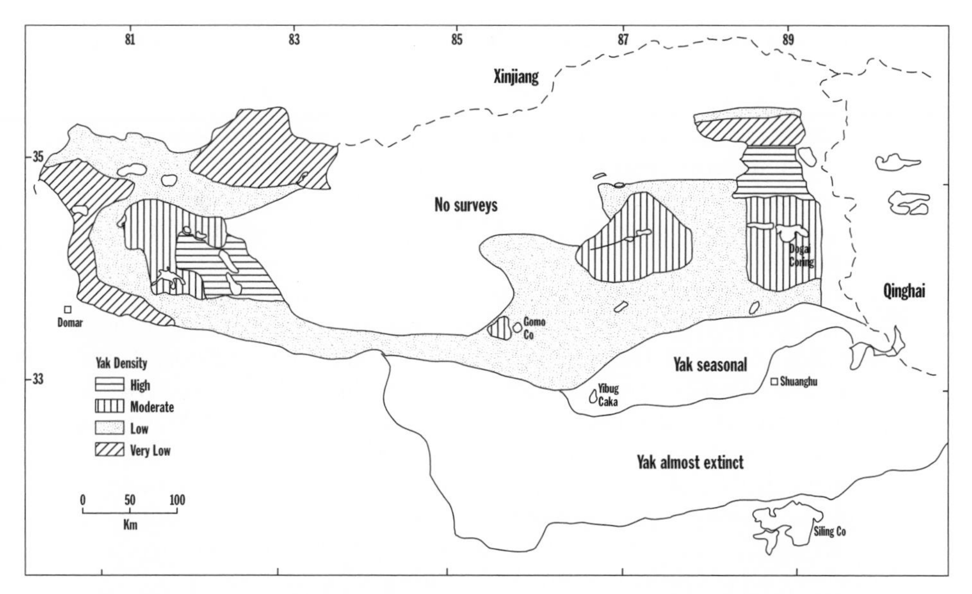

7.3. Relative density of wild yaks in the Chang Tang Reserve. (Adapted from Schaller and Liu 1996.).

others make clear, yak herds were usually encountered at intervals, droppings and trails indicating that they ranged widely. Today the reserve can be divided into three zones with respect to yak distribution (fig. 7.3).

The yaks in the southern quarter (24%) have in recent years been exterminated except for stragglers and two small populations, which are said to contain hybrids. The status of the species was quite different a century ago. Southeast of Shuanghu, "there were enormous numbers of yak about, and the Tibetans, galloping after a herd, succeeded in singling one out and riding it down" (Bower 1894). Also in the Shuanghu area, Hedin found that "yaks and kulans were remarkably numerous. Occasionally we counted them by the hundreds" (1903). Few pastoralists inhabited the steppes north of Siling Co before the 1960s, when Shuanghu was established as an administrative center. Yaks were still there in number during the early 1970s (B. Li, pers. comm.) in spite of a large influx of nomads, but by the late 1980s herds no longer frequented the vicinity. Ma noted in 1986 that Shuanghu officials "have two special tasks: gathering fuel (firewood and yak dung) and hunting wild yaks" (1991). They "drive into the barren wilds to hunt yaks for ten to fifteen days of primitive living." By the early 1990s, trucks from town went as far as 250 km north and northwest to kill yaks. Hunters from the town of Gerze drove north a similar distance to the Aru Basin. Within 30 years the yak in the southern part of the reserve were virtually eliminated.

In the eastern part between Shuanghu and Yibug Caka, an area comprising about 5% of the reserve, yaks are now mainly seasonal visitors. In December 1991 we surveyed 17,500 km^2, including a major part of this area (see table 6.1), and tallied 13 yaks, all bulls. According to local people, yaks from the north visit briefly in autumn; in confirmation of this, we saw herds there only four times and all in September. The usual distribution of yaks suggests that some of these herds had traveled south for 50 km or more.

Yaks occupy the rest of the reserve (71%) at varying densities. The distribution of large herds with females provides a better idea of relative yak abundance than the mere presence of solitary males and small herds. A total of 39 herds with 15 or more yaks were observed in the reserve (fig. 7.2). Three visits were made to the Aru Basin but only the 1990 data are used here. Excluding the yaks in the Aru Basin, the remaining 26 herds comprised 1093 yaks, or 80% of the total of 1373 animals tallied. An effort was made to count all yaks in the Aru Basin and within a 4000 km^2 block of terrain in the eastern part of the reserve. Otherwise our encounters with yak herds were largely accidental while traveling cross-country.

Sightings of large herds indicated two main yak concentrations (fig. 7.3). These two sites had several characteristics that made them attractive to yaks. There were partially glaciated mountains, the Aru Range and Rola Kangri, that provided freshwater streams and slopes with alpine meadow or luxuriant alpine steppe. The terrain enabled yaks to shift seasonally up and down slopes to select the best available forage. The volcanic hills east of Rola Kangri had greened by early June, some 2–3 weeks before the plains. With plants dormant from September to May, the availability of early high-quality forage may be critical to wildlife in that region, and yaks, chirus, and kiangs had concentrated there.

In 1990, 681 yaks were counted in the 1800 km^2 Aru Basin and 114 more in the small Yalung Basin on the west slopes of the Aru Range. Probably at least 1000 yaks inhabited the 5000 km^2 in and around the Aru Basin (Schaller and Gu 1994). We did little work west of the Aru area. A century ago, British hunters visited on occasion because wildlife was more abundant there than on the Indian side of the border. Phelps saw a herd of 200 yaks at Orba Co and during his stay there killed "7 Yak, 27 Antelope, and 2 Burrhel" [blue sheep] ([1900] 1983). Fifty-four yaks were seen during one brief survey by the Tibet Forest Bureau in the late 1980s. Information from local people indicated a population of moderate size. However, some animals there probably come from the Aru area. At the time of the surveys perhaps 1500 yaks were in that western area of concentration.

At the other end of the reserve, between the lakes Dogai Coring and Dogai Coring Qangco, we counted 437 yaks in about 4000 km^2 during a

June 1994 survey during which we were able to scan most of the terrain. To the south between Dogai Coring and the glaciated massif Purog Kangri was an area of moderate density in which we tallied 146 animals. In the same area during July 1901, Hedin observed that "the margin of the glacier was literally black with yaks. We counted considerably over 300, many of them quite little calves" (1903). Good yak habitat extends east to the Qinghai border. This eastern concentration area possibly contained 1000 yaks.

West of Dogai Coring are extensive arid flats and hills with little wildlife, but there is then an area with moderate yak numbers. The Tian Shui River flows east from the glaciated Zangser Kangri into several lakes. North of these lakes is the Yako Basin, also known as Pamachungtsong. We observed 45 yaks in the Tian Shui Valley in 1991 and 30 yaks in the Pamachungtsong area in 1994. However, the region had more yaks than our brief surveys indicated. People from Shuanghu drove there by truck to hunt yaks for meat in October 1991 and saw 300–400. The Pamachungtsong area, we were told, was used mainly by male yaks. We confirmed this three years later by observing only males there and counting a total of 55 skulls, of which 50 were from males. Possibly 500 yaks inhabited that region. About 12% of the reserve is composed of areas of moderate-to-high yak density, containing an estimated 3000 yaks, or about 13 km^2 per yak.

About 59% of the reserve is presumed to have low to very low densities. We were unable to visit all parts (see fig. 1.4 for routes), but based on our knowledge of terrain and vegetation, as well as on information from other expeditions, we doubt that more major concentrations occur. Few wild yaks and little spoor were encountered in low-density areas. For example, driving north past Margog Caka we saw 3 yaks in 143 km; going west from Dogai Coring no yaks were seen in 101 km, but then a herd of 39 was encountered; and traversing northwest of Yibug Caka past Gomo Co to the Zangser massif, 45 yaks were tallied in 218 km. These routes were devoid of pastoralists except for a few near Yibug Caka and around Gomo Co. If such low-density areas, which compose 39% of the reserve, averaged about 30 km^2 per yak, the total would be 4342 yaks.

In 1990, we made a survey over about 8000 km^2 of desert steppe north of the Aru Basin between the Toze Kangri massif and Heishi Beihu near the Xinjiang border. A total of 73 yaks were tallied, or 110 km^2 per yak. Similar habitat extends eastward along the northern edge of the reserve as well as north into Xinjiang. The western edge of the reserve also has few yaks because there is a major highway near which truck drivers hunt. An estimated 100 km^2 per yak in very low density areas extrapolates to about 668 yaks.

Based on these rough calculations, the reserve had about 8000 yaks, or about 42 km^2 per yak. This figure is somehwat higher than our earlier estimate of 7000–7500 (Schaller and Liu 1996) because our calculations are based on a reserve size of 334,000 km^2 rather than on the official figure of 284,000 km^2 used previously.

Estimates of total yak numbers

Recent published estimates of the number of wild yaks on the Tibetan Plateau range from 500 (Larrick and Burck 1986) and 15,000 (Miller, Harris, and Cai 1994) to 35,000 (Feng 1991b) and 20,000–40,000 (Lu, Li, and Ju 1993). The yak's last major refuge is in the remote northwestern part of the Tibetan Plateau. At most, an estimated 8000 animals were in the Chang Tang Reserve. Contiguously to the east in Qinghai, there were about 1200 in the Yeniugou and surrounding areas (Miller, Harris, and Cai 1994) and perhaps 1000–1500 more, judging by Feng's (1991a) survey. The Arjin Shan Reserve still had a substantial population in the late 1980s but hunting in the past few years has decimated it. The desert steppe west of the reserve has few animals. Possibly 2000–2500 yaks persist in Xinjiang. These figures total 12,200–13,200 within a block of about 400,000 km^2, an estimate based on fragmentary data, but, I believe, on the correct order of magnitude.

I have no information upon which to judge numbers elsewhere except to suspect on the basis of local interviews that populations are small. In Tibet, possibly 1000 yaks persist outside the Chang Tang Reserve. If so, Tibet's wild yak population would be about 9000, similar to the estimate of 7156–8758 given by Liu and Yin (1993). The size of the population in northern Qinghai is also unknown, but perhaps as many as 1000 survive there.

The calculated total number of wild yaks is thus approximately 14,200–15,200, but given the lack of precise information a preliminary estimate of around 15,000 seems appropriate.

Population and herd dynamics

Solitary bulls and bull herds were often on gentle terrain, whereas herds with cows and calves tended to be in or near high hills, often on the upper slopes, as also noted by Miller, Harris, and Cai (1994). Such habitat selection may partly be the result of human persecution in that hills provide escape: animals usually retreated high up on slopes when disturbed. However, the availability of more nutritious forage on hills than on the plains may be a principal reason why cows select rugged ground. Some bulls are with the cows at all seasons, but many are scattered widely, sometimes near cows but often far from them. Such sexual segregation made it difficult to determine the sex ratio of adults. Any yak two years and older was considered adult even though females first conceive at the age of 3–4 years and bulls "have the strongest sexual desire at the age of four to six years" (Zhang 1989), at least in domestic yaks.

In July–August 1990, 7.2% (n = 586) of yaks along the west side of the Aru Basin consisted of solitary bulls and bull herds, and in 1992, 10.1% (n = 414) consisted of such bulls. Outside the Aru Basin, the figure was 15% (n = 273) during August–September 1990, 28% (n = 146) in July 1991, and also 28% (n = 731) in June 1994. Figure 7.4 shows a similar difference between bulls in the Aru Basin and outside it. Since cows are rarely alone or in herds with 5 or fewer members, most such animals are bulls, 6.4% in the basin and 23.4% outside. With yak density in the Aru Basin high, bulls may have joined cows because they were readily available. It is also possible that proportionately more bulls dispersed out of the basin. Many bulls did not associate with cows during the rut, which in domestic and wild yaks is said to be from July to September (Zhang 1989; Ward 1924) or in wild ones mainly in September (Prejevalsky 1876).

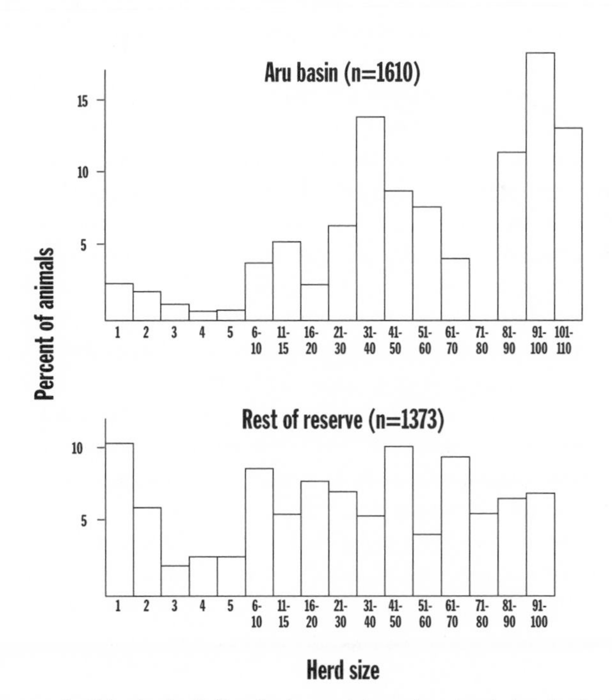

7.4. Percentage of wild yaks (including both sexes) in various herd sizes in the Chang Tang Reserve. The Aru Basin data include both 1990 and 1992.

I lack adequate data on percentages of bulls in mixed herds. In the Aru Basin bulls composed 14.8% of 12 mixed herds (n = 227); outside the basin bulls comprised 10.5% of 14 herds (n = 229). Mixed herds in the Yeniugou averaged 4.0% adult bulls (Miller, Harris, and Cai 1994). However, the Yeniugou population contained many young and yearlings, in contrast to the animals in the reserve, and this lowered the proportion of bulls.

The various calculations suggest that cows outnumbered bulls by 25–33%. Since bulls are often solitary and in open terrain, they are vulnerable to predation, especially by motorized hunters. Of 27 skulls found in the Aru Basin, 78% were bulls, and of 53 skulls in the Dogai Coring area 74% were bulls. Circumstantial evidence indicated that most had been killed by hunters.

In August 1990, 5.7% (n = 700) of the yaks in the Aru Basin and in the adjoining Yalung Basin consisted of calves; in July 1992, calves composed 1% (n = 315), and only 1 yearling was seen (Schaller and Gu 1994). Elsewhere in the reserve, calves composed 12.7% (n = 267) in August–September 1990, 2.7% (n = 146) in late June and July 1991, 3.6% (n = 111) in June and early July 1992, and 2.5% (n = 690) in June 1994. Domestic yaks give birth to single calves in alternate years (*Le yak* 1976; Zhang 1989) but in areas of poor grazing a few may give birth only once every three years according to our local informants. Wild yaks probably have a similar calving interval. Given this reproductive rate the expected percentage of calves would be at least 10–15%. With a rut from July to September and an average gestation period of 258–270 days (Bonnemaire 1976; Zhang 1989) most births would be expected to occur from April through June. I observed 6 calves during the first week in June, which probably had been born in late May, and a newborn on 7 June. Most calves in the Chang Tang were probably born between late May and the end of June, placing the main rut from mid-August through September. Several of our calf counts were made during June, when the birth season was not yet over. However, of the counts made between July and September only one, the 12.7% in 1990, reached an expected level. D. Miller (pers. comm.) calculated a calf population of 15% in the Yeniugou during summer 1991. The years 1991 and 1992 were essentially reproductive failures for the Aru population, because either cows failed to give birth or calves soon died.

One can only conjecture about the reasons for such low reproductive success. Wolf predation is one possible cause. Yaks have various defenses against wolves in addition to size and dangerous horns. When harassed, the yaks may "all rush together and remain thus with their heads toward the threatened danger" (Rawling 1905), or they may flee with the calves in the center (Prejevalsky 1876). One bull retreated into a river when chased by dogs (Hedin 1903). Five bulls merely faced a wolf with lowered horns (Zhang 1984). The wolf packs in the Aru Basin no doubt preyed on calves, but it is unlikely that they killed almost all. An analysis of wolf droppings revealed yak as only a minor summer food item (see table 11.4).

Brucellosis, caused by the bacterium *Brucella*, is a widespread livestock disease in Tibet that could also affect reproduction in wild yaks. Fetal abortion is the most striking aspect of the disease. Species of *Brucella* have distinct host preferences, and just as cattle in North America infected bison (Meagher and Meyer 1994), so domestic yaks could infect their wild relatives. Several bacterial diseases (*Salmonella*, *Endamoeba*, *Chlamydia*) may also affect calf survival in domestic yaks on the Tibetan Plateau (Cai and Wiener 1995) and presumably have an impact on wild yaks as well. Whatever the reasons for the poor calf survival, the Aru population cannot sustain itself, even if all poaching were to cease, unless the reproductive rate improved soon after our counts.

Hunting by humans aside, causes of adult mortality include wolf predation and disease. Przewalski reported mange in wild yaks (Prjevalsky 1876). According to Bema Tsandö (pers. comm.), the chief veterinarian of Nyima County, diseases of domestic yaks include brucellosis and a deadly bleeding disease called *dotu*, both of which also affect wildlife. Other virulent diseases include anthrax and bovine pleuropneumonia (Cai and Wiener 1995). Malnutrition after heavy snowfalls, such as occurred in Qinghai in 1985, causes periodic mortality. Huc and Gabet related an unusual form of death:

> As we were passing the Mouroui-Oussou, on the ice, a singular spectacle presented itself. We had previously, from our encampment, observed dark, shapeless masses, ranged across this great river; and it was not until we came quite close to these fantastic islets that we could at all make head or tail of them. Then we found that they were neither more nor less than upwards of fifty wild cattle, absolutely encrusted in the ice. They had no doubt attempted to swim across the river, at the precise moment of the concretion of the waters, and had been so hemmed in by the flakes as to be unable to extricate themselves. Their fine heads, surmounted with great horns, were still above the surface; the rest of the bodies was enclosed by the ice, which was so transparent as to give a full view of the form and position of the unlucky animals, which looked as though they were still swimming. ([1850] 1987)

Miller, Harris, and Cai (1994) accurately determined the ages of six yak skulls from the Yeniugou by cementum analysis of incisors. Four bulls were 6, 10, 15, and 16 years old, and two females were 10 and 11 years old, indicating that yaks have potentially a life span approaching 20 years.

Przewalski noted that bulls may occur singly or in herds of 2–5 and exceptionally 10–12 and that females may gather in herds of several hundred (Prejevalsky 1876). Large concentrations are now almost a thing of the past. However, Miller, Harris, and Cai (1994) saw 368 yaks scattered in one valley and also one herd of 245. In 1992, one herd in the Aru Basin contained at least 200 individuals but no accurate count was possible; the following day we tallied 236 yaks in the area but they were in five distinct herds.

As described earlier, bulls are often separated from the females. Of 507 such bulls tallied, 36% were alone, 43% in herds of 2–5, 13% in herds of 6–10, and the rest in herds with up to 19 individuals (fig. 7.5). One herd of 40 in the Aru Basin contained at least 31 bulls. Most solitary bulls were fully adult, but bull herds occasionally had two-year-olds and yearlings in them.

A solitary cow was observed once, and on another occasion a cow was alone with her newborn. However, Zhang (1984) watched a cow giving birth at the edge of a herd consisting of 4 bulls, 15 cows, and 4 calves. Small herds with up to 15 individuals sometimes contained either just cows or also young, yearlings, and bulls. Most cows were in large mixed herds with up to 100 or more animals of all ages and usually both sexes (fig. 7.6), as the composition of 10 such herds shows (table 7.1). Herds were not stable units: they readily split or two or more joined. Some large bulls

7.5. Wild yak bulls are often in small bachelor herds.

7.6. A herd of wild yak cows and yearlings alertly observes my approach.

Table 7.1 Composition of selected, large, mixed wild yak herds

Month/year	Adult males	Adult females	Yearling males and females	Young	Total
6/1991	8	39	3	0	50
6/1994	2	14	2	0	18
7/1991	4	25	4	2	35
8/1990*	1	17	0	0	18
8/1990*	7	23	2	0	32
8/1990*	2	26	4	4	36
8/1990*	0	32	7	9	48
8/1990*	7	26	3	4	40
9/1990	1	12	3	5	21
9/1990	3	28	3	14	48

*Observed in Aru Basin.

seemed to shift from life in a mixed herd, to a peripheral existence during which they remained within 1 km or so of a herd, and to a period when they were alone or just with other bulls. Herds in areas with few yaks probably maintained considerable cohesion, similar to African buffaloes (Sinclair 1977), in contrast to herds in areas with a high yak density.

Herd sizes (both sexes combined) are summarized in figure 7.4. The Aru Basin is treated separately from the rest of the reserve because herd size in that large population differed somewhat from other areas. There were proportionately few solitary bulls and small herds in the Aru Basin and many animals in herds with 80 or more individuals. In the Aru Basin, mean herd size (n = 64, excluding solitary individuals) was 24.5 and the median was 58.1; elsewhere in the reserve, mean herd size (n = 109) was 11.3 and the median 26.5.

Conservation

"In these inhospitable wastes, in the midst of a desolate nature, yet far removed from pitiless man, the famous long-haired ox roams in unrestricted freedom" (Prejevalsky 1876). "Pitiless man" has, of course, intruded. Yaks may occasionally defend themselves against such intrusion by chasing a car (Achuff and Petocz 1988), attacking a hunter (Hedin 1909), or injuring a pastoralist, but they cannot long endure. The range of the species has been reduced by more than a half during this century. Except for scattered populations with a tenuous future, yaks are now confined to one large tract. They have lost most of the best alpine meadow and steppe habitat to pastoralists. Their decrease is not due to habitat degradation or destruction, as with many endangered species, but due to hunting.

Most surviving yaks are in the Chang Tang and Arjin Shan Reserves, and a few frequent Gansu's Yanchiwan Reserve at certain seasons. However, China has placed yaks in its Class I protective category, which indicates that it may be killed only with a federal permit, and the species is also in Appendix I of the Convention on International Trade in Endangered Species (CITES). Even though two of the reserves are huge, yaks remain under threat within them. Hunting is difficult to suppress in such remote areas without a mobile patrol force, as the recent decimation of wildlife in the Arjin Shan Reserve has shown. Furthermore, wild yaks, more than other wild ungulates in the Chang Tang, must have large tracts of wilderness in which pastoralists are sparse or absent. Domestic and wild yaks readily hybridize and the pure wild strain will vanish with frequent contact. Disease transmission from domestic to wild yaks is also a constant threat.

Currently pastoralists are moving farther north into the Chang Tang to settle the last areas of good grazing. The drastic decline in the number of wild yaks during this century has shown that the animals do not long persist in areas with many people and much livestock. The Chang Tang Reserve north of about 33°30′ N and from the Qinghai border west to and including the Aru Basin should be closed to pastoralism (see chapter 15). That area is marginal for livestock production and still remains largely uninhabited. The wild yak will survive in its last stronghold only if it can roam, in Przewalski's words, with "unrestricted freedom."

8
White-lipped Deer

> We got into camp about 4 P.M., and as the natives said there were bears about, Dr. Thorold went out to look for one; he saw no bears but was lucky enough to find a herd of six Shoa-u-chu [white-lipped deer] stags, and killed one and wounded another. . . . Dr. Thorold had a long day's tracking in the snow before coming on the wounded one he was after, but he got it all right. I trudged about the whole day over hills covered with bushes and a foot of snow, without even seeing the tracks of one; so I fancy they are pretty rare.
>
> Hamilton Bower (1894)

ELEVEN DEER species occur on the Tibetan Plateau, most of them confined to the forested eastern parts. Eight of these deer are small, including three musk deer species, three muntjac species, tufted deer, and roe deer; and three are large, the sambar, red deer (with three subspecies), and white-lipped deer. Certain of these species venture from forest into open habitats. For example, the musk deer *(Moschus sifanicus)* usually inhabits forest and scrub, but in some places I saw it on mountainsides broken by cliffs and boulders. Southeast of Lhasa, Tibet red deer *(Cervus elaphus wallichi)* frequent high, rolling hills covered with grass and patches of low scrub (Schaller, Liu, and Wang 1996), and Sichuan red deer *(C. e. macneilli)* occur sparsely on the vast alpine meadows in Qinghai, where I encountered 16 animals in three herds as far west as 34°20′ N, 94°15′ E in November 1985. However, the white-lipped deer favors open habitats more than any of the other species, and because its range extends to the eastern edge of the Chang Tang, I include here a summary of its habits.

Endemic to the plateau, the white-lipped deer is morphologically distinctive, with affinities both to rusine deer such as the sambar and rucervine deer such as the Indian swamp deer (Groves and Grubb 1987). One of the cossacks on Przewalski's third expedition shot two white-lipped deer in June 1876 in the Dang He drainage, an area that is now within Gansu's Yanchiwan Reserve. One of these animals became the type specimen of the species, *Cervus albirostris.* In December 1891, W. G. Thorold shot two stags of the same species in eastern Tibet (about 32° N, 94° E), as related at the beginning of this chapter. Since Przewalski's description of the white-lipped deer was not widely known, these animals were given a new name, *Cervus thoroldi*, which later was shown to be a synonym of *albirostris*. I encountered this large and striking deer occasionally during our surveys and add here my limited observations to those of Cai (1988), Kaji et al. (1989), Miura et al. (1989), and a few other reports.

Description

White-lipped deer are large and robust. Adult stags stand about 120–140 cm tall at the shoulder and weigh 180–230 kg, whereas females are about 115 cm tall and weigh usually less than 180 kg (Cai 1988; Miura et al. 1989; Sheng and Ohtaishi 1993). The antlers are smooth and somewhat flattened, and those of large stags have a brow tine and 4–5 other tines clustered at the distal half of the 90–130 cm long main beam (Jaczewski 1986).

The pelage is coarse and stiff and the hairs are hollow as in blue sheep, providing an insulating layer of warm air. The dark hairs have a light-colored band near the tip, giving the animal a grizzled appearance. As their name implies, the deer have a white upper lip and tip of muzzle, and they also have white inside the ears, on chin and throat, and on the inside of the legs. The belly is cream to grayish white. The body is light buff or grayish brown to dark brown, pelage color varying somewhat with sex, season, and possibly locality. Some stags become dark during the rut and the hair on the neck elongates. It was also my impression that the deer in Shule Nanshan of northeastern Qinghai had a lighter-colored winter coat than those in eastern Tibet (figs. 8.1 and 8.2). The rump patch is rust-colored and surrounds the tail.

Walking deer make a clicking sound that emanates from their hooves, a noise similar to that made by caribou and Père David's deer (Schaller and Hamer 1978). The hooves are exceptionally broad and high, almost resembling those of cattle (Flerov 1952).

Status and distribution

White-lipped deer occur in Tibet from the vicinity of Lhasa eastward into western Sichuan, and in the eastern two-thirds of Qinghai, roughly east of 93° E, into Gansu (fig. 8.3). Within that area they inhabit high hills and mountains covered with a mosaic of forest, rhododendron, willow, other shrubs, and meadows, including those above timberline. They also occur in arid, treeless ranges, as in the Qilian Shan, and on alpine meadows west to the fringe of the Chang Tang. They are basically deer of high, open pastures, usually in hills above an elevation of 3500 up to the limit of vegetation above 5000 m. White-lipped deer have a highly discontinuous

8.1. A white-lipped deer herd is on a ridge in the Qilian Shan of northeastern Qinghai. The rut is in progress, and a stag tends a female closely. (October 1985)

8.2. These white-lipped deer are in a fenced pasture in northeastern Tibet. They appear to have a darker pelage than those in the Qilian Shan. (October 1988)

distribution, created in part by the terrain but in part also by hunting pressure, which has eliminated populations in many localities. The distributional information in figure 8.3, which is based on our data, the work of Kaji et al. (1989, 1993), and a few other sources in the literature, is incomplete in that surveys have so far covered only a part of the deer's range, with eastern Tibet in particular needing more work.

In northeast Qinghai, Zheng et al. (1989) observed 59 whitelipped deer in Gansu's Yanchiwan Reserve. Nearby, in the Shule Nanshan, we tallied 176 deer in our 610 km^2 survey block, a density of $0.3/km^2$. The deer occur scattered in the Burhan Budai Shan as far west as the Yeniugou, west of the Lhasa-Golmud highway (Harris and Miller 1995). They are

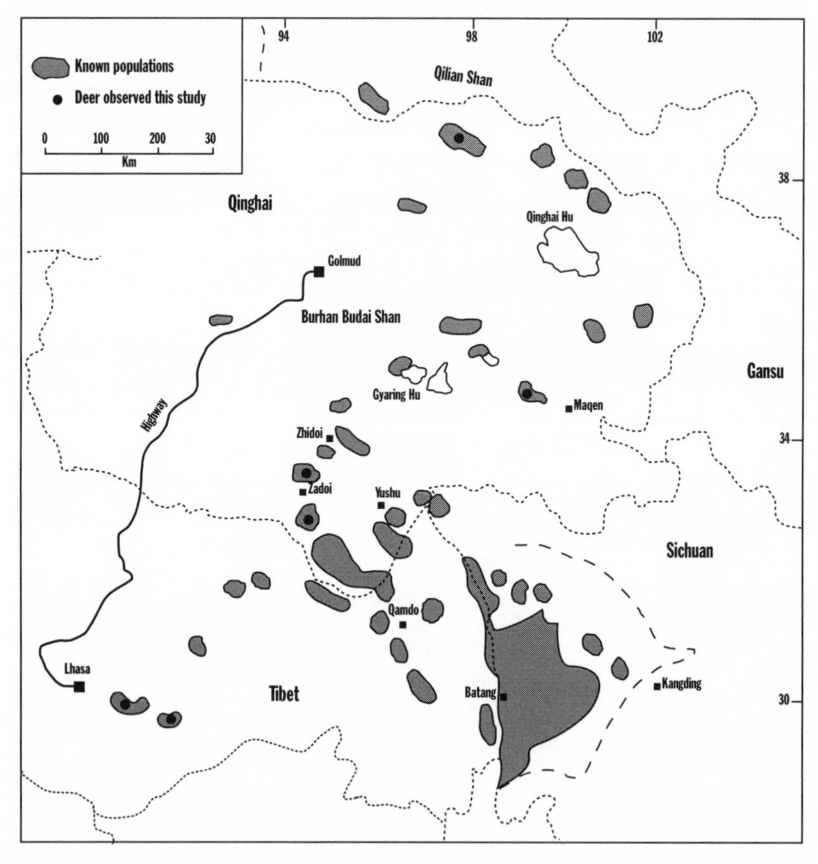

8.3. The known distribution of white-lipped deer, based on surveys by this project and by Cai (1988), Kaji et al. (1989), Koizumi et al. (1993), and a few other literature records. The eastern limit in Sichuan (dashed line) is from *Expedition* (1977). Eastern Tibet has not been adequately surveyed, and distribution is more extensive than indicated.

also found north of Donggai Co, according to local informants, and in the Anyemaqen Range, where we saw two stags (possibly escaped from a deer farm). Fewer than 100 deer survived along the west side of Gyaring Hu in 1985 (Miura et al. 1989). The October blizzard that year appeared to have caused heavy mortality because in 1986 only 25 animals could be found, and by 1988 only 10 survived (Kaji et al. 1993). Cai (1982) reported the species west of Gyaring Hu, at the junction of the Tuotuo and Dam Qu Rivers (34° N, 93° E). Around Zhidoi we found only a cast antler. Miura et al. (1989) observed a herd of 31 in that area, and to the northwest near Qumaleb, Kaji et al. (1993) counted 101 deer in 80 km^2, a density of $1.3/km^2$. To the south, in the vicinity of Zadoi, we censused several mountain blocks, totaling 405 km^2, and tallied 54 white-lipped deer, a density of $0.1/km^2$. A main stronghold of the species is in Nangqen County, in the southeast corner of Qinghai. We were told that herds with 100 or more animals could still be seen there, and Harris (1991) counted a herd of 200 in Baizha Forest, where the animals were concentrated in their wintering area.

The distribution of the deer in Tibet is only imperfectly known. I was told of several areas where they occur and in October 1995 checked on their status in the mountains between the Lhasa and Yarlung Tsangpo Rivers. There they occurred in the upper parts of several tributary valleys, usually on steep slopes covered with a mosaic of alpine meadows and various shrubs, including willow (*Salix*) and rhododendron. Only remnants survived in some valleys but in others they remained moderately common, according to local people. Near the village of Layu in Sangri County we tallied 75 deer in a two-day search.

In Sichuan, the species extends east to the vicinity of Kangding (*Expedition* 1977). Dolan wrote that "they were seen by us to the south of Lidang around Batang, and to the northwest of Jyekundo [Yushu], and tracks were found in the Tachienlu Mountains. However, they have been so persecuted that the stand is down to nothing in many localities" (1939). Kaji et al. (1993) observed them in several areas, especially in Serxu County, where they counted 1152 deer in 24 herds, often concentrated near monasteries, where they are protected. In one area of 155 km^2 there were 684 deer, a density of $4.4/km^2$ (Koizumi et al. 1993).

As Roborovsky and Kozlov's expedition to northeast Qinghai already noted in 1894–1895, "The white-faced *maral (Cervus albirostris)*, also discovered by Prejevalsky, is too much hunted to be numerous" ("Central Asian Expedition" 1896). Given the intensive hunting pressure for meat and antlers, much in demand for the medicinal market, it is surprising that populations remain widespread, though fragmented and seldom large. The deer have the highest level of legal protection (Class I) in China, but laws often have little local impact. A few herds have informal protection in the vicinity of monasteries (Harris 1991). Small deer farms are found throughout the range of the species—1500–2000 white-lipped deer are in such farms (Sheng and Ohtaishi 1993)—and they may help to maintain the animals as long as the price of antlers remains high and the operations remain profitable. However, deer farms also have a negative impact in that they capture young deer to stock farms or replenish existing captive herds. A number of farms keep both white-lipped and red deer. The two species readily hybridize. This negates any conservation role of the farms, and when animals escape, as they often do, hybrids may contaminate wild populations. Livestock disturb the deer and compete with them for forage, with the result that the deer move into marginal habitat (Miura et al. 1989). An estimated 50,000–100,000 deer survive (Sheng and Ohtaishi 1993).

Population and herd dynamics

We classified 175 white-lipped deer in the Shule Nanshan during late summer and autumn. The ratio of males to females was 45:100 (table 8.1). Kaji et al. (1989) sampled 90 animals in the Maniganggo area of Sichuan and found a ratio of 29 males to 100 females. Around Zhenda, Sichuan, Miura et al. (1993) tallied 684 deer, with a ratio of 59 males to 100 females. Because these data were collected just before and during the rut, when the males and females are most likely to be in the same area, the disparate sex ratios probably reflect the actual situation, due in part to selective hunting for antlers.

The ratio of yearlings to adult females in our sample was 36:100, and the ratio of young to adult females was 50:100. Kaji et al. (1989) found a young-to-female ratio of 27:100, and Miura et al. (1993) found 32:100, but they combined adult and yearling females. Since most females probably

Table 8.1 Composition of mixed white-lipped deer herds

Location	Month/year	Adult male	Yearling male	Female (adult and yearling)	Young	Total
Shule Nanshan	10/1985	7	0	9	2	18
		3	1	11	5	20
		10	5	51	20	92*
		14	1	23	7	45
Zadoi	9/1986	0	1	6	6	13
Sangri	10/1995	1	0	4	3	8
		3	1	1	1	6
		9	2	29	11	51

*Six animals were not classified.

do not give birth until 3 years of age, as is typical of large deer, their ratios include a number of reproductively inactive animals. Still, there were relatively few young only 3–4 months after the birth peak. On two deer farms only about a third of the females had young at heel, and the birthrate of fully adult females (3 years and older) was only 30–52% (Kaji et al. 1989).

Antlers are cast in April and May and the new ones reach full growth in September, when stags shed the velvet (Kaji et al. 1989); young males may retain velvet into October. The rut on deer farms may vary somewhat with elevation and locality. In the Qilian Shan at 3300 m the duration of the rut is from mid-September to early November; in the Anyemaqen Shan at 3900 m, from late September to mid-November; and in Zhidoi at 4300 m, from early October to mid-November (Zheng et al. 1989). On a deer farm near Serxu in Sichuan, Kaji et al. (1989) heard the characteristic rutting roar of stags during their 6–20 October visit. According to Miura et al. (1989, 1993) and Zheng et al. (1989), adult stags eat little during the rut and spar often, and each attempts to collect a harem of females.

At 1345 hours on 7 October 1985, I came upon a herd of 45 animals scattered on a grass ridge in the Shule Nanshan at an elevation of 3700 m:

> The herd comprises 3 small stags, 6 medium stags, and 4 large stags and many females and subadults; a fifth large stag joins the herd later. Each of the 4 large stags is in a different part of the herd. Three have a cluster of at least 5 females each, and 1 stag is with just 1 female. The small and medium-sized stags are dispersed throughout the herd without being focused on any females; 2 spar. One large stag sniffs the anal area of a female, then lifts his muzzle and curls up his lip. The largest stag raises his head and with open mouth emits a loud gurgling roar, a call that is reminiscent of that given by Père David's deer during the rut. After I have observed the deer for 45 minutes, they sense me. A female barks and the herd moves slowly to another ridge. There, several stags give their gurgling roar and one emits a short, whiny bugle. Three large stags each tend a female within the herd. The stag stands by her side. When she moves ahead, he cuts in front of her, stands broadside, and cuts off her retreat. Rather than attempting to be with a small harem, as observed an hour earlier, stags now seem to claim possession of a single female. On a sedge meadow in the vicinity of the herd but out of sight is a large stag standing in a mud wallow about 2 × 3 m in size. The stag's muzzle is muddy and his neck wet.

On 28 October 1995 we briefly observed a herd of 51, including 9 adult stags, near Layu in southeastern Tibet. Two of the three largest stags were each close to a female, and one twice attempted to mount.

Deer show considerable plasticity in their social structure during the rut in that in such species as red deer and caribou the stags may either maintain harems or tend individual females. The white-lipped deer may have a slightly different system, one that resembles the Indian swamp deer (Schaller 1967), in which stags in a herd tolerate each other but the largest stags lay claim to the females, either forming small harems within the herd or tending them singly.

Most young are born between late May and the end of June (Zheng et al. 1989) or early July, with a few as late as August (Kaji et al. 1989). Zheng et al. (1989) give a gestation period of 7.5–8.3 months; Kaji et al. (1989) give 7–7.5 months.

For part of the year the stags may wander alone or in small male herds with up to 8 members, but by the onset of the rut in September, most are in mixed herds with many members of both sexes and all ages. One to 8 very large stags are usually with a mixed herd during the rut (Miura et al. 1993). In September and October eight mixed herds ranged in size from 6 to 92 (table 8.1). In western Sichuan average herd size in October, at the height of the rut, was 51.1, with as many as 165 and 169 individuals together (Miura et al. 1993), and herds with 200–300 animals have been reported (Kaji et al. 1989; Harris 1991). Miura et al. (1989) observed a herd of 25 that retained its composition from 21 to 30 August along the margin of Gyaring Hu. During this period it ranged over about 35 km^2, traveling on an average of 3.5 km per day, including trips to several offshore islands that the deer could reach only after a 25-minute swim.

9

Wild Bactrian Camel

Such a journey as the one we have just mentioned, besides its geographical interest, would finally set at rest the question of the existence of wild camels and horses. The natives repeatedly told us of the existence of both, and described them fully.

According to our informants wild camels are numerous in North-western Tsaidam, where the country is barren, the soil being clay, overgrown with *budarhana*, and so destitute of water that they have to go seventy miles to drink, and in winter are obliged to satisfy their thirst with snow.

The herds are small, averaging five to ten in each, never more than twenty. Their appearance is slightly different from the domesticated breed; their humps are smaller, the muzzle more pointed, and the colour of the hair grey.

Nikolai Prejevalsky (1876)

As Przewalski crossed the Qaidam Basin during his first expedition in 1872, he was told by the inhabitants that wild camels were in that area. At the time the existence of truly wild camels was still in doubt, such animals being generally dismissed as feral domesticated camels. But, as H. Yule in his introduction to Przewalski's (1876) book makes clear, references to wild camels had existed in the literature for at least 450 years. He notes that D. Forsyth in his Report of a Mission to Yarkand and Kashgar in 1873 was told that "the wild animals of Lob are the wild camel. . . . It is a small animal, not much bigger than a horse, and has two humps. It is not a tame camel; its limbs are very thin, and it is altogether slim built." Yet even after Przewalski killed several camels in March 1877 during his second expedition, he still posed the question: "Are the camels found by us the direct descendants of wild parents, or are they domesticated specimens which have wandered into the steppe?" (Prejevalsky 1879).

It is now generally agreed that *Camelus (bactrianus) ferus* is the wild progenitor of the domestic form (Bannikov 1976). The two differ genetically (see chapter 13), suggesting that domestication occurred long ago. Camel bones found in Neolithic sites over 6000 years old cannot be ascribed with certainty to domestic or wild animals because proportions and form of bones and teeth are similar in the two (Olsen 1988). Among the oldest art that depicts camels and people is a relief on a black obelisk in Iran that shows Bactrian camels and their handlers bringing tribute to Shalmaneser III in 841 B.C. (Olsen 1988). The wild form of the one-humped dromedary has been extinct for about 2000 years (Köhler-Rollefson 1991), making wild Bactrian camels the last of an extraordinary genus that has roamed Asia at least since the early Pleistocene when *Camelus knoblochi* existed in Inner Mongolia and elsewhere (Olsen 1988). Because camels were once found along the northern part of the Tibetan Plateau and a few still frequent the Arjin Shan, I include them in this report.

During wildlife surveys in Xinjiang, I visited the fringe of camel habitat around the Gobi and Taklimakan Deserts, but the Lop Nur region, the animal's principal habitat, was closed to foreigners at that time for military reasons. Information from the old and recent literature (Prejevalsky 1879; Hedin 1898; Gu, Gao, and Zhou 1991; Hare 1995, 1996, 1997) and from local informants provided an overview of camel distribution. Mongolia's Great Gobi National Park is the only place outside China with wild camels, and I studied the animals there for several months between 1989 and 1993 (Tulgat and Schaller 1992; Schaller 1995).

Description

The domestic Bactrian camel is a familiar animal, impressively large and bulky with a shoulder height up to 2 m and with two distinctively large humps consisting of fibrous tissue and fat. The stout legs have broad two-toed feet cushioned by pads. The long curved neck is surmounted by a small head with a convex muzzle and slack lips. The slitlike nostrils can be closed during desert storms and the long eyelashes protect the deep-set eyes from dust. Males have well-developed occipital glands at the back of the head below the nuchal crest and these exude a coffee-colored liquid. The pelage is brown to dark brown, short in summer and long and woolly in winter, with the hair darkest and longest on the neck, top of head, humps, and elbows, especially in males (fig. 9.1).

Unlike the ponderous domestic form, the wild camel is relatively small, lithe, and slender-legged, with very narrow feet and a body that looks laterally compressed. The Mongolians call the animal *havtagai,* which means "flat." The pelage on the upper legs, neck, top of humps, and tail is elongated and dark brown, much as in domestic animals, but the rest of the body is more a sandy grayish brown, especially in the short summer coat. The wool is shorter and sparser than that of domestic animals. The tuft of hair on the top of the head is small, and, instead of the large irregularly shaped humps of domestic camels, the wild ones have small, pointed, conical humps (fig. 9.2). The forelegs lack callosities on the knees (Bannikov 1976). Early in 1995, Atkins (1995) measured the shoulder height of wild

9.1. We ask direction in Mongolia's Gobi desert from a herdsman leading a domestic Bactrian camel. The animal has characteristic large humps, and it is in its thick, woolly winter coat. (November 1992)

9.2. This subadult male Bactrian camel was caught as an infant and reared by a domestic foster mother. He shows the typical small, conical humps of wild camels.

camels that had been captured as young between 1987 and 1991. Seven females had an average shoulder height of 171 (167–180) cm; and two males, 172 (164–180) cm. On the basis of three measurements multiplied by a known factor, as determined by her research on dromedaries, Atkins (1995) calculated the average body mass of these females at 446 (377–517) kg, and the males at 394 (367–422). The males were born in 1990 and 1991 and at 4–5 years of age had probably not reached their full size.

Status and distribution

The Bactrian camel is adapted to arid plains and hills where water sources are few and vegetation is sparse, often little beyond some drought-resistant shrubs. The range of the species in historic times extended from about the great bend of the Yellow River at 110° E westward across the deserts of southern Mongolia and northwest China to central Kazakhstan (Heptner, Nasimovic, and Bannikov 1966). Much hunted for its meat and hide, the camel had by the 1850s vanished from the western part of its range and persisted only in remote desert tracts of the Gobi and Taklimakan. When Przewalski discovered the camel for science in 1877 (Prejevalsky 1879), its distribution was apparently still continuous between the Gobi and Taklimakan but populations had become fragmented by the 1920s.

In the Gobi, camels still occurred as far east as Gaxun Nur (*nur*, or *nor*, means "lake" in Mongolian) in Inner Mongolia during the 1920s and stragglers went north to Boontsagaan Nur in Mongolia as late as the 1960s (Zhirnov and Ilyinsky 1986). Today the animals are confined to about 28,000 km^2 of Mongolia's Great Gobi National Park, which was established in 1976 for their protection (fig. 9.3). Herds occasionally leave the park, crossing the border into northern Gansu (Hare 1995) and the adjoining parts of Xinjiang and Inner Mongolia. Some males wander up to 100 km south of the border during winter, where they may join domestic herds.

The size of the camel population in Great Gobi National Park has been estimated at over 300 in 1943, 400–500 in 1960, about 900 in 1974 (Bannikov 1976), 400–700 in 1976 (Dash et al. 1977), and 500–800 in 1980–1981 (Zhirnov and Ilyinsky 1986). Between 1982 and 1989, annual transects gave estimates of 480–555 (Tulgat and Schaller 1992). In the opinion of park staff, camel numbers declined between 1990 and 1995 to about 300–400 as a result of emigration to China and wolf predation (Tulgat 1995), but no comprehensive censuses have ever been conducted. In August 1993, I traveled 650 km in the western half of the park and the only fresh sign of camels was the tracks of a herd of 8. During an aerial strip census of the Gobi park in March 1997, R. Reading (pers. comm.) saw 314 camels and calculated a total of at least 900, casting doubt on the accuracy of the earlier figures.

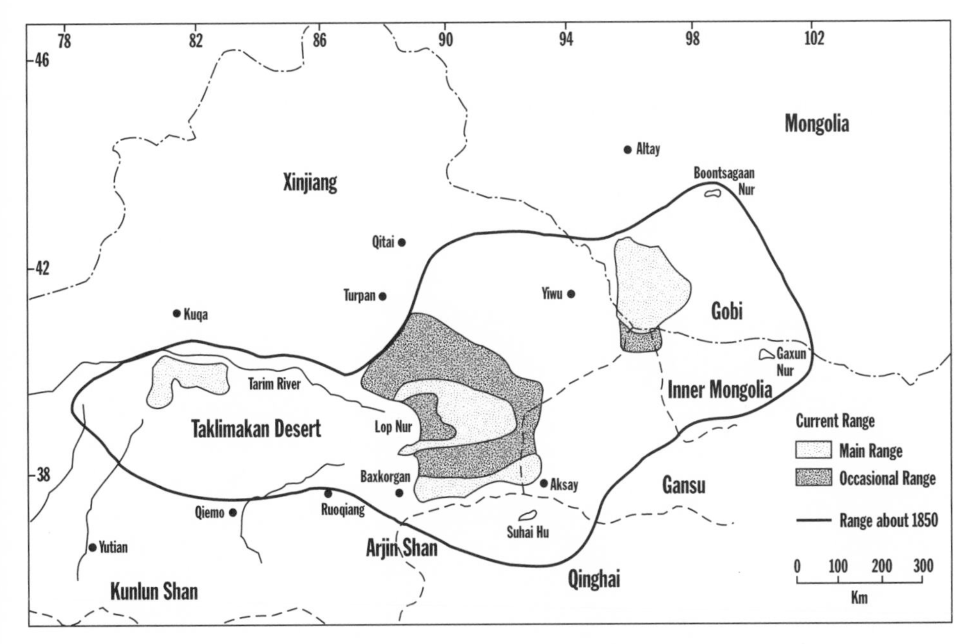

9.3 The past and current distribution of wild Bactrian camels in Mongolia and China. (Adapted from Tulgat and Schaller 1992.)

In China, camels disappeared from north of the Tian Shan sometime after Przewalski reported them from near the town of Qitai (Prschewalski 1884), but the animals persisted south of the Tian Shan. Their main stronghold now is around Lop Nur, a lake and marsh that dried up after the waters of the Tarim River were diverted for irrigation. Hedin (1903) hunted camels in the Kuruk Tag northeast of Lop Nur and Przewalski hunted them to the east (Prejevalsky 1879). In 1980 and 1981, Gu and Gao (1985) observed 98 camels in 21 groups in this region and estimated up to 200 there. By 1995, only about 80–100 survived, a decrease of 50% or more, due mainly to hunting (Hare 1997). Much of the Taklimakan Desert west of Lop Nur consists of shifting dunes and rock-littered plains. On a flight low over the area, I saw huge tracts devoid of water and almost without vegetation. However, several rivers flow north from the Kunlun Shan into the desert, where they vanish underground, their course marked by intermittent stands of *Phragmites* reeds, *Tamarix* shrubs, and *Populus* trees. Przewalski heard of camels along the lower Qiemo River (Prejevalsky 1879). Farther west, Hedin (1898) observed camels along the Yutian River in the middle of the Taklimakan, 250 km north of Yutian. North of Hedin's route near the Tarim River, E. Murzayev found camels in the late 1950s (Zhirnov and Ilyinsky 1986) and a population continued to persist there according to Gu, Gao, and Zhou (1991) and our local informants. A road was built through the camel's habitat south to Khotan in the 1990s and major oil development began. Perhaps only 50–80 camels persist in that region (Hare 1997).

Most explorers reported camels along the Arjin Shan or Altyn Tag. In 1901, Hedin found that "the wild camel was very common in the neighbourhood of Gashun-gol. We frequently saw troops of 15 to 20 individuals, sometimes on the right of our march, that is to say, on the outermost slopes of the mountain next [to] the desert" (1903). In May 1893, Littledale discovered "on the north side of the Altyn Tag, the fresh tracks of a camel. . . . After this we occasionally saw more tracks, but we never beheld the animal in the flesh until the 16th day; we were then in the most forlorn country imaginable, not a blade of grass for our animals was to be had, and water was only procurable about every second day" (1894). And Przewalski noted that "during the excessive heats in summer, the camels are attracted by the cool temperature of the higher valleys of the Altyn-tagh, and make their way thither to an altitude of 11,000 feet (3355 m), and even higher" (Prejevalsky 1879). In 1995, Hare (1997) observed a total of 49 camels in the Lapeiquan Spring area northwest of Baxkorgan, and he estimated that 250–300 persist along the Arjin Shan.

Camels were also found on the southern slopes of the Arjin Shan in the Qaidam Basin. The Russian explorers Roborovsky and Kozlov reported "great numbers of wild camels, khulans, and antelopes" around the lakes Khuitin Nur, Sukhain Nur (Suhai Hu), and Bagasirtin Nur ("Central Asian Expedition" 1896), and Przewalski also noted them there (Prschewalski 1884). I have no evidence that camels are now resident in the Qaidam Basin. However, a few are north of the Arjin Shan in Gansu (Cheng 1984) and the animals still appear east of Baxkorgan near the Qinghai-Xinjiang border, indicating that camels might still enter the basin on occasion.

Tulgat and Schaller (1992) thought that fewer than 1000 wild camels survive in Mongolia and China, and Hare (1997) presented a figure of 730–880. Given the results of the 1997 census in Great Gobi National Park, perhaps at least 1500 may persist. For comparison, some 2.5 million domestic camels occur in central Asia (Mukasa-Mugerwa 1981). Only a remote and desolate habitat and extreme shyness have enabled the animals to persist. Over a century ago, Przewalski wrote that "twenty years ago, wild camels were numerous near Lake Lob, where the village of Chargalik [Ruoqiang] now stands, and farther to the east along the foot of the Altyn-tagh, as well as in the range itself. . . . With an increase of population at Chargalik, the hunters of Lob-nor became more numerous, and camels scarcer" (Prejevalsky 1879). Today motorized hunters penetrate the desert, and herders and miners occupy oases, preventing wildlife access to water.

In 1980, a geological survey team shot 40 camels (Hare 1995). Since wild and domestic camels readily hybridize, contact between the two leads to contamination of the wild stock, especially since wild males tend to drive off domestic females during the rut. The Gobi park authorities have shot several hybrids in recent years. Although fully protected by law and ostensibly safe in the Gobi (44,190 km^2), Annanba (3900 km^2), and Altun Shan (15,120 km^2) Reserves (see fig. 1.3), wild camels have a tenuous existence as humans penetrate ever more persistently into their last inhospitable haven.

Population and herd dynamics

Great Gobi National Park has a continental climate of winters with temperatures to −30°C and below, summers to 40°C and above, strong winds, and precipitation usually of 100 mm or less annually. Only about 30 permanent springs were in the park at the end of a drought that had lasted through the 1980s. Away from the few oases, with their *Phragmites*, *Achnatherum*, and other tall grasses and a few *Populus* trees, the vegetation consists of a thin cover of shrubs such as *Haloxylon*, *Caragana*, *Reaumuria*, and *Salsola*, which serve as the camel's principal food; grasses and forbs are scarce. Dry watercourses, the *sairs*, cut the sun-baked and barren terrain. Several mountain massifs, precipitous and with narrow gorges, protrude several hundred meters from the plains.

The behavior of camels in the Gobi was described by Tulgat and Schaller (1992). Herds move widely, their distribution linked to water. The animals tend to concentrate in and around the massifs, such as the Atas Ula, because most springs are there and snow on the slopes may provide the only moisture in winter. They also shift to areas where a local shower has created a flush of green. Concentrations of up to 100 camels occur near the mountains especially in October and November toward the onset of the rut. In the Gobi, as in the Lop Nur region (Prejevalsky 1879), the rut begins in November and peaks from late January to late February. A male attempts to collect a harem, sometimes with as many as 10–20 females. With the sex ratio at about 1 male to 1.6 females, competition between males is fierce, occasionally "terminating in the death of one or other of the combatants" (Prejevalsky 1879). Usually, however, males display several visual and olfactory patterns to express dominance. A male may rub his occipital gland against his anterior hump and grind his teeth; and he may spread his hindlegs and lower his rump as he flaps his tail audibly, urinating with his posteriorly directed penis to saturate his tail, hindlegs, and rump (Wemmer and Murtaugh 1980). Most births are in March and early April, though Hare (1995) reported a newborn as late as

Table 9.1 Percentage of young in the wild camel population of Great Gobi National Park, September–December 1981–1994

Year	Sample size	% young*
1981	420	10.7
1982	189	6.3
1983	98	4.1
1984	255	4.7
1985	157	3.8
1986	62	3.2
1987	440	2.5
1988	322	9.3
1989	183	3.2
1990	81	3.7
1991	243	5.3
1992	112	1.8
1993	251	8.0
1994	217	3.7

Sources: Data for 1981: Zhirnov and Ilyinsky 1986. Data for 1982–89 and 1992: Tulgat and Schaller 1992. Data for 1990, 1991, and 1994: Tulgat 1995. Data for 1993: Atkins 1994; G. Neumann-Denzau, pers. comm.
*Excludes 22 young taken into captivity 1987–91.

12 May. The average gestation period is 406 days in domestic animals (Novoa 1970).

The number of young in herds has been recorded intermittently by park wardens, biologists, and visitors in the Gobi park since 1980. Most samples are small and potentially biased. For instance, in August–September 1989, I tallied 106 camels and there were no young. In October–December that year, a sample of 183 camels contained 3.1% young. Nevertheless, the data from 1982 onward showed a consistent pattern: the percentage of young was low. Counts during September–December, when young were at least 5–6 months old, showed an average of 5.0% (1.8–10.7%) young from 1981 to 1994 (table 9.1). In China, Gu and Gao (1985) counted 13.8% young in 80 camels, and Hare (1995) counted 24.8% in 50 camels. Females generally mate first at the age of 4 years and the long gestation period enables them to have a young only once every two years (Novoa 1970). With such a slow reproductive rate, a population would have a low percentage of young, perhaps 10–15%, but not as low as the camels in the Gobi Park.

The scarcity of young was the result of poor survival rather than poor production, as shown by a steady decrease from March–April to June–July to October–December in 3 sample years: 1986—8.2%, 4.1%, and 3.2%; 1987—13.8%, 6.2%, 2.5%; 1988—15.0%, 11.5%, 9.3%. Tulgat (1995) also noted a marked decrease in numbers between spring and autumn in

1990, 1991, and 1994. Camels may live as long as 35 years (Jones 1993), and an occasional poor reproductive season can be absorbed by a population. But young survived adequately only during two years between 1982 and 1994. Given the annual mortality from various causes, the population, whatever its precise size, cannot maintain itself with such a low percentage of young.

A total of 89 camels were found dead between 1984 and 1989: 61% killed by wolves, 2% by snow leopard, 12% in fights between males, 9% by poachers, and 16% of causes unknown. Of the wolf kills, 31% were camel young; of 5 wolf stomachs examined in 1987–1989, all contained camel young, and 1 also contained red fox and goitered gazelle. I examined a site where wolves had killed a camel young, and also observed an aborted hunt:

> At 1800 hours a herd of 16 camels, including at least 3 adult males, rests on barren ground near the tall grass of an oasis. Seven animals lie and the others stand. Suddenly the herd flees tightly bunched about 200 m. They halt a few seconds and then walk off rapidly in single file. As the camels run, I hear two wolves howl. The wolves trot into view, angle up a low ridge across which the camels have fled, and sit on top.

With wolves obviously having an impact on the camel population, the park authorities began a wolf control program in 1987. A total of 118 wolves were killed between 1987 and 1989 and a few after that. Wolves have coexisted with camels for millennia, and it is not clear why predation manifested itself so strongly during the 1980s. However, the decade of drought, during which mean annual precipitation was only 50 mm, may have had an influence. The water table dropped, causing springs to vanish. The Tsagaan Bogd Range had 9–10 springs during the 1970s, according to a longtime resident, but only 4–5 by the early 1990s. Without rain, shrubs failed to sprout, and tender, nutritious forage was scarce. In 1989 the camels were in poor condition—lean, their ribs prominent. Herds became localized around the few oases that had not only water but also tall green grass and poplar leaves. This possibly made the camels vulnerable to wolf packs, which focused on these localities. The drought also caused an apparent decline in argalis, kulans, and goitered gazelles, and with these species low in number the wolves may have concentrated even more on camels. The summer rains were reasonably good in 1990, providing a test of this explanation. With wolf numbers reduced by hunting and herds presumably dispersed, survival of young should have improved. Yet it did not that year nor in the two following years. The 1993 rains were the heaviest since 1938. The desert was green with forbs and sprouting seedlings where in previous years the ground was bare. Survival of young was better than

Table 9.2 Size distribution of wild camel herds in Great Gobi National Park, 1982–1989

	Rut (Dec.–Feb.)	Birth season (Mar.–Apr.)	Rest of year (May–Nov.)
Total number of herds	76	236	363
Total number of animals	451	1246	2370
Mean herd size	5.9	5.3	6.5
Herd sizes			
1	3.1%	5.2%	3.2%
2–5	28.6	25.5	18.4
6–10	25.7	23.5	27.3
11–15	24.8	19.7	16.8
16–20	7.1	10.4	12.9
21–25	0	8.9	6.8
26–30	0	4.2	8.4
31+	10.6	2.6	6.1

Source: Data from Tulgat and Schaller 1992.
Note: Mean herd size during the birth season differs significantly ($P < 0.05$) from that during May–Nov.

9.4 A herd of wild Bactrian camels files across a scrub-covered valley toward an oasis in Great Gobi National Park. The animals are in their summer pelage. (August 1989)

average that autumn but dropped again in 1994 for unknown reasons (table 9.1).

Average herd size was 6.0 in the 675 herds tallied during the 1980s (table 9.2) and it remained similar in the 1990s (Tulgat 1995). A few individuals (3–5%) of both sexes were solitary, but most (63–79%) were in herds with 2–15 members (fig. 9.4). Herd size was smallest during the birth season because females apparently became solitary to have their young and they remain alone for about two weeks, according to park rangers. The largest herd in the sample contained 48 individuals, but temporary aggregations were excluded. For example, 98 camels were observed at an oasis in October 1988. In China, Hedin (1898, 1903) observed herds with up to 6 animals and considered herds with 12–15 rare; Littledale (1894) saw a solitary female, a herd of 2 twice, and a herd of 9 once; Gu and Gao (1985) recorded lone animals and herds with up to 23 individuals and an average of 4.7. Przewalski was told by a local hunter that in the past it was possible "to see some dozens, or even a hundred animals together" (Prejevalsky 1879). The small average herd size in the past 100 years reflects not just the desperate aridity of the environment but heavy hunting pressure. All camel populations are for one reason or another severely threatened, yet little intensive research and conservation effort has focused on them. If present trends continue, they will not survive the next century.

Captive breeding

With the last wild camel populations in decline, every captive must become part of a well-managed captive-breeding effort to ensure the survival of the wild form of the species. Like Przewalski's horses, the animals might then someday be returned to the wild. In the mid-1990s, nine camels were captive in four institutions in China, with four females and two males in Wuwei, Gansu, where two young were born in 1995 (Hare 1995; D. Miller, pers. comm.).

Between 1987 and 1991, a total of 22 camel young (10 males, 12 females) were captured in Great Gobi National Park ostensibly for captive breeding (fig. 9.5). Nine of the young soon died, but the rest (6 males, 7 females) were raised by free-ranging domestic camel females around Bayan Tooroi, the park headquarters. When these camels matured, the park authorities failed to segregate the wild and domestic animals. One female, born in 1987, was bred by a domestic male and gave birth to a hybrid in 1992 and a second hybrid in 1994. During the winter rut of 1992–1993, a wild male became aggressive toward humans and mated with domestic females. Although the local people have handled camels for centuries and well know how to deal with irascible males, they made no effort to manage the wild animals, such as by restricting them to corrals during the rut. A male that came near a yurt was shot in the leg. Another male mauled a domestic male while in quest of females; he suddenly vanished and no doubt was killed by a herdsman. When a male injured a person in 1994, he, together with another male, was taken far into the park and released.

9.5. The authorities of Great Gobi National Park captured wild Bactrian camel young for a captive-breeding program. This animal is about 2 months old. (June 1990)

The naive animals could not have known the location of water sources, and their fate is unknown.

Concerned about the lack of management, the UNDP Mongolia Biodiversity Project funded Atkins (1994, 1995) to establish a breeding program, and she spent several months during the winters of 1993–1994 and 1994–1995 at Bayan Tooroi. A corral was built to confine the camels during the rut. Atkins tamed the animals so that they could be handled, and she supervised feeding, mating, and other aspects. Cooperation from the park staff was minimal, and the program lapsed after she left. However, her efforts were successful in that at least one of the two young born in 1995 and 1996 appeared to be purebred. I mention this sad endeavor to show how an important breeding program would have wholly failed through negligence and disinterest on the part of the authorities except for the brief intervention of one dedicated person. Mongolia should not so carelessly doom these magnificent animals.

10

Kiang (Tibetan Wild Ass)

Everywhere the tracks of kyangs were seen and a good many of wild yaks; a few of the latter animals were grazing in a side valley. The kyangs had been rare so far. Now they appeared in six different herds in the extensive valley. We know too little of high Tibet to be able to draw maps of the occurrence of big game and its wanderings with the seasons. On a journey like this one gets, however, a very strong impression of the fact that in some regions there is no kind of big game at all, in others the yak is common, in others the kyang or both.

Sven Hedin (1922)

KIANGS ARE still quite common in some areas of the Chang Tang, where they add a vibrant presence to a landscape that often seems desolate and lifeless. At times they galloped over the golden pastures, tasseled tails streaming in the wind and hooves trailing feathers of dust until they suddenly wheeled like well-trained cavalry and lined up to observe us pass. One strives to protect the Chang Tang in part to guard such memories. The kiangs are also a delight to field biologists because, unlike other species, they are often inquisitive, standing and watching, instead of fleeing, most elegant in their russet and white coats. Such behavior was not always appreciated by foreign travelers. Their "ill-timed curiosity often spoils a stalk after nobler game" (Bower 1894) by alerting others of potential danger. Furthermore, "their apparent tameness is often deceptive, enabling them to draw quite close to the unwary traveller, and then with a sudden dash seize him by the stomach" (Landor 1899), a comment unique in the annals of kiang behavior.

The Asian wild asses of the species *Equus hemionus* have declined drastically during this century. Remnant populations of a few hundred to a thousand or two persist in Iran, Turkmenistan, and India, but only in southern Mongolia and a small part of bordering China does the species survive in numbers that possibly exceed 8000. The kiang, *E. kiang*, is an exception among Asian wild asses in that its distribution remains extensive and its numbers so large that there is no immediate threat to its existence. The animals actually appear to be increasing in some areas—to the consternation of pastoralists, who assert that they deprive livestock of forage. But a reasonably secure current status does not eliminate concern for the ani-

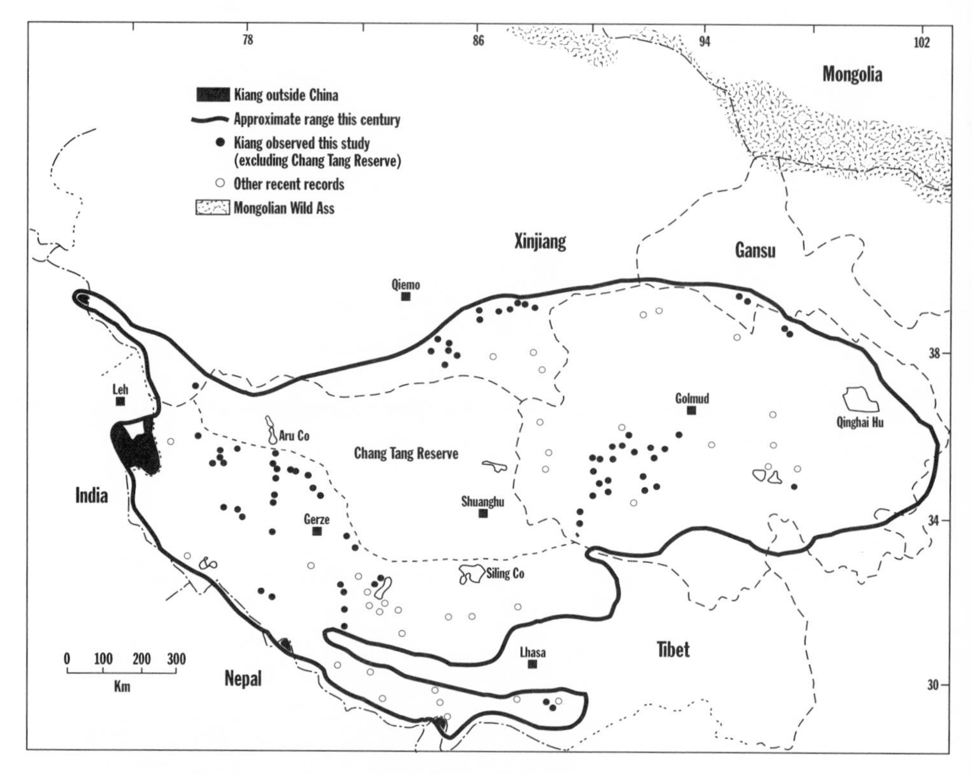

10.1. The distribution of kiangs (Tibetan wild asses). Locality observations from this project outside the Chang Tang Reserve are indicated, as are selected literature records. The current range of kulans (Mongolian wild asses) in China and Mongolia is also shown.

mal's future. Kiangs have in recent decades been decimated or eradicated from large tracts, and this trend will continue as pastoralists and their livestock increase. The species will need careful management, but first it requires detailed study, which has not yet been attempted.

Taxonomy

The kiang has often been considered to be a subspecies of *Equus (Asinus) hemionus: E. h. kiang*. On the basis of such morphological features as shape of rump, angle of upper incisors, and color pattern, Groves and Mazák (1967) placed the kiang into its own species: *E. kiang*. Studies of mitochondrial DNA indicated that kiangs "are discernibly discrete and there is no evidence of recent gene flow" between *kiang* and *hemionus* (Ryder and Chemnick 1990). Kiangs may well have been isolated on the Tibetan Plateau for several thousand years, and today 350 km separate them from the nearest *hemionus* in northern Gansu and Inner Mongolia (fig. 10.1).

Groves and Mazák (1967) proposed three subspecies of kiangs based on coat color, body size, and skull proportions. The subspecies with the largest body *(holdereri)* is said to occur in the eastern part of the plateau; a second subspecies *(kiang)*, intermediate in size and with a very dark russet coat, is found in the western part; and the third subspecies *(polyodon)* is "very small in size" and is found just "north of Sikkim." Groves (1974) mapped these subspecies as geographically distinct. Actually kiang distribution is, or until recent decades was, continuous across the rangelands of the plateau, including an extension along the foothills of the Himalaya (fig. 10.1). I have observed the three supposed subspecies and noted no marked difference in size or color. Although slight regional variation in kiangs may exist, an acceptance of subspecies seems premature.

Description

The kiang is the largest of the wild asses, a robust yet trim animal up to 142 cm high at the shoulder (Groves 1974; *Research on Flora and Fauna* 1979; Feng, Cai, and Zheng 1986) and with an estimated body mass of 250–300 kg, with some stallions up to 350–400 kg (Schäfer 1937d), although, like in all equids, sexual dimorphism is slight. One captive stallion weighed 223 kg at death and one mare 214 kg (Crandall 1964). The shoulder height of three wild stallions ranged from 132 to 142 cm (*Research on Flora and Fauna* 1979). Newborns may weigh 30 kg (Pohle 1991) and stand 90 cm tall at the shoulder (Hedin 1903). The kiang's head is large, the muzzle blunt, and nose convex. The mane is upright and relatively short. The coat is a rich chestnut color, darker brown in winter and a sleek reddish brown in late summer after the animal has shed its woolly pelage (fig. 10.2). The summer coat is about 1.5 cm long and the winter one over double that length (Groves 1974). The legs and undersides, including the ventral part of the neck, are white, as are the insides of the ears and end of the muzzle. A dark dorsal stripe extends from the mane to the end of the tail, which has a tuft of stringy black hairs. The tips of the ears and a narrow band along the margin of the hooves are also black.

Status and distribution

The kiang is an animal of open terrain, of plains, basins, broad valleys, and hills, wherever suitable forage, especially grass and sedge, is abundant. It reaches its highest densities on the vast alpine meadows and alpine steppes. However, it also occurs in desert steppe and other arid habitats, such as in the Qaidam Basin as low as 2700 m in elevation, and in mountainous terrain up to about 5300 m. The species confines itself to the pla-

10.2. These female kiangs are shedding their winter coats. (July 1991)

teau except at the western end, where it has penetrated the dry Yarkant and Oprang Valleys between the Kunlun and Karakoram Mountains as far west as the Shimshal Pass at the Pakistan-China border. Figure 10.1 indicates where I observed kiangs and also shows some recent records from others to delineate the range of the species.

Outside China

Kiangs occur in only a few small areas outside China. The westernmost kiangs enter the Shimshal Valley of Pakistan, where as many as 20–25 have been reported (Rasool 1989). In eastern Ladakh, India, Ronaldshay found that "galloping about in every direction were great herds of *kiang* or wild horse; and as I crossed the plain I must have seen hundreds of these ugly fiddle-headed creatures" (1902). Hunting has reduced the species to such an extent that now only about 1500 animals survive in Ladakh (fig. 5.4), at a density of at most 0.25/km^2 (Fox, Nurbu, and Chundawat 1991a). Farther east, kiangs occasionally cross from Tibet into the Dolpo and Mustang areas of Nepal (R. Jackson, pers. comm.). Similarly, kiangs enter northern Sikkim from Tibet at certain seasons, but numbers are low, ranging from 10 to 120, because suitable habitat is limited to no more than 30 km^2 (Lachungpa 1994; Shah 1994).

Qinghai and Gansu

At one time, kiangs were abundant in eastern Qinghai, except for parts of the Qaidam Basin "probably owing to the crust of salt on the surface of the ground there" (Prejevalsky 1876). In the Qilian Shan and adjoining ranges, "the wild donkeys . . . are very numerous, and go to the lower valleys as well" ("Central Asian Expedition" 1896). During a survey of a 610 km^2 area in that area in 1985, we tallied 13 kiangs, and in the nearby Yanchiwan Reserve in Gansu we saw 58, all on the Yema (Wild Horse) flats. Kiangs once occurred near Qinghai Hu in "herds of several hundred" (Prejevalsky 1876), but they are now rare in that region. Farther south on the great rangelands around the headwaters of the Yellow River, the species was once abundant. Going west from Donggi Co, at the western end of the Anyemaqen Range, Rockhill (1891) saw "at least a thousand" kiangs in 120 km. More recently, Kaji et al. (1993) counted 186 kiangs in a month of fieldwork there. In 1935, Schäfer (1937d) found kiangs in northwest Sichuan and tallied over 1000 kiangs in one day in the upper Yalong Valley across the border in Qinghai. The animals were intensively hunted throughout this part of the plateau after a highway was constructed during the 1950s and especially from 1958 to 1961 when during the Great Leap Forward severe food shortages existed (Zhang 1984).

Kiangs remain widespread in southwest Qinghai, as surveys along the eastern edge of the Chang Tang have shown. In the Yeniugou, west of Kunlun Pass, Harris (1993) estimated a population of 843 in 1051 km^2, or 0.8/km^2. In one census block of 2100 km^2 near Wudaoliang on the Lhasa-Golmud road, 213 kiangs, or 0.1/km^2, were seen, and 1500–2000 were estimated within an area of 20,000 km^2, including that block (Schaller, Ren, and Qiu 1991). Toward the Tibet border, Feng (1991b) censused seven blocks totaling 2736 km^2 and tallied 510 animals, or 0.19/km^2. For the whole region of 75,000 km^2, he estimated 1000–1500 kiangs, a figure that may be somewhat conservative even though large tracts have few or no animals. For example, we did not observe a single kiang in 170 km of the Tuotuo Valley upstream from Tuotuohe in November 1993.

Xinjiang

In Xinjiang, we drove along the broad Tula Valley, which traces the southern margin of the Arjin Shan, and in about 300 km saw 56 kiangs. In a week's survey in the Yüsüpalik Tag, a subrange of the Arjin Shan, we tallied 40 kiangs, a third of them solitary. Traversing about 4000 km^2 of the plateau on camels west of the Arjin Shan Reserve, we counted 108 kiangs, a minimum density of 0.03/km^2. A survey of 23,000 km^2 in the western half of the Arjin Shan Reserve showed that most kiangs were concentrated in about 5795 km^2, at a density of 0.08/km^2 (Achuff and Petocz 1988); the rest of the survey area was almost devoid of kiangs, and average density in the whole 23,000 km^2 probably did not exceed 0.02–0.03/km^2. Kiangs were more abundant in the eastern half of the reserve, where Butler, Achuff, and Johnston (1986) saw over 1000. Feng (1991a) apparently counted a concentration area in the reserve when he noted 770 kiangs in 1030 km^2, or 0.75/km^2. The estimate of 41,262 for the whole reserve (Gao and Gu 1989) was no doubt much too high.

In the Aksai Chin region of Xinjiang, Shaw ([1871] 1984) reported 100 kiangs at the head of the Karakax River. Kiangs still occur in the area, but no surveys have been conducted there.

Tibet

In Tibet, kiangs persist in fragmented populations along the foothills of the Himalaya. Those in the eastern part are separated from the northern populations by agriculture and the many villages in the valley of the Yarlung Tsangpo. At one time, kiangs were abundant on these southern rangelands. Traveling south and east of Gyanze, Bailey (1911) noted that "this animal may be seen anywhere." During a wildlife survey in October 1995 we attempted to delineate current distribution between the Bhutan border and the Yarlung Tsangpo. According to local people, kiangs were exterminated in most areas between the 1960s and 1980s. In that eastern part of their range we found kiangs around Chigo Co (28°40′ N, 91°40′ E) in three populations totaling probably no more than 200 animals (fig. 10.3); and others persist just south of Yamdrok Co (W. Liu, pers. comm.). Scattered populations survive also along the Himalaya west of Bhutan, west of about 89° E. There are some animals just north of Sikkim (Shah 1994), and about 200–300 in the Qomolangma Reserve, particularly near Pegu Co (R. Jackson, pers. comm.).

The eastern Chang Tang, east of a line from Nam Co to Siling Co, is now almost devoid of kiangs. But west and north of Siling Co, including the whole Chang Tang Reserve, the animals are widely distributed and moderately common in places, and they even remain visible along the road that crosses the Chang Tang north to Coqen and west to Gerze and Shiquanhe. But a few quotations from Sven Hedin's books indicate just how abundant kiangs were a century ago. Not far from Dogai Coring, he noted that "a little higher up the hillside was dotted over with yaks, and there were more kulans [kiang] and antelopes than we could count"; and north of the Yako Basin, at about 36° N, 87° E, in an area we did not visit, Hedin noted that "the region we were now travelling through swarmed with khulans" (1898). Traveling west of Nyima in 1906, Hedin recorded: "Kyangs are very numerous; we had never seen so many animals of this kind gathered on so small an area" ([1922] 1991). And northwest of Gerze in Febru-

10.3. Kiangs are now rare in much of southern Tibet. This herd was observed in the Chigo Co area south of Lhasa. (October 1995)

ary 1908, Hedin found kiangs in "enormous numbers . . . at least one thousand were seen at a time" ([1922] 1991).

Numbers in the Chang Tang Reserve

To census kiangs, we counted all animals within about 1 km on each side of our travel routes. As described for gazelles (chapter 6), travel distances and kiang numbers were recorded on a grid and densities were calculated in animals/km^2 for each quadrat (fig. 10.4). Kiangs tended to concentrate in valleys and basins. Because vehicle routes also followed such terrain, our counts were probably not wholly representative. Kiangs tended to avoid the vicinity of main roads because of disturbance, including shooting, and this also biased counts. Figure 10.4 shows that kiangs were widespread though of patchy distribution inside the reserve (including quadrats that straddle the border), with densities seldom over 1.0/km^2. A density was calculated for each of the three vegetation zones, excluding the 11.6% of the reserve that consists of glaciers, lakes, and other barren terrain (see fig. 2.7). The *Stipa* zone (130,260 km^2, at a reserve size of 334,000 km^2) had a mean density of 0.39 kiangs/km^2 (table 6.2). A concentration of 806 kiangs in four quadrats south of Yibug Caka in October 1993 skewed this mean. Without these quadrats, the density was 0.13/km^2, a more realistic

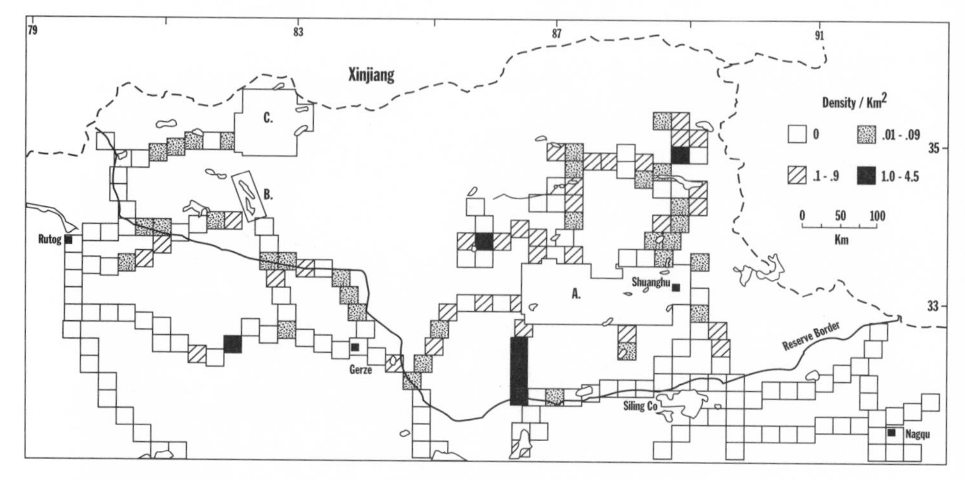

10.4. Mean density (per km²) of kiangs in and adjoining the Chang Tang Reserve, based on vehicle transects, showing the patchy distribution of the species. The three blocks marked A, B, and C indicate areas in which total counts were attempted.

figure. Counts within a block of terrain between Shuanghu and Yibug Caka gave densities of similar magnitude (table 6.1). Transects in the *Stipa-Carex* zone (78,490 km²) gave a mean density of 0.15/km², again a result that appears to be somewhat high. The Aru Basin had an estimated 250 kiangs within its 1800 km², or 0.14/km², and I considered that the area had a higher than average density. Kiangs were rare in the *Carex-Ceratoides* zone (86,840 km²), as judged by transects and a census of an 8000 km² block of terrain; the density probably was about 0.01/km². Thus densities in the *Stipa* and *Stipa-Carex* zones seemed to lie between 0.10–0.13/km². Based on these figures, the number of kiangs in the reserve would be 21,743–28,006, given a reserve size of 334,000 km² (or 18,488–23,813 at the official size of 284,000 km²).

Numbers on the Tibetan Plateau

Kiangs have no doubt decreased markedly in number during this century, especially in areas with many nomads and their livestock. The species is, however, still found within much of its former range of 1.5 million km², and in some areas it appears to be on the increase. The actual number of kiangs is at present impossible to estimate accurately but, based on our surveys, I will make some guesses.

In Qinghai, the southwestern part comprises 100,000 km², and this area has the highest kiang density in the province. At an average density of 0.1/km², as in much of the Chang Tang Reserve, there would be 10,000 animals, possibly too high an estimate. Kiangs also extend at low densities east to Ngoring Hu and north across the Qaidam Basin to the Qilian Shan, a vast tract for which no estimates can be made. Perhaps 15,000 kiangs exist in Qinghai.

The kiang's range in Xinjiang consists mostly of desert steppe with a low density of animals, but there is good habitat in the eastern part. My estimated total is about 4500–5500 kiangs.

In Tibet, about 22,000–28,000 kiangs are calculated for the Chang Tang Reserve, and judging by distribution and relative abundance, another 15,000–20,000 exist outside the reserve, for a total of 37,000–48,000, which is somewhat lower than the 54,000–62,500 proposed by Piao (1994). Thus, the Tibetan Plateau, including parts of Ladakh and Gansu and other small areas, may have around 60,000–70,000 kiangs. Gao and Gu (1989) gave an estimate of 200,000.

Population and herd dynamics

Stallions and mares are so similar in appearance that it is often difficult to distinguish the sexes at a distance, especially when the animals stand in heat waves or are in flight. On encountering a herd, I was usually able to do little more than make a count and note the number of foals; however, on occasion the sex of all adults could be ascertained. The data are divided into two periods, May to August and September to December, based on the kiang's reproductive calendar. The birth season is from about mid-July to mid-August. The first young of the year in Xinjiang was seen on 22 July and in Tibet on 28 July, and one newborn was noted as late as 18 August. Hedin (1903) reported newborns in late July and Rawling (1905) as late as 19 August. Since the kiang's gestation period is around 355 days (Groves 1974), females without young come into estrus at about the same time as others give birth. I observed mounting on 28 and 30 July. Schäfer (1937d) placed the onset of the rut in late July and a peak in the second half of August. With such a reproductive cycle, a mare would give birth only once every two years. *E. hemionus* mares first breed at 2 or 3 years and have their first offspring at 3 or 4 years (Heptner, Nasimovic, and Bannikov 1966); kiangs are probably similar.

Young were classified as yearlings when, in August, they were a year old. By late autumn yearlings were so large that they could not easily be distinguished from adults. The percentage of foals in the population as a whole between birth and year's end averaged about 11% (table 10.1). There then seemed to be heavy winter mortality, for by the following summer the old foals and new yearlings, now aged 10–15 months, composed only 4–5%. Except for one foal found dead of unknown causes (there were

Table 10.1 Percentage of young and yearlings in the kiang population of the Chang Tang

Location	Month/year	Sample size	%
YOUNG (0–12 mo.)			
Reserve	8–10/1990	1367	12.9
Reserve	9/1988	228	13.6
Reserve	10/1993	1942	11.3
Reserve	12/1991	1063	7.9
Qinghai	11/1985	138	13.0
Qinghai	11/1986	363	8.5
Reserve	6/1994	425	4.9
Reserve	6–7/1991	262	3.1
Aru Basin only	7/1992	181	4.4
YEARLING (12–15 mo.)			
Aru Basin only	8/1990	212	5.7
Reserve	8–10/1990	929	6.9
Reserve	8/1992	218	2.3

Table 10.2 Age at death of kiangs, based on upper incisor eruption and wear

Age (yr.)	Unsexed	Male	Female
<0.6	2		
0.6–1	1		
1–2	3		
2.5–3	4		
3.5–4		3	4
5–6		7	5
7–9		17	13
9–11		16	7
11–13		9	4
12–14		3	4
13–16		6	4
15–18		3	7
17–20	—	—	1
Total	10	64	49

Note: Age criteria based on plains zebra (Klingel and Klingel 1966).

hemorrhages in the walls of its digestive tract) in October, we obtained no data on foal mortality.

The upper incisor teeth of skulls found in the field were aged by tooth eruption and wear criteria developed for East African plains zebra (Klingel and Klingel 1966), a species whose food habits resemble those of kiangs. Of 123 skulls found, three were of foals and three of yearlings (table 10.2). Stallions outnumbered mares in the sample by a ratio of 131:100, suggesting higher mortality in males, but without data on the ratio in the living population the extent of the disparity is unknown. Most kiang mortality (41–52%) occurred in the prime age classes (7–11 years) in both sexes.

Hedin found that "the skeletons and skulls of kulans and orongo antelopes were scattered all over the locality; these must have belonged to animals which died a natural death, or were killed by wolves, for the Tibetans never meddle with kulans" (1903). These deaths were probably caused by starvation after a blizzard such as the one in October 1985, during which kiangs also died.

Hunting was and continues to be a major cause of mortality. Hedin came upon a woman with three children: "She had arrived from Gertse [*sic*] seventeen days before with her two husbands, who had returned to Gertse after they had filled the tent for her with wild ass meat. She owned a few yaks and a small flock of sheep, and would live for the next three months on game—yaks, kiangs and antelopes" (1909). Some pastoralists still hunt the species for subsistence, as I saw in Qinghai and Tibet, but most do not eat kiang meat. Kiangs were also shot around Shuanghu to feed a temporary influx of Chinese construction workers, and meat finds a ready sale in towns. Kiang remains were often near roads and some of the animals had been shot. Since stallions seem to roam more than mares and are often less shy, they may be more vulnerable to such hunting.

Kiangs are usually alone or in small herds but at times they congregate, forming herds of 300–400 (Schäfer 1937d) and even 500–1000 (Rockhill 1895). A total of 8668 kiangs were tallied in Qinghai and Tibet, of which 513 were solitary and the rest in 877 herds. From May to August 13.6% of the kiangs were alone, compared to 2.5% between September and December, a significant difference that is probably related to the rut. Of 109 solitary individuals sexed, all but 3 were stallions. It is possible that some stallions are seasonally territorial, behavior also noted in Grevy's zebra, African wild asses, wild asses, and feral domestic asses (Moehlman 1985; Neumann-Denzau 1991; Feh, Boldsukh, and Tourenq 1994). For example, on 7 August we counted 31 kiangs along 20 km of Aru Co. Of these, 15 were stallions, 9 mares, 1 yearling, 2 foals, and 4 unsexed solitary adults. Twelve of the stallions were alone, well-spaced and reluctant to leave their sites, and they chased intruding stallions and herded passing mares, all patterns characteristic of territorial animals. One stallion (recognizable by his missing tail) remained at one site for several days. But given the social plasticity of equids (Moehlman 1985), territorialism by a few stallions does not preclude the existence of small, nonterritorial harems consisting of one stallion and one or more mares with their offspring. Such associations would be temporary, unlike those of feral horses and plains zebra (Berger 1986). After the rut, most solitary individuals rejoined herds.

Herds were smallest in summer during the disruptive time of mating and giving birth. At that time 56.3% of the animals were in herds of 2–15, compared to 39.6% from September onward (fig. 10.5). Starting in about mid-September kiangs often amalgamated to form herds with 50 or more individuals. In summer, 8.3% of the animals were in such large herds, whereas later in the year the figure was 19.1%. The largest well-defined herd comprised 112 animals. Mean herd size, excluding solitary individuals, in summer was 6.8 and in winter 10.9; median herd size in summer was 9.5 and in winter 19.9 (fig. 10.6). In Xinjiang, mean herd size, excluding lone animals, has been calculated as 9.4 (Feng 1991a) and 6.4 (Achuff and Petocz 1988), with a range of 2–62. Piao (1994) gave a mean of 7.0 for Tibet.

Herds often congregated on good pastures during autumn and winter. Many animals were then in a limited area, a few alone and the rest in unstable herds of various sizes. The largest of such aggregations consisted of 131, 140, 172, 200, 236, and 261 animals. There is no evidence that kiangs migrate, but these large aggregations suggest considerable local movements in some areas. Repeated censuses of the same area near Garco also showed much seasonal variation in numbers (table 3.2).

Like the other Chang Tang ungulates, kiangs associated in male herds, female herds, and mixed herds containing adults of both sexes. As noted, stallions, including at times yearlings, were often solitary. All members of herds could be sexed on occasion, and these included 24 male herds. Two stallions were together on 12 occasions; three on 5 occasions; four on 2 occasions; five, six, and eight once each; and eleven on 2 occasions. Mares were seldom alone, but a mare with a foal or yearling, and two or three mares with their offspring, were commonly noted. Mares in reproductive

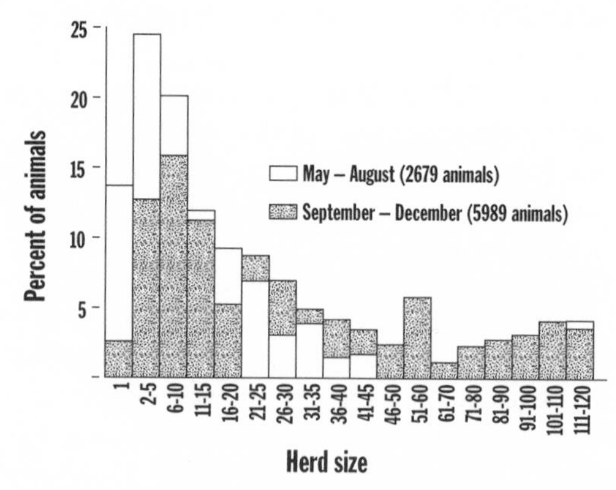

10.5. Herd sizes of kiangs in spring and summer and after the rut in autumn and winter.

10.6. A herd of kiangs on *Stipa* pasture in a lake basin, a typical winter habitat of the species. (December 1991)

synchrony tended to band together. Herds consisting solely of 5, 6, and 10 pregnant mares were observed, and in one herd of 11 at least 6 mares would give birth within a month. One herd of 22, observed in December, consisted solely of mares, none with foals. Small female herds seemed to retain considerable cohesion, especially when apart from others. Stallions often associated with mares in small herds. Thirty-two of 38 such mixed herds contained only 1 stallion, and the number of mares varied from 1 to 9. Of the remaining mixed herds, five had 2–4 stallions each, and one herd contained 6 stallions and 6 mares, 2 of them pregnant. Stallions often trailed behind such small herds, and in several instances they then veered aside while the others traveled on.

Comparisons with Mongolian wild ass (kulan)

Two subspecies of kulan occur in Mongolia: *E. h. hemionus* in the north, now possibly extinct; and *E. h. luteus* in the south across the deserts and semideserts of the Gobi east to about 110° E. The southern kulans are large, up to 130 cm tall at the shoulder (Groves and Mazák 1967), with pale tan upper parts blending into light buff to white undersides; like ki-

10.7 A solitary male kulan stands on a rise just above an oasis in Great Gobi National Park.

angs, they have a short erect mane and a dark dorsal stripe (fig. 10.7). I observed kulans mainly in August and September, or after the birth and mating seasons, which begin in late May and extend to the end of July (Zhirnov and Ilyinski 1986; Feh, Boldsukh, and Tourenq 1994), starting at least a month earlier than kiangs.

In 1994, 535 kulans were tallied in South Gobi Province, and of these, 22.8% were young, a high figure. Farther east, in East Gobi Province, 775 kulans were classified, and of these, 11.5% were young, an average figure. South Gobi Province had a severe drought in 1992, during which many mares probably failed to conceive or lost their young. This would bring a large number of mares into estrus in 1993 with the resulting high birthrate.

Most kulans are in small herds with 2–15 animals (Zhirnov and Ilyinski 1986) but occasionally large aggregations of over 1000 are formed (Andrews 1933). Excluding two large aggregations, maximum herd size in the eastern Gobi during August 1994 was 25. Of 642 animals, 14.5% were alone, 53.7% in herds of 2–10, 27.9% in herds of 11–20, and 3.9% in a herd of 25. Mean herd size was 5.5 excluding lone individuals and 3.3 including them. In addition, two large aggregations, with herds joining and parting as they fled, contained 234 and about 500 (411 were counted) individuals.

Of 29 solitary individuals, all but 1 were stallions. Female herds consisting of 1–5 mares with their foals were common. Three mixed herds were accompanied by 1 stallion, and the largest of these had 6 mares and 4 foals. One mixed herd contained 2 stallions; another contained 3 stallions. As with kiangs, stallions seemed to be transient members of such herds. On one occasion, on 7 August, 3 stallions harassed female herds:

> Two stallions are in a dry streambed pawing at a crater in the sand to obtain seepage water, while a third stands on the bank. When the third stallion sees a mare with foal approaching, he walks up to her and follows closely. She retracts her ears and lashes out with her hindlegs. He then tries to herd the two, cutting in front of them, but both trot off without drinking. The stallion stands until another mare and foal arrive. As soon as he advances, they veer aside and leave. The stallion then joins the others in the streambed. A herd of 5 mares and 5 foals arrives and two of the stallions gallop to them, scatter the herd, and then each tries unsuccessfully to round up mares. Later the stallions drift away, one alone and two together.

These observations on herd composition differ from those of Zhirnov and Ilyinsky (1986), who reported that one stallion is typically found with several females. Similarly, Feh, Boldsukh, and Tourenq, working in the western part of Great Gobi National Park between June and October, found that "there was no indication of territorial males and no group consisted of only females and offspring" (1994). Typical herds observed by them contained one stallion and two mares with their offspring in summer, and two stallions and four mares in autumn, indicating a seasonal increase in herd size as in kiangs. They surmised that bonds between stallions and mares persisted, behavior reminiscent of plains zebras and horses.

Herd sizes, seasonal aggregations, solitary stallions, and some other aspects of herd dynamics are similar in kiangs and kulans, but possible differences, such as territorialism in kiangs and the seeming persistence of a bond between a stallion and mares in kulans, require additional observations.

11
The Carnivores

I dreamed that to the east beyond this high majestic mountain
A colossal pillar was standing.
At the top crouched a great lion.
His mane of turquoise flowing everywhere,
He spread his claws upon the snow,
His eyes gazed upward,
And he roamed proudly on the vast whiteness.
I tell this to the Lama Buddha of the Three Ages.

I took it as a happy omen
And rejoiced at this great fortune.
I wish you to tell me its meaning.

Milarepa, eleventh-century Tibetan poet and saint

THE UNGULATES of the Tibetan Plateau evolved various behavior patterns partly or wholly in response to predation pressure. Fleetness of foot and other escape methods, habitat selection, herd sizes, and movement patterns were probably influenced by wolves and snow leopards, the two principal large predators. There are in addition other predators, from brown bears and lynx to sand foxes. I endeavored to learn something about these predators and their impact on prey populations, but my encounters with them were infrequent and usually brief. Shot and trapped for their pelts and in defense of livestock, predators are scarce and elusive. I never saw a snow leopard on the Tibetan Plateau and only encountered a Eurasian badger and a polecat once, a manul cat twice, and lynx three times. Dholes are also said to occur in some parts according to local people.

With direct observations difficult to make, I concentrated on the food habits of several carnivores by examining the contents of droppings. Prey remains were identified by the distinctive color and texture of hairs and the presence of bones, claws, and other material. The percentage of each food item was estimated. Even cursory examination of droppings soon indicated that, in addition to ungulates, small mammals such as pikas and marmots contributed importantly to the diet of predators. I give a brief overview of such prey species first to place the predators into their ecological context.

Rodents and lagomorphs as prey species

The principal small rodents in the Chang Tang are *Alticola stoliczkanus*, *Pitymus leucurus*, and *Cricetulus kamensis* (Feng, Cai, and Zheng 1986). All are usually uncommon, their colonies widely scattered and small. Himalayan marmots, too, have a highly local distribution, but their large size makes them a favored prey of wolves and brown bears. Woolly hares, though sparse in most places, are widespread. The black–lipped pika is at times so abundant on the alpine meadows and steppes that it probably has the highest mammalian biomass in many areas and is a key link in the food chain for most predatory birds and mammals.

Marmot

Only one species of marmot (*M. himalayana*) extends across the Tibetan Plateau. In the Karakoram and western Kunlun Mountains its place is taken by *M. caudata* and in the Tian Shan and Mongolia by *M. bobac.*

Himalayan marmots favor alpine meadows with lush vegetation in the vicinity of their burrows. Since they hibernate six months or more, they need to store ample fat reserves during the short growing season. They frequently dig a burrow system with two or more entrances in rolling terrain and on slopes, often at the base of boulders. Rocky sites tend to be well drained and also difficult for bears to excavate. Although marmots are usually found in hilly to mountainous terrain up to an elevation of 5000 m, they readily occupy level plains that have suitable forage, such as east of Zhidoi in Qinghai. In the Chang Tang, with its arid steppes, marmots occurred sporadically in small colonies along seepages, in mountain valleys, at the bases of glaciated ranges, and in other moist sites. The Aru and Jangngai Ranges were typical marmot habitat. I saw no marmots during one month in desert steppe near Toze Kangri, and Hedin ([1922] 1991) also commented on their scarcity in the northwestern Chang Tang. Even in alpine steppe we often traveled 100 km or more without noting marmots, an indication of scarcity because the animals are diurnal and their den mounds conspicuous. Marmots begin hibernation in late September or early October, depending on elevation; my latest marmot sighting in the Chang Tang Reserve was 7 October. Pastoralists told me that the animals emerge during April.

Marmots are dark tan with black, especially on the crown, and they weigh 4–8 kg (Feng, Cai, and Zheng 1986). Average litter size is 4.8–6.9 (Huang et al. 1986), although I seldom observed more than 3–4 around a den. Of 101 marmots tallied during September 1986 in eastern Qinghai, 53% were born that year. The young are especially vulnerable to predation when they first venture from their burrows. Once, in late June, I observed

11.1. A Tibetan woolly hare attempts to remain inconspicuous by sitting close to lava boulders.

a herder's dog hunt marmots *(M. caudata)* intermittently for five hours and kill 4 young and an adult male. Marmots are much hunted for their hides in some areas (Tibetans usually do not eat marmot meat). In Qinghai the animals are also poisoned because they are thought to compete with livestock for forage. Marmots are susceptible to the flea-borne sylvatic plague, a bacterial *(Pasteurella pestis)* disease, which may greatly reduce local populations and also affect people (Li Xing, Xinjiang Local Disease Prevention Research Unit, pers. comm.).

Woolly hare

The woolly, or gray-rumped, hare is confined to the Tibetan Plateau. Of the five subspecies recognized, *L. o. oiostolus* occurs in the Chang Tang (Feng, Cai, and Zheng 1986). These hares weigh about 2.0–3.0 kg (*Research on Flora and Fauna* 1979), have a yellow to grayish brown coat with white undersides, a conspicuously gray rump, and a white tail with a central dark stripe on the dorsal side (fig. 11.1). Hares are widely distributed on the plateau, extending up to 5200 m in elevation. Exceedingly adaptable, they occur in desert steppe so arid that even pikas cannot live there and on pastures around villages where livestock and people are abundant; they occur on steep mountainsides as well as on plains. But they prefer terrain with cover, such as boulder fields and small thickets.

Usually only one or two hares were seen in a day's travel, but at times they were abundant. In one valley of the Shule Nanshan, I tallied 17 hares within 250 m and flushed 7 hares from a brush patch; in the Chang Tang up to a dozen hares may inhabit a small rocky knoll or hillside. Hares are easy to overlook because they typically crouch unless closely approached and then they hop away only to crouch again. In open terrain they dig scrapes, about 25–40 cm long, shallow at one end and to 15 cm deep at the other, into which they tuck their rumps and lie still. Marmot burrows may also be used as retreats. One hare young lived in a pika burrow at Memar Co. Over a six-day period, it appeared at the den entrance between 1030 and 1230 hours and remained crouched there during daytime; after dark it presumably foraged nearby. One hare fecal sample from the Yako Basin contained 54.3% *Ceratoides*, 19.4% *Stipa*, 16.7% other graminoids, and the rest legumes and *Saussurea*.

Black-lipped pika

A number of pika species occur on the plateau, nine in Tibet alone (Feng, Cai, and Zheng 1986), but only the black-lipped, or Plateau, pika is ubiquitous on alpine meadows and steppes. This pika, a typical chunky, tailless lagomorph, has a brown to reddish tan coat with light gray undersides (fig. 11.2); adults weigh about 100–200 g (Suo 1964; Kaiser and Gebauer 1993). Colonies occur almost wherever the terrain is flat to gently sloping, well drained, and with a silty or sandy soil devoid of rocks. Pikas greatly favor alpine meadows, especially areas where the turf is broken into blocks beneath which they can burrow. They may be found in barren desert steppe or up to the limit of vegetation at 5300 m, where only scattered *Ceratoides* clumps persist. On the lush alpine meadowland of Qinghai densities vary from about 12 to 380 animals per hectare (Kaiser and Gebauer 1993).

Pikas live in burrows, which they leave for several hours in daytime to forage. Since they do not hibernate, daily activity cycles vary with the seasons. In midyear they may appear above ground by 0600 and remain intermittently active until 2100, except that few are out in early afternoon. In winter, they leave the burrows after sunrise but spend little time in the cold. Once, on 5 November, sunup was at 0830 but clouds obscured the sun, and pikas remained in their burrows until 0930, when direct rays reached the colony. The following day they emerged at about 0830. Predators have to adapt their hunting schedules to conform to such activity changes (see Wei et al. 1996).

Pikas live in families consisting of 1 adult male, 1 adult female, and 5–10 juveniles from at least two litters (Smith et al. 1986; Smith 1988). Each

11.2. A black-lipped pika is by its latrine, which is tucked against a rock.

family has a burrow system with many entrances and a number of aboveground latrines. I excavated one such system near Donggi Co in Qinghai (fig. 11.3). The burrow consisted of 833 cm of tunnels 20–40 cm below the surface, considerably below the 10 cm thick turf layer. There were three entrances to the tunnel system and 14 latrines tucked into niches along the tunnels. At one place only 2 cm of earth separated the burrow system from a neighboring one. High pika densities such as this are possible only when food is abundant. In the desert steppe, with its sparse resources, pikas appear to have a less cohesive social system, judging by their burrows. Each of seven burrows we excavated near Toze Kangri consisted of a single nonbranching tunnel 70–233 cm long angling downward, usually with one bend, to a depth of 30–60 cm. Six tunnels ended abruptly but one had its end enlarged into a chamber 30 cm wide and 20 cm high containing 435 g of *Stipa*. A system of scattered burrows enables pikas to forage widely and scuttle to the nearest hole in times of danger. However, simple burrows make pikas vulnerable to being dug out by bears. Single entrances provide no alternate escape routes, and, with permafrost less than 1 m below the surface in places, pikas cannot dig deeply.

Pikas eat a great variety of plant species. On the alpine meadows of eastern Qinghai, I casually recorded 11 graminoids and 37 forbs as food plants (table 11.1). Jiang and Xia (1987) found that about half of the pika's food in the Qinghai region consisted of graminoids and half of forbs. By contrast, one small pika colony in alpine steppe had only *Stipa* and *Leontopodium* available near the burrows, and one in desert steppe appeared to subsist primarily on *Ceratoides*. In the latter, on 3 July, one pika visited 37 *Ceratoides* shrubs in four hours, using three burrows as retreats within an area of about 300 m^2. On alpine meadow the pikas made hay piles in summer and autumn, tucking the drying vegetation along erosion terraces and into burrow entrances and other protected sites. Such hay piles were only rarely observed in alpine and desert steppe. Hay piles are perhaps difficult to maintain in the Chang Tang because winds may blow them away, social structure is less cohesive, and ungulates may eat the harvest. Unless food is taken underground, pikas must forage daily, making them vulnerable to inclement weather and predation.

Pikas can have considerable impact on vegetation in the vicinity of burrows. *Stipa* culms may be dug up and *Ceratoides* clipped down to a bristle of branches. In alpine steppe there are occasional bare areas 10–15 m in diameter with abandoned burrows that look as if pikas had eliminated the vegetation and then moved elsewhere. In alpine meadow habitat, with densities of 100 pikas or more per hectare, the amount of food they eat is considerable, and pastoralists feel that pikas compete with livestock for forage. Ungulates, both domestic and wild, and pikas overlap greatly in

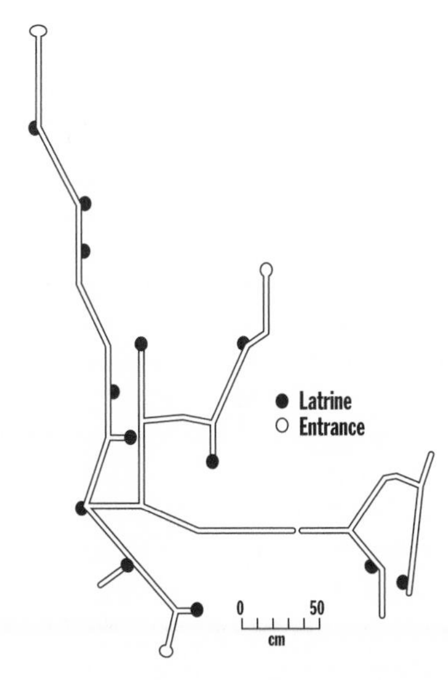

11.3. The tunnel system of a black-lipped pika family in alpine meadow habitat, Qinghai. One tunnel almost joins the tunnel of a neighboring family.

Table 11.1 Food plants of pikas in the alpine meadow areas of Madoi and Zhidoi, Qinghai

Graminoids	
Carex inanovai	*Kobresia rumulis*
Carex przevalskii	*Littledalea tibetica*
Elymus atratus	*Poa sinatteruta*
Elymus durus	*Stipa regeliana*
Kobresia robusta	*Trisetum spicatum*
Forbs and Shrubs	
Aconitum gymnandrum	**Meconopsis horridula*
Ajania tenuifolia	*Melanchum glandulosum*
Ajuga lupulina	*Morina chinensis*
Anaphalis flavescens	*Oxytropis ochrocephala*
Androsace mariae	*Oxytropis* sp.
Artemisia nanschanica	*Pedicularis anas*
Artemisia sp.	*Pedicularis curvituba*
Aster glaccidus	*Pedicularis kansuensis*
Corydalis sp.	*Pleurospermum* sp.
Delphinium trichophorum	*Polygonum sibiricum*
Gentiana straminea	*Polygonum sphaerostachyum*
Gentianopsis paludosa	*Polygonum viviparum*
Heracleum millefolium	*Potentilla saundersiana*
**Iris songarica*	*Saussurea arenaria*
Lagotis brachystachya	*Saussurea brunneopilosa*
Lamiophlomis rotata	*Saxifraga tangutica*
Leontopodium longifolium	**Stellaria chamaejasme*
Leontopodium nanum	*Thalictrum rutaefolium*
**Ligularia virgaurea*	**Thermopsis inflata*
Lonicera rupicola	

*Usually not eaten by livestock.

their food habits (see chapter 12), but the long-term impact on the rangelands of this dependence on the same plant species is difficult to measure. Jiang and Xia (1987) found that pikas at high densities competed with livestock but that at low densities they concentrated on different foods. Pikas consume certain plants that livestock find unpalatable (table 11.1), and they may thus help slow the spread of such species. The prairie dog, North America's analogue to the pika, also lives in large colonies, and these sites attract bison, pronghorns, and elk apparently because of the more nutritious forage there (Whicker and Detling 1988). Nutrient enhancement, partly based on droppings and on cycling of soil by digging, may be a benefit of pika activity too, as may an increase of plant species diversity (Huntly and Reichman 1994).

Pikas are also blamed for causing serious hill erosion. Since pikas prefer to inhabit places where the turf layer is broken and eroded, they are sometimes particularly abundant at such sites, as, for example, near Maqen and Zhidoi in Qinghai. There is, however, no evidence that pikas initiated the damage. Sheet erosion with turf slumping and sliding to expose bare soil and rock often appears to be caused by solifluction, by human and livestock activity, or by both. Anything that cuts the turf layer, such as roads, livestock trails, and yak wallows, can cause erosion. Pikas could then contribute further to erosion by their digging. But in some seriously damaged rangelands, as around Amdo, pikas are not abundant and the people realize that poor livestock management is responsible. Nevertheless, pikas have been poisoned over large tracts of eastern Qinghai with zinc phosphide. This has reduced pika numbers, and presumably also pika predators; I found a dead polecat in one such poisoned colony. The situation is reminiscent of the campaigns against prairie dogs and the virtual extermination of one of their principal predators, the black-footed ferret. A marked reduction of pika burrows could also affect Hume's ground jays and several species of snow finches, which nest in them.

Many predators in the Chang Tang and other parts of the plateau subsist primarily on pikas. The two most common large raptors in the region, upland hawk and saker falcon, were often perched in a pika colony. Beneath one upland hawk nest with a large young were 10 pellets, all with pika remains, and one half-eaten pika. A sample of 25 pellets from beneath a saker falcon nest with two young contained 90% pikas and 10% birds. Pellets in a colony of about 50 black kite nests in southwest Qinghai contained mainly pikas (probably *O. macrotus*) and dung beetles. Manul cats probably prey mostly on pikas, as do polecats. On 17 July 1985, near Donggi Co in Qinghai, I observed a polecat, presumably a female, with two half-grown young in a pika burrow:

> At 0700 hours she dashes after a pika, grabs it over the shoulders, and carries it into a burrow. Two minutes later she emerges and enters three entrances in succession. From the third burrow she appears with a pika, which she takes to her young. She reappears within 30 seconds and slides into yet another burrow. When she comes out with a pika a herder's dog rushes up. She retreats into the burrow, having dropped the pika, which the dog gulps down. At 0712 the polecat resumes her hunt, entering a burrow but without catching anything. She then joins her young in their burrow and has not reappeared by 0745. At 1000, the two young are at the entrance.

In addition to predation, other factors may cause extensive pika mortality. I saw a large larva, probably of warble fly, emerge from a dying pika. During the blizzard of October 1985, dead and dying pikas, heavily infested with fleas, were found on the snow. Wang and Smith (1988) also commented on high mortality after a snowstorm. In 1994, large tracts north of Shuanghu were almost devoid of pikas, but the many silent burrows attested to their recent presence. Mummified pikas lay scattered along the saline Yupan Co. Disease may have devastated the populations over several thousand km^2. Given high death rates, high reproductive rates are essential. Litters are large, containing 4–6 young, females may have two litters a year, and some females reproduce in the year of their birth (Smith 1988).

Fox

Two fox species occur on the plateau. The sand fox is small, weighing 3.0–4.5 kg (Suo 1964), and it has sandy to pale rufous back and sides, white belly, gray on neck, thighs, and rump, and a dark gray tail with a large white tip. The red fox is larger than the sand fox, with a body mass of 4.6–5.3 kg (Suo 1964). Its coat is conspicuously rufous, and its legs, back of ear, and white-tipped tail are dark gray to black. Sand foxes are found in high plains, so characteristic of the Chang Tang, whereas red foxes favor mountains, including forested ones. However, there is a broad zone of overlap in the Chang Tang, with red foxes the scarcer of the two, in part because they are much hunted for their pelts.

Red foxes were observed 10 times within the limits of the Chang Tang, 5 times outside the reserve—for example, at Tanggula Pass, at Kunlun Pass, and near Coqen. Within the reserve, they were noted once at Memar Co, once north of Yibug Caka, and 3 times in the volcanic hills north and south of Dogai Coring. By contrast, sand foxes were recorded 90 times. Though widely distributed, they were seldom common. For instance, on desert steppe in June 1992 we tallied 5 sand foxes in 1848 km of driving and an undetermined distance of walking. In the more benign habitat of the upper Tuotuo River in Qinghai, where pikas were abundant, 15 foxes were seen in 367 km, an unusually high density.

Droppings from three mountain areas in eastern Qinghai revealed that red foxes preyed heavily on pikas and small rodents, followed by marmots and hares; ungulates, probably scavenged, contributed a little to the diet. Sand foxes often deposited their droppings in conspicuous places, such as on the earth mound by a pika burrow and on top of *Ceratoides* shrubs. The droppings revealed that the foxes subsisted mainly on pikas, augmented by occasional small rodents, birds, and chiru remains (table 11.2). One dropping contained *Ephedra gerardiana* berries. In the Aru Basin, a subadult female (weight 2.8 kg) dashed from her burrow and under the wheels of our truck. Her stomach contained a whole pika. On several occasions I observed a sand fox trot or stalk in a crouch among pika burrows. One fox hovered near a domestic sheep kill which a wolf had just vacated.

Since pikas are diurnal, sand foxes appear to be active mainly in daytime as well. Of the foxes tallied, all except six pairs were solitary, indicating

Table 11.2 Food items in red fox and sand fox droppings, expressed as percentage of total content in sample

	Red fox (Qinghai)			Sand fox (Chang Tang Reserve, Tibet)	
Item	Shule Nanchan 9–10/1985 n = 42	Zadoi 8–9/1986 n = 17	Anyemaqen 10/1986 n = 72	Northwest 6–7/1992 n = 34	East 6/1994 n = 79
Blue sheep	5.2	—	4.0	—	—
Chiru	—	—	—	2.9	2.5
Musk deer	—	6.5	2.1	—	—
Hare	9.5	—	5.6	—	—
Pika or small rodent*	67.0	57.9	50.5	94.1	93.5
Marmot	16.1	28.8	33.1	—	—
Bird	—	5.6	2.2	—	2.7
Insect	—	—	1.1	—	1.0
Vegetation	0.2	1.2	—	2.9	0.3
Soil	1.9	—	—	—	—

*Red fox droppings often contained small-rodent remains; sand fox droppings rarely did.

that most hunted alone. Dens were usually in hummocks or low on slopes and they consisted of one or two entrances 25–35 cm in diameter; one den site was on a sandy mound with four entrances.

On four occasions a sand fox meandered around and through a chiru herd and once a gazelle herd. Although the fox was at times within 3 m or less of the animals, they gave no response beyond a brief look. The reason for this seeming association was unclear.

Wolf

Wolves remain widespread on the high uplands of the plateau. Adaptable and far-ranging, they frequent all types of terrain, ascending as high as 5300 m. I observed wolves or their spoor most commonly in areas sparsely settled by pastoralists or in uninhabited places, an indication that as human presence increases, the predators not surprisingly decrease. Tolerance for wolves by pastoralists is low. The wolf is the only species in the Chang Tang Reserve without legal protection. Wolf carcasses sometimes lie near roads, the animals shot from vehicles, and at Rongma three bodies were in the village dump with jaws wired shut.

Most wolves were brown to grayish brown, some with black along the back (fig. 11.4), but there were also reddish ones, particularly in the Aru Basin, and a few were almost white or black. They were often seen alone or in groups of 2–3 and occasionally as many as 8 (fig. 11.5). However,

11.4. A wolf investigates a piece of yak skin.

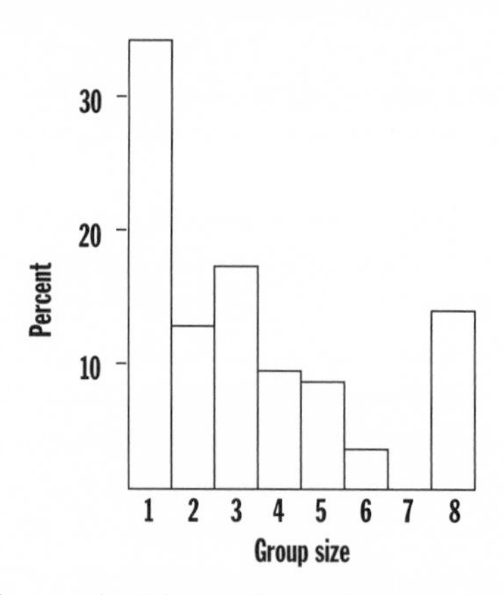

11.5. Group sizes of wolves on the Tibetan Plateau, based on 172 animals.

group size may bear little relation to total pack size because all pack members may not necessarily be seen together. For example, a pack at Memar had at least 8 members but only once did I encounter them in one group. Once near Garco in December, I noted three wolves at rest in a dip of the plains. After a few minutes a fourth wolf came across the steppe, alertly looking around until it sensed the others. Trotting closer in a crouch, ears

Table 11.3 Food items in wolf droppings from the Tibetan Plateau in Qinghai and Xinjiang, expressed as percentage of total content in sample

Item	Shule Nanshan Qinghai 9–10/1985 n = 29	Kunlun Pass Qinghai 12/1986 n = 21	Anyemaqen Qinghai 9/1986 n = 19	Kunlun Shan Xinjiang 5–6/1987 n = 115
Blue sheep	24.1	45.7	47.4	2.6
Tibetan argali	—	—	—	2.6
Chiru	—	—	—	9.6
White-lipped deer	3.4	—	—	—
Musk deer	—	—	5.3	—
Tibetan gazelle	—	9.5	—	—
Yak*	—	4.8	—	—
Domestic sheep/goat	3.4	9.5	—	—
Marmot	61.2	28.6	47.1	80.0
Pika or small rodent	—	—	—	trace
Hare	6.6	—	—	4.3
Bird	0.7	1.9	—	—
Vegetation	0.5	—	0.3	—

*Domestic or wild yak.

laid back, it met them in a melee of tail-wagging, rolling on the back, and mouth-nibbling. Still later a fifth wolf joined, and the group, composed of two adults and three young-of-the-year, moved out in single file.

Judging by size, pups leave dens in June. One den in the Aru Basin consisted of two entrances, each about 45 cm in diameter, dug 8 m apart into an old lake terrace. The pups already hunted with the adults in August, but they still visited the den site alone or together to lounge around.

Wolf howls were heard only twice during the study, on 14 August and 24 September.

Wolves often deposited feces at latrine sites. These sometimes had 20 or more fecal piles and usually were at a landmark such as a pond, the base of a river bluff, an isolated hillock, a mountain spur, or a yak skull. Contents of droppings revealed that wolves preyed on mammals ranging in size from pikas to yaks (tables 11.3 and 11.4). Prey availability varied with the seasons. Often a third or more of the wolf's summer diet consisted of marmots, but these rodents hibernate for half the year. During the chiru migration, the wolves around Toze Kangri subsisted mainly on this species, but at other times only pikas remained abundant there. That wolves will prey heavily on pikas when ungulates are scarce is shown from the Natigan dropping sample. Blue sheep are local in distribution but they are a principal prey item when available. Kiangs were not noted in droppings, a surprising omission given their general availability in the Chang Tang. Livestock were at most a minor food item, in contrast to some areas in Pakistan and Nepal (Schaller 1977b; Oli 1994b). Wolves in the Karakoram

Table 11.4 Food items in wolf droppings from the Chang Tang Reserve, expressed as percentage of total content in sample

Item	Aru area 8/1990 $n = 58$	Toze Kangri 6/1992 $n = 24$	Natigan 10/1990 $n = 28$	Dogai Coring 6/1991, 1994 $n = 90$
Blue sheep	21.0	—	10.9	4.9
Tibetan argali	3.4	—	—	—
Chiru	12.1	65.6	—	21.9
Tibetan gazelle	5.2	—	—	—
Yak*	3.4	—	3.6	5.5
Domestic sheep/goat	3.4	—	—	—
Marmot	35.0	—	11.6	41.2
Pika	7.1	25.0	66.8	16.9
Hare	1.7	4.2	—	6.3
Unidentified ungulate hair	4.9	—	7.1	2.1
Bird	—	4.2	—	—
Vegetation	0.7	1.0	—	1.1

*Domestic or wild yak.

and Tian Shan also preyed heavily on marmots and, with blue sheep absent, on ibex (Schaller et al. 1987, 1988).

Several times I encountered wolves poised near marmot burrows as if waiting for an unwary animal; one wolf left a small colony carrying a marmot it had just killed. Once 36 kiangs fled from our car and inadvertently pounded past a resting group of eight wolves, two of which loped after the herd for about 300 m. But actual hunts were observed only twice:

> A herd of 40 wild yaks grazes and rests on the flats in the Aru Basin nearly 1 km from the hills. At 1130 the herd suddenly bunches tightly and runs toward the hills. Bounding beside and behind the herd are five wolves. The yaks soon lose their cohesion, and the wolves mingle with the herd but without attacking any animal. Instead of defending themselves communally, individual yaks threaten the wolves with lowered horns and lunges. After fleeing for several hundred meters the herd halts. And the wolves gather, stand around briefly, as if undecided about further action, and then trot toward a sixth wolf which has appeared on a rise nearby.

The attack seemed desultory, as if the wolves had merely checked the herd for a calf or other vulnerable animal.

> At 1430 hours about 50 chiru females and yearlings run tightly clustered up a sloping stream bank with a large wolf close behind. The herd races in an arc onto a low ridge and then along it, the wolf lagging. However, when the herd dips slightly over the crest, the wolf takes a shortcut and with a burst of speed narrows the gap. One female veers abruptly from the bunched herd and stands. The wolf passes her by, focused on a second female which

> is fleeing at an angle away from the herd. With a final sprint, the wolf pulls her down in a cloud of dust after a chase of about two km.

Examination of the chiru showed a 4 cm cut through the skin on the side, apparently where the wolf had first grabbed her. Bites through the top of the skull and around the throat killed her. She was pregnant, the fetus weighing 1.8 kg and the placenta another 1.5 kg.

Ten fresh ungulate kills were examined, and of these, one was a half-eaten large gazelle young, one a just-killed domestic sheep, and the rest were chirus (six adult females, a yearling female, and an adult male). The fat content of the bone marrow of six of the animals was checked and all were in good condition. Two females, killed in September and October, were lactating. Four of the eight chirus were especially vulnerable to predation: two females were pregnant, one with a full-term fetus (3.1 kg); one female was old, with all molars worn flat; and the yearling had a cataract in one eye (a raven had eaten the other eye).

We observed an association between wolves and brown bears on four occasions during June 1994. Three times a wolf merely rested within 30–100 m of a bear as it excavated pika burrows, but once the two predators were close to each other:

> Two bears, apparently a courting couple, are on a hillside at 1950 hours. The larger of the two is presumably the male. The female digs for pikas, her forequarters deep in a hole, while the male lies 50 m away. A wolf stands 3 m from the digging bear watching her alertly. She then ambles along poking her muzzle into five burrows, as if testing for the presence of a pika by scent, while the wolf walks 1.5 m behind or beside her. As the female digs up the sixth burrow, the wolf facing her from as close as 1 m, the male bear walks up and reclines beside her. Apparently no pika is caught and all three then walk for several hundred meters in single file, the female in the lead, trailed closely by the wolf, and finally the male. Both bears dig again at 2030 hours, the wolf first standing close to one and then the other. The wolf is obviously waiting to snatch any pikas that escape the bears' jaws but none do. At 2045 the two bears cross a glacial stream and the wolf stays behind.

Tibetan brown bear

The brown bears on the Tibetan Plateau belong to the subspecies *Ursus arctos pruinosus.* In the Himalaya west of Nepal, the Karakoram, and western Kunlun another subspecies, *U. a. isabellinus,* occurs, but it is not known where a contact zone, if any, exists. *Isabellinus* extends north across the Pamirs to the Tian Shan and then east along it. Within sight of the eastern end of the Tian Shan are small ranges in Great Gobi National Park, Mongolia, where the desert-dwelling Gobi brown bear occurs as a small remnant population. This bear has been variously referred to as *U. pruinosus* or *U. a. pruinosus* on the assumption that it is similar to the bears on the plateau, or as *U. gobiensis* on the premise that it is specifically distinct (Schaller, Tulgat, and Navantsatsvalt 1993).

The bears on the plateau are medium-sized animals—one male shot in October weighed 109 kg (Wallace 1913)—with a shaggy coat and sometimes a conspicuous ruff. Their pelage color is distinctive though highly variable. Adults usually are dark brown to black except that the face is rust brown to tan and a white collar extends from the shoulders to the chest, becoming broader below. The ears are also black and at times so hairy that they seem tasseled. Two courting pairs revealed the variation in pelage color. In one pair, the male, presumed to be the larger of the two, was dark except for the collar, whereas the female had a light tan saddle. In the other pair, the male was straw-colored on neck and sides, whereas the female was dark. Photographs and descriptions of adult bears from the eastern edge of the plateau in Gansu, Sichuan, and Qinghai in the reports of Wallace (1913), Burdsall and Emmons (1935), and Dolan (1939) reveal coat colors similar to those I saw in the Chang Tang. Subadults are generally lighter colored than adults. Although the legs, face, and hump are dark, the rest of the coat may range from pale brown to such a light tan that the animals appear almost white at a distance.

U. a. isabellinus has a tan, rufous, or dark brown coat, sometimes with a silver tinge, and a few animals, especially cubs, may have a white collar (Sterndale 1884). The Gobi bear also has a tan to rufous coat, the legs at times somewhat darker than the rest (Schaller, Tulgat, and Navantsatsvalt 1993; Schaller 1995), a pelage quite different from that of the Tibetan bear. The affinity of the Gobi animals thus lies with those to the west and north, not with those on the Tibetan Plateau, and it is inappropriate to designate them as *pruinosus*. Chestin (1996) noted that the skulls of Tibetan bears were significantly different from others in central Asia.

The Tibetan brown bear is now rare on the steppes, where they mainly persist in or near mountainous terrain. Yet they were once common, particularly on the alpine meadows of Qinghai. Rockhill (1891) found that "bears are very numerous around the Yellow River," and as late as the 1930s E. Schäfer saw as many as 14 in a day (Dolan 1939). Tibetans tend to consider the bears "the most terrible of animals," in Rockhill's (1891) words, and they often kill them. Consequently I saw few bears and only occasionally found their spoor, such as tracks, feces, and burrows excavated in pursuit of pikas and marmots (fig. 11.6).

The bears hibernate from October to April according to local informants. In the Anyemaqen Shan, I recorded fresh tracks in snow on 21 September. In the Shule Nanshan, I followed tracks to a freshly dug den

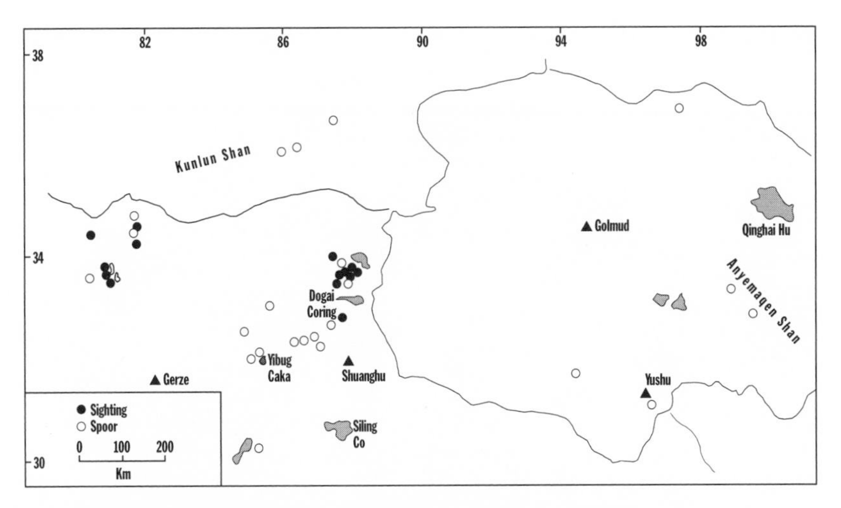

11.6. Spoor and sightings of Tibetan brown bear observed during the project.

on a north-facing slope on 27 September. Since the bear was probably inside, I did not measure that den. Another den, facing southwest, near Zadoi had an entrance 60 cm high and a tunnel 2 m deep. On 17 December 1991, a male bear came to a herder's tent west of Yibug Caka and killed a dog. A woman fled the tent, which the bear then entered and ate the hindquarters of a sheep. The bear was shot later that day about 1 km away (fig. 11.7). His canines and incisors were worn to stumps and the body lacked fat, no doubt the reason why the animal hunted in midwinter; his stomach contained mutton.

I observed bears in only two areas of the Chang Tang Reserve: in the northwestern corner and the eastern part. In 1990, 5 different bears were observed in the Aru Basin: 2 adults, 1 subadult, and 2 small cubs. Two years later only 1 adult was seen. Four bears were tallied near Toze Kangri in 1992: a pair on 6 June and 2 other adults. In June 1994, we encountered at least 11 different bears in 12 sightings in the eastern part between Purog Kangri and Rola Kangri, including two courting pairs and a female with a small cub. Nine of these bears were in high hills just east of Rola Kangri, the only area in the region with an ample supply of pikas.

The bears usually fled as soon as they perceived a vehicle, making it difficult for us to observe undisturbed animals. One female charged and pursued our car when it inadvertently approached her two cubs, which were out of sight in a swale, and a bear charged two members of our camp staff to within 30 m when they suddenly came upon it. Both members of a courting pair retreated a short distance on seeing our car, but when the female continued her escape, the male cut broadside in front of her four

11.7. A male Tibetan brown bear entered this tent in search of food in December and was shot. His hide shows the typical white pelage on neck and shoulders of this subspecies.

Table 11.5 Food items in Tibetan brown bear droppings (n = 48) in the Chang Tang Reserve, expressed as percentage of total content in sample

Item	%
Blue sheep	9.1
Argali	0.6
Chiru	2.6
Yak	0.4
Pika	59.3
Marmot	1.8
Bear	trace
Insect	trace
Grass	20.0
Roots	6.0

times as if to bar her way. Each time she trotted around him and he finally followed, watched by a yak bull and a wolf from nearby hills. On three occasions bears excavated pika burrows. Muzzle in the hole, they shoveled the loose soil back alternately using each foreleg or just one foreleg (see section on wolves).

A total of 48 droppings were collected in the reserve (table 11.5), most of them at excavations and rest sites. A bear sometimes pawed a shallow resting hollow in the shelter of a cliff or boulder or at a vantage point such as on a river bluff. The edge of the hollow sometimes had one or more fecal deposits. Pikas contributed 59% of the diet. Marmots were seldom caught, judging by our sample, probably because of the effort needed to dig them out of their deep, rocky burrows. Ungulates, most likely scavenged, composed 13% of the fecal content. Green and dry grass and roots were also an important food at 26%. At one site near Yushu, a bear had dug up and eaten the pencil-thin taproots of blue poppies (*Meconopsis*), discarding leaves and stems. Nine droppings of a bear in the Shule Nanshan showed that it ate a varied diet of marmot, blue sheep, grass, and *Tamarix* stems. Hedin (1903) found a marmot and several forbs in the stomach of one bear. And Kozlov reported from the northern part of the plateau that "the stomachs of all seven bears we shot contained nothing but pikas . . . there were 25 of them in the stomach of one bear killed while it was still hunting" (1899, quoted in Zhirnov and Ilyinsky 1986).

Lynx

Lynx are widely distributed on the plateau, where they are much hunted for their pelts and also because they prey occasionally on domestic sheep and goats. In the Chamdo region of eastern Tibet, 870 lynx pelts were sold to the government in a three-year period (1968, 1970, 1971), more

11.8. A lynx, probably a male, rests on a rock outcrop.

than for any other of the 13 carnivore species except for two kinds of marten (Feng, Cai, and Zheng 1986).

Lynx were observed on only three occasions in the Chang Tang. On 6 June 1987, at 1550 hours, in the hills just west of the Arjin Shan Reserve, we came upon a dead male marmot 2 m from its hole. Fresh blood on throat and nape indicated that it had just been killed. Two lynx ascended a slope at a fast walk 50 m away. North of Toze Kangri, a hillock with a cliff projects from the plain. On 20 June 1992, a large lynx, probably a male, rested near the top of the cliff within 10 m of a niche that contained two downy saker falcon young. The lynx was remarkably tolerant of us, merely watching us at 20–30 m (fig. 11.8). A female lynx with a large cub sat on a knoll north of Yibug Caka at 1015 hours on 16 October 1993.

On 17 December 1991, a truck driver saw a lynx feeding on a chiru kill on the road between Garco and Tsasang. He drove over the cat, killing it. The lynx was a male with a weight of 22.3 kg, and his kill was female and weighed 29.5 kg, including about 1.5 kg he had eaten along the side of the neck and one thigh. Her bone marrow was in good condition. Phelps ([1900] 1983) observed a lynx stalking a chiru young.

Snow leopard

Snow leopards inhabit some of the most remote and highest ranges on earth. Their luxuriant smoke-gray coats with black rosettes and their long, lush tails are evocative of snow and immense solitudes. Rare and elusive, the cat has become symbolic of the wildness and wilderness of central Asia's mountains. We searched for them but they remained little more than an imminence—a track, a dropping. Although the species has a wide distribution, covering over 2 million km^2, and extends into twelve countries, from eastern Afghanistan and northern Pakistan east over the Himalaya and the Tibetan Plateau and northeast through the Pamirs and along the Tian Shan and Altay Mountains as far north as the Sayan Mountains near Lake Baikal in Russia, much of its range lies within China. Little about the cat in China was known when we began our surveys in 1984. In the early stages of the project, we traveled widely to determine status and distribution in Qinghai, Gansu, and Xinjiang (Schaller et al. 1987; Schaller et al. 1988; Schaller, Ren, and Qiu 1988) and later also in Mongolia (Schaller, Tserendeleg, and Amarsanaa 1994). Surveys in Tibet focused on the Chang Tang. Except for a brief, intensive study of a snow leopard population in the central part of the Altay (at 45°40′ N, 96°50′ E) in Mongolia (fig. 11.9), work on the cat consisted of surveys. Detailed studies include those of Koshkarev (1989) in Kyrgystan, Fox et al. (1991) and Chundawat and Rawat (1994) in India, Jackson and Ahlborn (1988, 1989), Oli, Taylor, and Rogers (1993, 1994), and Oli (1994a, 1994c) in Nepal, and T. McCarthy (in prep.) in Mongolia.

Distribution

Since western China and Mongolia comprise over 4 million km^2, we could do little more than sample this vast area for snow leopards. In determining distribution, we first perused topography and habitat to delineate potential range and then inquired about the cat from local people. Snow leopards typically inhabit rugged terrain such as steep slopes with bluffs, ridges broken by outcrops, and valleys interrupted by cliffs (Jackson and Ahlborn 1988). Such topography is characteristic not just of the high Himalaya, where the cats tend to stalk our imagination, but also of the desert massifs of the Gobi, where the animals occur as low as 1000 m in elevation. Though partial to a broken landscape, snow leopards will traverse expanses of flat to hilly ground. The cats generally inhabit treeless terrain, but they may enter sparse forests of conifer and oak, especially in winter (Schaller 1977b).

To obtain direct evidence of snow leopards we made cross-country transects, usually on foot but also on horseback, in search of spoor. Raking their hindpaws, the cats make characteristic scrapes on the ground at mountain passes, the base of cliffs, the confluence of streams, by prominent boulders, and other conspicuous places; the abundance of scrapes, as well as tracks and droppings, provides a rough index of relative numbers. However, counts of scrapes as a measure of abundance must take into account

11.9. A Mongolian herdsman and I (on right) hold a sedated female snow leopard after radio-collaring her in the Altay Mountains, Mongolia. (October 1992)

differences in use of terrain by snow leopards from area to area. Cats in the Shule Nanshan traveled mainly along the bottoms of ravines, whereas in the western Kunlun Shan they mostly traced the base of hills flanking broad valleys. In the limestone massifs of southeast Qinghai the cats had no well-defined linear routes (fig. 11.10). Travel patterns presumably reflected the probability of encountering prey.

11.10. Cliffs and steep slopes are favored snow leopard habitat, here in southeast Qinghai. (August 1984)

The detailed results of the surveys in Xinjiang, Qinghai, Gansu, and Mongolia can be found in our previous publications. I provide here an overview of these areas and add a few new data from Tibet and Inner Mongolia (fig. 11.11).

XINJIANG. Snow leopards occur in all major ranges, totaling about 10.6% of Xinjiang's 1,603,774 km^2. We sampled three areas in the Tian Shan and two in the Kunlun Shan. In the Tian Shan no sign was found at the eastern tip, the animals were rare in the central part, but in the western part, in the Tomur Feng Reserve (3000 km^2) adjoining Kyrgystan, a viable population persisted at that time. East of the reserve, in the Kokosu area, local hunters killed 12 snow leopards during the winter of 1985–1986, and southwest of the reserve, 11 snow leopards were killed during the winter of 1984–1985, according to informants. Two months of surveys in and around the Taxkorgan Reserve revealed that the cats were generally scarce, though in one locality, around Mariang, they seemed quite common. No surveys have been made along the Kunlun and Arjin Shan, which stretch for over 1500 km along the northern rim of the Tibetan Plateau. We checked the area just north and west of the Arjin Shan Reserve and found no spoor, although the Arjin Shan Reserve itself is said to contain a few snow leopards (Butler, Achuff, and Johnston 1986). Przewalski found the cats "very rare" in the Arjin Shan (Prejevalsky 1879).

QINGHAI. Snow leopards are found continuously, or almost so, along the three mountain chains that cross Qinghai longitudinally. Along the northern edge of the plateau is the Arjin Shan, which becomes the Qilian Shan and its subsidiary ranges; in the central part is the Kunlun Shan, which extends east as the Anyemaqen Shan; and along the border with Tibet is the Tanggula Shan. In addition many small massifs harbor snow leopards, especially in the southeast, where the cat's distribution is highly fragmented (Liao 1985; Schaller, Ren, and Qiu 1988; Yang 1994). The range of snow leopards encompasses about 65,000 km², or 9% of the province. Densities varied from rare to moderately abundant. A four-day search in the Burhan Budai sector of the Kunlun Shan south of Dulan revealed only one track and one scrape. In parts of the Anyemaqen Shan and various

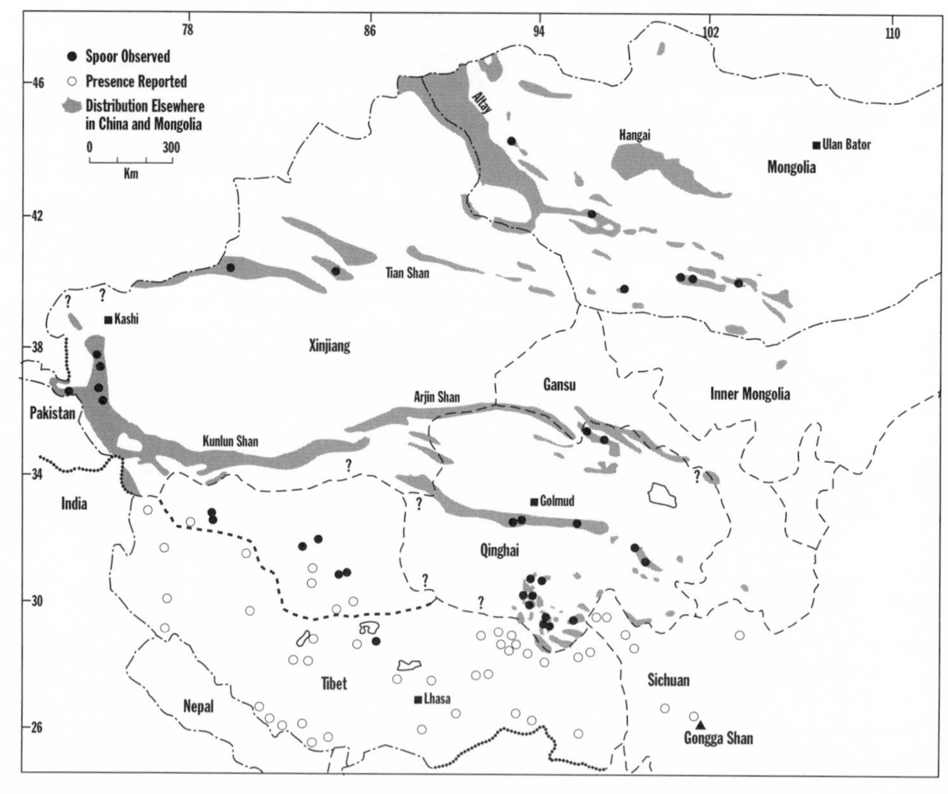

11.11. The approximate distribution of snow leopards in China and Mongolia. No surveys have as yet been conducted in much of Tibet and western Sichuan.

ranges around Zadoi we found spoor frequently. And in one part of the Shule Nanshan spoor was abundant—we tallied 170 scrapes and 91 droppings—even though 12 snow leopards (3 cubs, 6 subadults 1–3 years old, 3 adults) had been trapped in the area for the Xining zoo during the previous two years. We were told that herdsmen killed at least 11 cats there during the winter following our survey—probably reflecting a trend relevant to snow leopard numbers in many parts of China since our surveys in the mid-1980s.

TIBET. The status and distribution of snow leopards remain unknown for much of Tibet. My records, reports from local people, and information from other sources indicate that the cat occurs at least sporadically over the whole region (fig. 11.11). Snow leopards no doubt extend along the whole length of the Himalaya. In the Qomolangma Reserve (33,910 km^2), they "are widely but sparsely distributed" (Jackson et al. 1994). Their status in reserves farther east along the Himalaya, in the Medog (676 km^2) and Zayu (1015 km^2) Reserves, has not been determined. In other parts of Tibet, populations are much fragmented not only by unfavorable habitat such as plains and forest belts but also by areas devoid of cats because of extensive hunting. One of our surveys, covering 40,000 km^2 from the Lhasa Valley south to the border of Bhutan and India during October 1995, revealed that snow leopards have almost been exterminated there during the past quarter century. "About 200 skins were purchased in 1977 in Tibet in spite of being listed as a protected species" (Wang and Wang 1986).

In the arid western half of Tibet, snow leopards generally seemed scarce in the Gangdise, Nyainqentanglha, and other ranges, and many of the small ranges were devoid of them even though blue sheep, a principal prey, were available. Either the cats had been exterminated, or blue sheep were too scarce and the ranges too small to maintain viable cat populations. In 1990, we were told that five snow leopards had been killed near Mount Kailas during the previous two years. South of Siling Co is a range of low rounded hills broken by a few cliffs on which blue sheep were sparse. The area looked marginal for snow leopards, yet I found spoor and examined a domestic goat that had been killed that day. The site emphasized that the species could survive in many areas in which they do not now occur.

Snow leopards are rare and local in the Chang Tang Reserve, even in areas uninhabited by pastoralists, largely because of unsuitable terrain and scarcity of resident prey. We were told of snow leopards along the limestone hills southwest of Lumajangdong Co, in the mountains west of Yibug Caka, and in a cluster of peaks and ridges northwest of Siling Co. I saw snow leopard spoor in only three localities. In many days of hiking in the Aru Range, I found three scrapes. South of the Tian Shui River are high hills broken by cliffs. One valley, flanked by bluffs and outcrops, penetrates these hills for 4 km before opening into more expansive terrain. This valley is a major travel route for wildlife. A total of 37 scrapes were found in the sand and gravel along the margins of the valley. The third locality was the Jangngai Shan, a rugged range with a few glaciers, where I tallied 20 scrapes on the north side of the crest. In addition, one skull was found in a low range north of Gomo Co.

Other areas in China. In Sichuan, snow leopards inhabit various areas above timberline, but a distributional survey has yet to be made. Liao and Tan (1988) listed counties in which the species is said to occur, including Baoxing, along the eastern edge of the plateau. This county contains the Qionglai Shan, which are partly in the Wolong Reserve. The eastern limit of snow leopards in Sichuan, as well as in Gansu, lies at about 104° E.

In Gansu, snow leopards extend along the Arjin Shan and Qilian Shan. Populations in the southern part of the province may have been exterminated except for a few animals in the Die Shan. The cats occurred in the Mazong Shan of northern Gansu until the 1980s, according to information obtained during our 1996 survey in the region.

Snow leopards were once found in most of the large desert ranges in the western half of Inner Mongolia, as determined by our interviews with local people. These ranges included the Dongda Shan, Yabrai Shan, Ulan Shan, Daqing Shan, and Helan Shan on the Inner Mongolia–Ningxia border, and Longshou Shan on the Inner Mongolia–Gansu border, but the cats on all of them have been exterminated since 1940. Except for an occasional straggler on the China-Mongolia border, the only snow leopards now in Inner Mongolia are a few animals in one part of the Lang Shan (41° N, 106°35′ E).

Mongolia. Mallon (1984) presented a general map of snow leopard distribution, and our surveys (Schaller, Tserendeleg, and Amarsanaa 1994) added detail, emphasizing especially the fragmented populations in the isolated desert massifs of the Gobi (fig. 11.11). The cats extend sporadically westward from about 106° E across the Gobi and then along the Altay and several of its subsidiary spurs to the Russian border. North of the Altay they occur at low to very low densities in the Hanhöhiy, Harkhyra, and Hangai Ranges, and a few may persist in the northernmost part of the country west of Hövsgol Lake. The total range of the snow leopard probably does not exceed 90,000 km^2, or 14.9% of the country. Snow leopards have legal protection in several reserves, including the Great Gobi (54,120 km^2) and Gobi Gurvensaikhan (20,000 km^2) Reserves in the Gobi, the Khokh Serkh Nuruu (660 km^2) and Khasagt Khairkhan Uul (270 km^2)

Reserves in the Altay, and the Otgon Tenger (951 km^2) Reserve in the Hangai Mountains.

Status

A number of estimates of snow leopard density have been published and these vary considerably, depending on the size of the survey area and intensity of research. Relatively small tracts yielded high densities largely because the site was selected for study based on a substantial population. In one area of Nepal, there were 5–10 (excluding small cubs)/100 km^2 (Jackson and Ahlborn 1989), and in another, 4.8–6.7/100 km^2 (Oli 1994c). In Mongolia, Schaller, Tserendeleg, and Amarsanaa (1994) estimated at least 3.6/100 km^2 within one area of 275 km^2. Average density estimates for large areas include 1.0–2.0/100 km^2 in 15,000 km^2 of Ladakh (Mallon 1991), 0.8/100 km^2 in 8200 km^2 of the Dzungarian Alatau in Kazakhstan (Annenkov 1990), and 1.0/100 km^2, with a variation of 0.8–4.7/100 km^2, depending on survey site, in 65,800 km^2 of the Tian Shan in Kyrgystan (Koshkarev 1989).

Our surveys were made mostly over a decade ago and they were too cursory to provide a basis for reliable population estimates. This makes me reluctant to offer figures. However, even a well-considered guess has validity when evaluating status of endangered species. Snow leopards retain much suitable habitat, but killing for pelts and bones (used in traditional medicines) and in defense of livestock has reduced many populations to isolated remnants. Furthermore, as hunters decimated the snow leopard's prey, cat densities either declined or livestock became a major food source—with predictable consequences for the cats.

An average density of about 1/100 km^2 over large tracts appears to be typical of three estimates quoted earlier. Xinjiang has 170,000 km^2 of potential habitat but I doubt that 1700 snow leopards survive. Aside from heavy hunting pressure, extensive areas of the arid Kunlun and Arjin Mountains normally support only low wildlife densities. Schaller et al. presumed that "there may be no more than 750" (1988). With 65,000 km^2 of snow leopard habitat, Qinghai would have 650 cats, a figure that may be of the correct order of magnitude. Several areas, such as the southeast corner of the province, had substantial populations, estimated at 3–4/100 km^2 in places. It is premature to estimate populations in other parts of China. Given the total range of the species, numbers in the country as a whole probably exceed 2000. The former Soviet Union had an estimated 1000–2000, most of them in what is now Kyrgystan (Braden 1982); Mongolia possibly had as many as 1000 (Schaller, Tserendeleg, and Amarsanaa 1994); and India, around 500 (Fox et al. 1991)—to give figures for three areas with moderately high populations for comparison and to indicate that China harbors a substantial percentage of the world's snow leopards. Fortunately China has a number of reserves containing snow leopards (fig. 1.3). However, some are relatively small, and the two largest—Arjin Shan and Chang Tang—contain few animals, largely because of unsuitable habitat. Fox (1994) estimated that about 4500–7350 snow leopards persist in the wild throughout their range.

Food habits

Snow leopards are medium-sized cats about 190–210 cm long, of which almost half is tail. Two adult males weighed 49.5 kg and 52.5 kg (Dang 1967; Suo 1964); two subadult males, 28 kg and 34 kg; and an adult female, 39 kg (Jackson and Ahlborn 1989). A cat of such size can prey on all ungulates within its realm except the largest, such as adult wild yak. Snow leopards are essentially solitary and hunt alone, but at times a pair travels together, especially during the winter mating period. A female may be accompanied by one to three cubs until they are a year or more old, and independent subadults may also associate intermittently (Schaller 1977b). Individuals tend to be most active from dawn to about 1000 hours and from late afternoon to midnight (Jackson and Ahlborn 1989; Schaller, Tserendeleg, and Amarsanaa 1994). In areas with ample prey, home ranges are relatively small. Average minimum home range size of five cats in Nepal's Langu Valley was 20.9 ± 4.9 km^2 (Jackson and Ahlborn 1989). Winter home range sizes of two females and a male in Manang, Nepal, varied from 12 to 23 km^2 (Oli 1994c). One male in Mongolia remained within an area of 12 km^2 for 41 days in winter before traveling into an adjoining drainage (Schaller, Tserendeleg, and Amarsanaa 1994).

Snow leopards prey on whatever ungulates are available, from wild pigs and gorals to Himalayan tahrs, markhors, takins, and argalis (Schaller 1977b), and in Mongolia on such animals as wild Bactrian camels, goitered gazelles, and kulans (Tulgat and Schaller 1992; Mallon 1984). However, their staple prey without which they could not survive in most areas is blue sheep and ibex. In addition, all forms of livestock are killed, and marmots are important among small mammals.

Dropping samples from each of four areas in Qinghai showed blue sheep (24–39%) and marmots (37–65%) as the most important prey in summer, supplemented with deer, hares, pikas, and an occasional bird (table 11.6). Similarly, blue sheep and marmots contributed over two-thirds to the diet of snow leopards in part of the Taxkorgan Reserve and the Manang area of Nepal (Schaller et al. 1987; Oli, Taylor, and Rogers 1993). In the Tomur Feng Reserve, where marmots do not occur, all 19 droppings consisted wholly of ibex; in our Mongolia study area the content of 22 droppings was 63% ibex, 18% marmot, 3% domestic yak, and the rest

Table 11.6 Food items in snow leopard droppings in Qinghai, expressed as percentage of total content in sample

Item	Shule Nanshan 9–10/1985 *n* = 91	Anyemaqen Shan 9/1986 *n* = 20	Yushu 8/1984 *n* = 46	Zadoi 8–9/1986 *n* = 36
Blue sheep	39.3	31.3	30.4	24.0
Cervus deer	1.1	—	2.2	—
Musk deer	—	—	—	13.8
Domestic sheep/goat	2.2	—	17.4	2.8
Domestic yak	—	—	4.3	—
Marmot	36.5	65.3	41.3	51.1
Hare	5.5	—	—	—
Pika or small rodent	2.2	—	—	—
Unidentified hair	2.1	—	2.2	—
Bird	0.1	—	—	—
Vegetation	11.0	3.4	2.2	8.4

Source: Adapted from Schaller et al. 1988.

vegetation. The snow leopards in the Langu Valley of Nepal differed from other populations in that they preyed heavily on pikas and small rodents, with up to a quarter of the fecal content consisting of the remains of such species (Jackson 1996).

I collected only incidental droppings in the Chang Tang Reserve. One dropping from Memar Co contained five hooves of blue sheep young; of four droppings from the Jangngaida Shan, three had blue sheep and one had marmot; and of three from the Tian Shui River area, one had chiru, one had marmot, and one had wild yak young.

Livestock represent a significant prey item in some areas. The percentage occurrence of livestock in all our dropping samples from China and Mongolia was less than 5% except for the Yushu area in Qinghai, where it was 22%. We collected droppings mainly in summer and autumn, but, as Oli, Taylor, and Rogers (1994) have shown, snow leopards are most likely to kill livestock in winter and spring. Elsewhere, Chundawat and Rawat (1994) had 15% occurrence of livestock in Ladakh, and Schaller (1977b) had 45% in a small sample from Pakistan. Most livestock kills consist of sheep and goats, but the cats also readily prey on horses, Bactrian camels, and yaks weighing up to 200 kg.

Vegetation may for unknown reasons occur in the droppings, with some wholly composed of twigs 2–4 cm long from such shrubs as *Tamarix* sp., *Salsola arbuscula*, and *Sibiraea angustata*. In Ladakh, *Myricaria germanica* and other plant material were found in 41.0% of the droppings (Chundawat and Rawat 1994).

Impact of predation

The lists of food items in droppings of large predators (tables 11.3, 11.4, and 11.6) reflect the importance of prey species but they do not provide data on biomass and number of animals killed. Such information has importance when assessing the impact of predation on wild ungulate populations, including the influence of marmots and pikas as buffers, and on livestock. Floyd, Mech, and Jordan (1978) developed a linear regression equation for converting wolf dropping data to number of prey consumed, which was later refined by Weaver (1993). Ackerman, Lindzey, and Hemker (1984) provided a similar equation for pumas, which is used here for snow leopards. Because small animals have a higher surface-to-volume ratio than large ones, they have a higher proportion of hair to flesh. Calculations of biomass and of numbers based on hair content in droppings will overrepresent biomass and underrepresent numbers unless the bias is reduced, as in these equations: for wolf, $y = 0.439 + 0.008x$, and for snow leopard, $y = 1.98 + 0.035x$, where y is the number of kilograms of prey per dropping and x the average weight of prey.

Effects of predation

A wolf requires at least 1.7 kg of food per day for basic maintenance (Mech 1970), or 620 kg per year. Inedible parts such as large bones and stomach contents average about one-third of an ungulate's total weight (Jackson and Ahlborn 1984) and about a quarter of a marmot's and hare's; pikas are totally consumed. If a wolf preyed solely on ungulates, it would have to kill about 930 kg per year, the equivalent of 23 adult blue sheep with an average weight of 40 kg or 31 chirus with an average weight of 30 kg. If a predator kills many young animals, which have a lower body mass than adults, then its impact would be greater than my calculations indicate.

In the Aru Basin, an area with a variety of resident prey, two-thirds of the biomass eaten by wolves consisted of ungulates and much of the rest consisted of marmots during the six-month season when these rodents were available. However, in terms of relative number of animals killed, ungulates totaled 8, marmots 22, and pikas 152, indicating that for small prey such as pikas even a low frequency of occurrence in droppings translates to large actual numbers (table 11.7). In the high, rather barren valley south of the Natigan massif, plains ungulates were sparse most of the year. There pikas provided over half of the biomass eaten and marmots another 12%. In a third area, the bleak desert steppe north of Toze Kangri, wolves preyed heavily on migrating chirus, which comprised 71% of the biomass eaten, and on pikas.

Medium-sized ungulates and marmots are the preferred prey of wolves,

Table 11.7 Diet of wolves during the six spring–autumn months in three Chang Tang Reserve areas with different prey availability

	Blue sheep	Argali	Chiru	Gazelle	Yak*	Sheep/ goat	Marmot	Pika	Hare
Estimated weight (kg) of prey	40	48	35♂†, 25♀‡	14	150	27	5.3	0.125	2.3
ARU BASIN (no. of droppings: 58)									
Frequency of occurrence	13	2	7	3	2	2	21	5	1
% of diet	27.5	5.5	14.0	4.6	9.1	3.7	28.2	6.1	1.3
No. of animals killed	3.2	0.4	1.9	1.5	0.3	0.6	22.0	152.1	2.3
NATIGAN (no. of droppings: 28)									
Frequency of occurrence	5	—	—	—	1	—	4	19	—
% of diet	24.1	—	—	—	10.4	—	12.2	53.2	—
No. of animals killed	2.8	—	—	—	0.3	—	9.6	1318.9	—
TOZE KANGRI (no. of droppings: 24)	—	—	—	—	—	—	—	—	—
Frequency of occurrence	—	—	17	—	—	—	—	9	1
% of diet	—	—	71.1	—	—	—	—	25.9	3.0
No. of animals killed	—	—	13.2	—	—	—	—	642.7	5.4

*Domestic or wild yak.
†Aru Basin.
‡Toze Kangri.

but when these are unavailable, the wolves readily switch to a small year-round food source, the pikas. Predation on pikas appeared to be inversely correlated with the abundance of larger prey: wolves consumed significantly more pikas at Natigan ($P < 0.001$) and Toze Kangri ($P = 0.007$) than in the Aru Basin, and more at Natigan ($P = 0.04$) than at Toze Kangri. Marmots reduce predation pressure on themselves when hibernating and chiru when migrating. Blue sheep, another favored prey, are local in distribution. Pikas are the most available and vulnerable food source for wolves and as such function as an important buffer, deflecting predation from other species. However, the prey base of pikas probably helps to maintain wolf numbers above a level they would otherwise have, and this in turn may have a significant impact on numbers of certain ungulate species.

Blue sheep and marmots are important prey of snow leopards as well, whereas pikas do not contribute significantly to the diet (table 11.8). Do snow leopards and wolves compete for resources when they inhabit the same areas? Large cats require 40–45 g of food per kilogram of cat per

Table 11.8 Diet of snow leopards in two areas of southeast Qinghai during the six spring–autumn months, showing differing predation rates on livestock

	Blue sheep	White-lipped deer	Musk deer*	Domestic yak	Sheep/ goat	Marmot
YUSHU (no. of droppings: 46)						
Frequency of occurrence	15	1	—	2	9	20
% of diet	35.9	4.5	—	10.2	18.7	30.7
No. of animals killed	3.7	0.1	—	0.3	2.8	21.2
ZADOI (no. of droppings: 36)						
Frequency of occurrence	9	—	6	—	1	21
% of diet	32.7	—	15.3	—	3.1	48.9
No. of animals killed	3.4	—	5.7	—	0.5	33.7

*Estimated weight, 11 kg; other weights as in tables 11.7 and 11.9.

Table 11.9 Comparison of diets of snow leopards and wolves during the six spring–autumn months in the Shule Nanshan, Qinghai

	Blue sheep	White-lipped deer*	Sheep/goat	Marmot	Hare
SNOW LEOPARD (no. of droppings: 91)					
Frequency of occurrence	38	—	2	39	5
% of diet	56.1	—	2.6	36.9	4.5
No. of animals killed	5.8	—	0.4	25.4	7.1
WOLF (no. of droppings: 29)					
Frequency of occurrence	8	1	2	19	2
% of diet	32.2	7.6	6.9	48.4	4.8
No. of animals killed	3.7	0.3	1.2	37.8	8.7

*Estimated weight, 125 kg; other weights as in table 11.7.

day (Emmons 1987). One snow leopard male ate a total of 66 kg of prey in 13 days between 12 November and 22 December, or an average of 1.7 kg per day (Schaller, Tserendeleg, and Amarsanaa 1994). His weight was 37.5 kg and he would thus require 1.5–1.7 kg per day (Emmons 1987), a figure similar to his consumption. At 1.5 kg per day a snow leopard weighing 35 kg would need to consume 548 kg a year and kill 822 kg, about as much as a wolf. However, digestive efficiency of felids may be somewhat lower than that of canids (Houston 1988). Even so, snow leopards and wolves select much the same prey and have similar food requirements, as shown in the Shule Nanshan (table 11.9). There the two predators showed no significant differences in predation on marmots, but blue sheep provided more biomass ($P = 0.026$) to snow leopards than to wolves. Overlap in diet was so great that competition was likely. With plains wildlife decimated by humans, wolves probably prey more on blue sheep now than they did in the past. In a competitive situation, the snow leopard is at a disadvantage because, unlike the wolf, it is not highly mobile.

A predation rate of about 10% of a population represents a "limiting equilibrium state for large predators and large mammalian prey" (Emmons 1987). Thus a blue sheep population with 150–200 animals and an annual increment of 15% could support one wolf or snow leopard if the population lacked other mortality, an unlikely occurrence. Even with pikas and marmots acting as buffers, it is probable that predation has a major impact on many blue sheep populations. Mortality is likely to approach and even exceed 10%, especially if human predation is added to the total. Marmot populations also appear to be significantly affected by predation pressure.

Predation on livestock

The occasional killing of livestock by wolves and snow leopards has brought pastoralists and predators into such conflict and become such a contentious issue that it affects the future survival of both species. As livestock numbers increase, herders penetrate ever higher into the mountains or into marginal pastures on the plains, bringing prey and predators into more frequent adverse contact. Local people in Nepal's Annapurna Conservation Area have such a negative attitude toward snow leopards that "most suggested that total extermination of leopards was the only acceptable solution to the predation problem" (Oli, Taylor, and Rogers 1994). In a subsistence economy even the loss of one or two sheep is of economic importance to a household, and there is a further incentive to kill predators for the pelts and bones that can be sold.

Losses to predation vary considerably from area to area and even between households of a community, depending upon where livestock is grazed and how well it is tended. Five households in the Anyemaqen Range owned a total of 2350 head, of which they had lost 6 sheep, 5 yaks, and 1 horse to snow leopards and 1 sheep and 1 horse to wolves during the past 12 months (total 0.6%). One family near Zadoi was treating a yak calf that had been attacked by a snow leopard a few days earlier (it had bite wounds on its back behind the shoulder), and 3 yaks and 3 sheep, or 2% of that family's livestock, had been killed the past year.

In the Qomolangma Reserve, annual livestock losses averaged about 1.2% of the total population (Jackson 1991). Officials in Rongma estimated that wolves killed each year about 0.7–0.9% of the livestock. Five households in the Aru Basin had lost about 4.5% of their animals to wolves during the past year.

Predation on livestock by snow leopards in the Mariang area of the Taxkorgan Reserve was heavy. Each household had an average annual loss of 3.3 sheep and goats and 0.3 large animals, principally yaks, or 7.6% of sheep and goats and 1.7% of large animals in the total population. Elsewhere in the reserve, nine households had lost 2.6% of their sheep and goats and 0.9% of their yaks to wolves (Schaller et al. 1987).

In the Manang area of Nepal, the average household had 26.6 head of livestock, of which 2.6% were killed in a two-year period (Oli, Taylor, and Rogers 1994).

In Mongolia, in 1990, our study area in the Altay Mountains had eight resident families with 3175 head of livestock. The total annual loss to snow leopards was 0.4% of sheep and goats, 11.9% of yaks, and 17.0% of horses, principally foals and yearlings. During December 1989, I visited a family in the Toost Uul, an isolated massif in the Gobi. Of 300 horses, snow leopards had killed 21 since the previous April and I was shown 3 injured foals, 2 with wounds on their shoulders and 1 with bites on its throat. These families grazed their livestock in high-density snow leopard areas. However, livestock losses to snow leopards and wolves were as a whole much lower, 0.34–0.38% of the total population in two districts of the Gobi and 0.13–0.14% in two districts of the Altay. Yaks and horses were killed in disproportionately large numbers because they roamed untended for days and weeks, whereas sheep and goats were guarded in daytime and brought close to homes at night. Unlike other peoples, Mongolians do not eat or otherwise use predator-killed carcasses, and the loss to a household is thus total. But, on the positive side, predators may be able to retain their prey, obviating the need to kill again immediately.

Predation is to some extent influenced by herding practices that make livestock vulnerable. Sheep and goats may be left unguarded, herder and dogs absent or asleep, and at night they are often not in secure enclosures. Jackson (1991) noted that children and women are less adept than men in protecting livestock and that dogs are generally ineffective. Horses and yaks cannot usually be stabled for long because fodder is too sparse to be gathered and stored. But better care of mares with foals would reduce predation. However, any change in herding practices would require certain new livestock grazing patterns, changes in the division of labor among family members, and other unwelcome adjustments.

Depletion of wild prey is likely to increase predation on livestock. Certain government policies encourage or once encouraged the reduction of prey populations. Both Tibet and Qinghai have poisoning programs to reduce pika populations. Qinghai also subsidized marmot hunting for pelts by providing local people with ammunition, and even guns—and these were used not just to hunt marmots. From the late 1950s to the late 1980s, 5000–10,000 blue sheep were shot annually in Qinghai for export as luxury meat to Europe. Such market hunting decimated or eliminated blue sheep on various ranges.

The Yushu area in southeast Qinghai appeared to have less wildlife and more livestock than the Zadoi area, and this is reflected in the amount of livestock killed by snow leopards (table 11.6). A third of the biomass eaten by the cats in each area was blue sheep, and marmot was also important. But in Yushu 22% of the diet was livestock, compared to 3% in Zadoi, a significant difference ($P = 0.015$).

Wolves, snow leopards, and other predators were also reduced directly with predator control programs. The government provided free or cheap ammunition and then bought the pelts. One such program in the Xigaze Prefecture of Tibet was terminated in the 1990s because the number of pelts had greatly declined and local people increasingly sold wildlife products for a better price privately rather than to the government (Miller and Jackson 1994).

I have presented information on the impact of predation on livestock in some detail to emphasize that the financial loss to households is at times so considerable that it may represent a significant proportion of annual income (Jackson 1991; Oli, Taylor, and Rogers 1994; Oli 1994b). Snow leopards and wolves can continue their existence in pastoral areas but only if the local people are willing to make certain concessions. At the same time, remedial action must be initiated to reduce the effects of predation. There is a need for better livestock husbandry, such as better herding practices and corrals, and for improved law enforcement to protect both predator and prey. In addition, a broad program of education, including religious aspects, and some form of community improvement are needed to provide economic benefits to the people living in and around reserves and other critical wildlife areas. Compensation for losses, tax incentives, shared tourist revenues, improvements in health and education services, and other programs specifically designed for an area are possible options that may help to promote the survival of predators. In an innovative pioneering effort, the International Snow Leopard Trust of Seattle and the Mongolian Association for Conservation of Nature and Environment are providing tea, noodles, clothing, and other much needed items to pastoralists in critical snow leopard areas in the Altay Mountains of Mongolia with the understanding that they will protect the wildlife. More such initiatives are essential if the large predators in central Asia are to endure.

12

Feeding Ecology of Ungulates

> But not withstanding their sterility and unfavourable conditions of climate, the deserts of Northern Tibet abound with animal life. Had we not seen with our own eyes it would have been impossible to believe that in these regions, left so destitute by nature, such immense herds of wild animals should be able to exist, and find sufficient nourishment to support life by roaming from place to place. But though food is scarce, they have no fear of encountering their worst enemy, man; and far removed from his bloodthirsty pursuit, they live in peace and liberty.
>
> Nikolai Prejevalsky (1876)

ONLY A FEW species of graminoids, forbs, and low-growing shrubs dominate plant production in the Chang Tang. Though seed stalks of *Pennisetum* and an occasional shrub may reach a height of 50 cm, most vegetation is less than 10–15 cm tall, and this in effect eliminates a vertical component to the landscape. Except for patches of alpine meadow at moist sites, plant coverage is sparse. Since most terrain lies above 4600 m and the vegetation limit lies at around 5200 m, available habitat is confined to a narrow band of plains and hills. Plant growth is dependent on precipitation and temperature. With winters long and precipitation scant, the growing season is short, from late May or June to September. Young growing plants contain much digestible protein and energy, but as they age the proportion of fiber increases and nutrient levels and digestibility decrease (Hudson and White 1985), sometimes below the maintenance requirements of ungulates. The long winters provide ungulates with only low-quality forage and it is occasionally covered with snow. Certain plants have secondary compounds that are toxic or inhibit digestion, further reducing the available forage. Some cushion plants also resist herbivory by their growth form. The ungulates in the Chang Tang thus live in an environment whose vegetation is structurally simple and nutritionally restricted. How then do the six ungulate species partition their resources to permit not just efficient use but survival?

This chapter discusses how the ungulates have adjusted to the availability and nutritional quality of forage both within and between species, a subject that has been much studied on Africa's savannah with its large and varied ungulate fauna, and I draw on this work for ideas. Vegetation is dependent on soil type, particle size, soil moisture, exposure, and the local impact of animals, including such features as the mounds of excavated soil at marmot burrows and the dust wallows of yaks. Because of local variation in these factors, vegetation consists of a mosaic of plant communities, even in the simple environment of the Chang Tang. Such vegetation patches promote coexistence in ungulates if they prefer different foods and forage at different patches (McNaughton 1983). In Tanzania's Serengeti National Park, with its 28 ungulate species, some species can exist on the same foods if they are of different size and are dispersed differently (Sinclair 1983). In the Chang Tang, as in the Serengeti, food selection is available to the animals in a hierarchy based on topography, vegetation formation, plant community, plant type, and finally certain species.

However, ungulate populations tend to be regulated less by food abundance than by quality (Fryxell, Greever, and Sinclair 1988), with winters and dry seasons being times of greatest stress. At those times, protein levels in plants are lowest, with nutrients stored below ground. Ungulates require 4-9% protein to maintain body weight (Sinclair 1975; Fryxell 1987; Koerth et al. 1984). Plants are then also more fibrous, making them less digestible. Digestibility of grasses may decline from 80% in green grass to 40% in dry grass (Sinclair 1975). Availability of moisture, whether from rain or snowmelt, provides a flush of growth and with it a renewal of nutrients, including minerals.

Shrubs have, on the average, deeper roots than grasses and forbs, and this enables them to tap nutrients at a greater depth. Consequently the quality of their forage tends to be higher and to fluctuate less seasonally (Boutton, Tieszen, and Imbamba 1988). In the Serengeti, McNaughton (1985) noted that when grazers concentrated on a sward, the frequent cropping prevented senescence of the grass tissues. This maintained the nutrient status of the grass by prolonging its growth and keeping a high leaf-to-stem ratio.

Different feeding strategies help reduce competition, with diet partitioned on the basis of plant species, growth stage, and plant part. On the Serengeti plains, zebra are the first to move into tall grasslands, where they subsist on coarse stems, sheaths, and some leaves, and they are followed by wildebeest, which prefer leaves, and then by Thomson's gazelles, which eat short grasses and forbs, a sequence of use termed a grazing succession (McNaughton 1985; Jarman and Sinclair 1979). The Grant's gazelles on the plains consume primarily forbs. In the wooded areas, buffaloes graze mainly on large-leafed grasses, and eland and impala graze in the wet season, when grass is nutritious, and browse in the dry season. During the dry season, forage quality declines to such an extent that it becomes a limiting resource (Sinclair 1975).

Grass leaves and stems grow from the intercalary meristem located at or below ground level, and they are, therefore, quite homogeneous in nutrients though often of rather low quality (Georgiadis and McNaughton 1988). By contrast, forbs and shrubs grow from the tips, and nutrients are concentrated there. An animal has to choose what is most profitable to eat in terms of nutrients, and such selection depends on body size and digestive physiology. To a herbivore, plants have two components: a cell content of soluble nutrients, including lipids, sugars, and protein; and the structural cell wall with its tough cellulose, hemicelluloses, and lignin (Van Soest 1982). Cell content is readily digestible once the cell wall has been broken down. In young tissues the cell wall is fragile, whereas in old ones it is hard. Herbivores lack the enzymes necessary to digest cell walls, and to break these down they must form a symbiotic relationship with microorganisms that can degrade cellulose and hemicelluloses by fermentation; lignin, the third structural component, is indigestible. Fermentation takes time and the rate depends on the amount of protein-rich food. Slow food passage, allowing time for digestion, and a fermentation site are two requisites for efficient forage use (Van Soest 1982; Hudson and White 1985).

Two types of digestive systems have evolved. Hindgut fermenters have an enlarged colon and a baglike caecum that stores, ferments, and absorbs nutrients. Kiangs and hares are of this type. Ruminants, from gazelles to yaks, have become specialized in that their foregut has become a multichambered vat where food is fermented before passing into the stomach for further digestion (Bunnell and Gillingham 1985). Equids lack the digestive efficiency of ruminants and they generally eat low-quality forage. However, the passage rate of food through their digestive tracts is twice as fast as that of ruminants. Though this enables them to process forage quickly, it also requires them to eat more to obtain the necessary nutrients (McNaughton 1985).

Body size affects food selection. A large species needs to eat more to meet nutritional requirements than a small species does, as can be expected, but because metabolic rate declines with size, the relationship between size and consumption is not linear. To consume enough, large species such as yak tend to be rather unselective feeders, often on an abundant but low-quality resource such as grass. Gazelles and other such species require small absolute amounts but their forage must have a high protein level to maintain a high fermentation rate, which in turn is directly correlated with metabolic rate. Small species, therefore, tend to be selective feeders of high-quality items (Sinclair 1983). Ungulates generally belong to one of three categories determined to a large extent by body size: grazers, browsers, or mixed feeders.

This overview of ungulate nutrition and factors influencing food selection provides a basis for analyzing the feeding ecology of the wild and domestic species in the Chang Tang. I describe first the available vegetation and its nutrients and then the food habits of species.

The vegetation

We sampled plant communities in alpine steppe to obtain a measure of species abundance and diversity. No transects were conducted in alpine meadows, with their small patches of turf densely covered with *Kobresia*, and in the barren desert steppe, which we visited only at the beginning of the growing season. Data on plant biomass were collected in alpine steppe, and protein analyses of plants and mineral analyses of soils were based on samples from a variety of sites throughout the Chang Tang.

Composition

Vegetation was sampled by two methods. In one, we used a 10-point frame with pins at an angle of 57°. A string was stretched along a selected transect within a plant community (tables 12.1 and 12.2), and either 10 or 20 samples, each with 10 points, were taken at 2 m intervals. Our total sample comprised 7300 points. When a pin touched a plant, it was recorded as a "hit." Calculations of species composition were based on one hit per stem or leaf; of vegetation cover, on a hit within 1 cm above ground. We also recorded whether the plant part was cropped by herbivores and, for graminoids, whether it was green or senescent. In the second method, my coworker Daniel Miller used a circular plot 0.25 m^2 in size (table 12.3). A transect consisted of 10 plots in 50 m, and percent coverage of each species was estimated. The sample comprised a total of 540 plots in a variety of plant communities in the eastern half of the reserve. The two methods produced similar results (tables 12.1–12.3, figs. 12.1–12.3).

One important characteristic of alpine steppe is that 72–92% (average 85%) of the ground is bare. Vegetation consists of isolated individuals, tufts, and cushions. Another characteristic is the predominance of graminoids, particularly *Stipa (S. purpurea, glareosa, subsessilifolia)*, *Kobresia (K. prainii, robusta, persica)*, *Poa (P. poiphagorum, pagophila, calliopsis, litwinowiana)*, *Elymus sibiricus*, and *Carex moorcroftii*, to mention the most important species. Graminoids often compose over 66% of the vegetation, although in some plant communities, such as those on gravel outwashes and unstable hillsides, they may compose 33% or less. Grasses and sedges, especially *Carex moorcroftii*, are also dominant in desert steppe, together with the dwarf shrub *Ceratoides compacta*. Forbs are a small component of alpine steppe, usually less than 15% in *Stipa* grassland (fig. 12.1) but at times more than 30% on mountain meadows (fig. 12.2), often in the form of

Table 12.1 Composition (in percentages) of plant communities in the Aru Basin, based on transects in August with a 10-point frame

	Stipa steppe	Gravel outwash	*Ajania* flats	Mountain meadow
No. of points	1200	400	200	600
% of bare ground	78.0	79.3	83.0	72.2
No. of vegetation contacts	1132	293	99	601
GRAMINOIDS				
Stipa sp.	33.0	15.4	19.2	9.8
Poa sp.	13.3	2.0	3.0	5.2
Elymus sibiricus	—	2.4	—	11.8
Pennisetum flaccidum	1.5	—	—	—
Kobresia sp.	21.4	6.8	—	36.3
Carex moorcroftii	17.9	—	—	3.0
Other graminoids	—	—	—	1.3
FORBS				
Biebersteinia odorata	1.9	25.9	—	—
Leontopodium pusillum	6.2	3.4	—	7.0
Saussurea stoliczkai	0.2	3.1	—	1.6
Arenaria pulvinata	—	—	24.2	—
Ajuga lupulina	0.1	—	—	1.0
Oxytropis glacialis	1.1	7.9	—	3.0
Oxytropis pauciflora	—	4.4	—	15.0
Oxytropis falcata	—	28.3	—	—
Astragalus heydei	0.3	—	—	0.5
Potentilla bifurca	2.0	0.3	1.0	4.3
Thermopsis inflata	0.2	—	—	—
Other forbs	0.9	—	—	0.2
DWARF SHRUBS				
Ephedra gerardiana	—	—	11.1	—
Ajania fruticolosa	—	—	41.4	—

cushions. A few forbs, such as *Potentilla bifurca*, *Saussurea stoliczkai*, and the mat-forming *Leontopodium pusillum*, are widespread, and legumes of the genera *Astragalus* and *Oxytropis* are among the most important, present in almost all transects. Most other forbs are local and sometimes uncommon. Variety was on the whole small, with fewer than 15 species and sometimes fewer than 10 species along transects in the various plant communities (tables 12.1–12.3). The *Stipa*-dominated steppes in the plains had fewer species than mountain meadows in the hills.

Dwarf shrubs are often only a minor component of the vegetation. The procumbent *Myricaria prostrata* and spiny *Hippophae thibetana* confine themselves mainly to lake and stream flats, and sage *(Ajania fruticolosa, Artemisia nanschanica)* is common only on some well-drained sites. *Potentilla* sp. (probably *parviflora*) is abundant on some hills. *Ceratoides compacta* is widespread and prominent in hilly terrain with unstable soils and on barren

Table 12.2 Composition (in percentages) of plant communities in alpine steppe, based on transects in July–August with a 10-point frame

	Stipa steppe	Hills north of Margog Caka	Tian Shui Valley	Hills south of Tian Shui
No. of points	3100	300	700	800
% of bare ground	92.2	84.7	91.0	85.0
No. of vegetation contacts	1770	158	156	236
GRAMINOIDS				
Stipa sp.	75.5	10.1	3.2	—
Poa sp.	0.8	4.4	3.2	8.5
Elymus sibiricus	—	6.3	—	—
Pennisetum flaccidum	0.1	—	—	—
Kobresia sp.	11.2	26.6	8.3	20.3
Carex moorcroftii	4.4	32.3	48.1	—
Festuca sp.	—	11.5	—	—
Other graminoids	0.6	—	1.9	—
FORBS				
Leontopodium pusillum	0.7	—	3.2	2.1
Saussurea sp.	0.5	—	—	1.7
Arenaria pulvinata	0.7	—	2.5	4.7
Draba sp.	0.2	—	—	—
Euphorbia tibetica	0.1	—	—	—
Rhodiola sp.	0.3	—	—	—
Ajuga lupulina	0.5	—	0.6	11.4
Oxytropis and *Astragalus* spp.	1.9	1.9	2.6	19.9
Potentilla bifurca	0.6	—	3.2	7.6
Heteropappus sp.	—	—	2.6	3.0
Androsace sp.	—	4.4	6.4	7.2
Ranunculus longicaulis	—	—	—	0.8
Other forbs	0.4	—	—	3.0
DWARF SHRUBS				
Ceratoides compacta	0.3	2.5	—	—
Potentilla sp.	1.2	—	14.1	9.7

upland flats. On *Stipa* grasslands, shrubs make up 5% or less of the vegetation composition and in hills 12% or less.

Phenology

When we arrived at Shuanghu on 28 May 1994, the long, cold, and dry winter was barely over, the vegetation still a muted sandy brown. However, a few green *Stipa* blades hidden in senescent yellow tufts were indicators of spring. Farther north, around Dogai Coring, a *Myricaria* was in bloom on an alkali flat on 4 June, and three days later there was a cluster of blue-flowered iris tucked among lava boulders. At that time, too, *Oxytropis*, *Draba*, *Allium*, and several other forbs had pushed up through the sandy soil. Grasses and sedges on a few slopes had a green sheen, and a week

Table 12.3 Composition (in percentages) of vegetation in alpine steppe, eastern Chang Tang Reserve, based on transects with a 0.25 m² circular plot

	Stipa (plains)	*Stipa* (plains)	*Stipa* (hills)	*Stipa-Carex*	*Stipa-Carex*	Cushion plant
Number of plots	180	100	60	40	100	40
Month/year	10/1993	6/1994	10/1993	10/1993	6/1994	6/1994
Average % of bare ground	84.5	87.6	76.4	89.4	89.8	90.2
GRAMINOIDS						
Stipa sp.	61.0	56.5	15.6	40.3	11.9	—
Poa poiphagorum	0.9	2.2	3.6	0.6	5.4	11.6
Poa sp.	—	—	4.3	1.8	—	—
Elymus sibiricus	—	0.1	0.1	—	—	3.2
Deyeuxia sp.	—	—	4.4	—	—	0.7
Roegneria thoroldii	—	—	—	—	0.1	—
Festuca sp.	—	—	—	—	—	1.7
Kobresia robusta	3.4	7.3	0.8	5.8	6.5	2.0
Kobresia macrantha	6.4	1.2	9.8	4.3	2.6	—
Kobresia pygmaea	0.6	0.3	6.9	—	0.7	10.8
Kobresia sp.	1.8	—	—	—	—	—
Carex moorcroftii	3.0	19.7	5.3	38.4	60.0	5.9
Other graminoids	—	—	1.8	—	—	—
FORBS						
Leontopodium sp.	3.5	1.3	14.9	1.6	2.4	3.1
Potentilla bifurca	4.2	4.1	2.2	1.7	0.6	4.4
Saussurea spp.	0.2	1.6	—	—	0.7	—
Oxytropis and *Astragalus*	5.7	1.9	4.7	0.9	4.0	4.6
Sibbaldia tetrandra	0.5	—	—	—	—	—
Draba sp.	1.1	—	—	—	0.2	—
Heteropappus sp.	3.3	—	0.2	—	—	—
Arenaria sp.	1.3	—	10.2	1.6	—	8.5
Androsace sp.	—	0.3	1.3	—	0.8	8.1
Thylacospermum sp.	—	0.3	—	—	0.6	25.9
Allium sp.	—	1.1	—	—	—	—
Other forbs	0.8	—	1.6	0.4	—	0.6
DWARF SHRUBS						
Ceratoides compacta	0.5	2.1	0.3	—	1.5	—
Potentilla parvifolia	1.3	—	11.9	2.6	0.8	8.9
Ajania sp.	3.6	—	—	—	—	—
Myricaria prostrata	trace	—	trace	0.1	—	—
Ephedra gerardiana	—	—	—	—	1.1	—

Source: From Miller and Schaller 1996.

later *Kobresia* turf in moist depressions was startlingly green in the somber landscape. Spring growth was earliest on slopes, probably because of the soil moisture from seepages and snowmelt. At one such site, the soil was moist 7 cm below the surface, whereas on the plain the soil was soft and dry for 12 cm and only slightly damp for another 15 cm. A car driving across the plains sinks about 5–7 cm into the soil. Such compaction of the

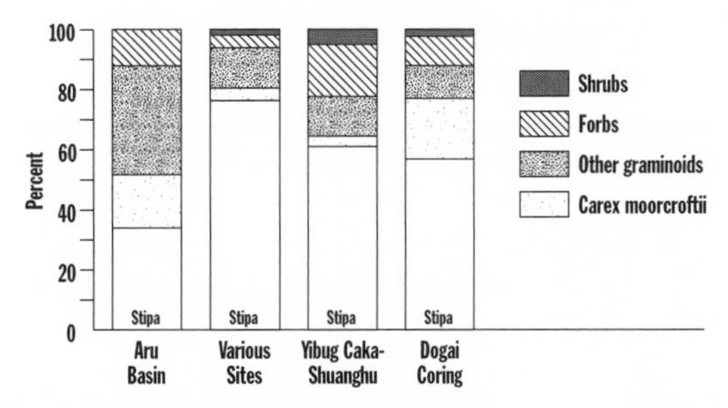

12.1. Composition of vegetation by plant form of alpine steppe communities on the plains in the Chang Tang Reserve during summer. (The vegetation was sampled with a 10-point frame.)

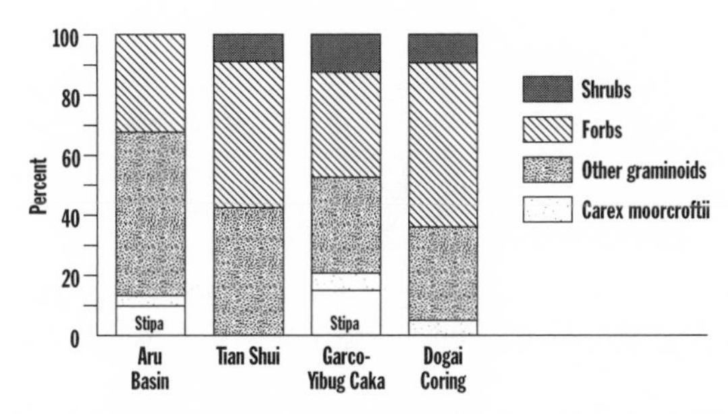

12.2. Composition of vegetation by plant form of alpine steppe communities on hillsides in the Chang Tang Reserve during summer. (The vegetation was sampled with a 10-point frame.)

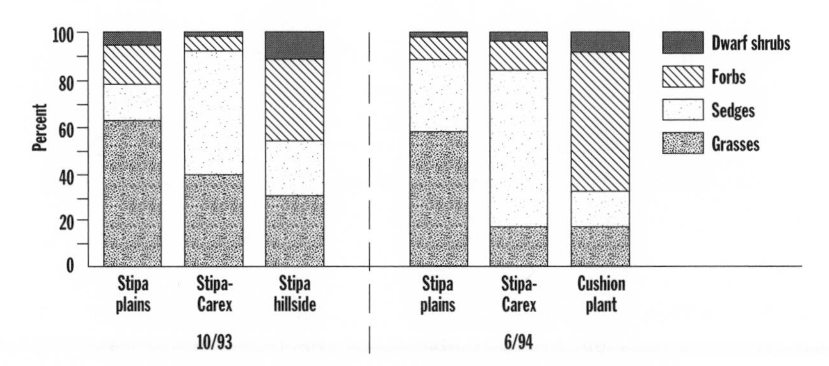

12.3. Composition of vegetation by plant form of alpine steppe communities in the eastern part of the Chang Tang Reserve during June and October. (The vegetation was sampled with circular plots.) (Adapted from Miller and Schaller 1997.)

soil apparently helps retain moisture, with the result that graminoids in tracks become green before the steppe nearby. Hillsides with the most abundant early growth were of volcanic origin, covered with black boulders that probably absorbed heat.

In desert steppe, between Toze Kangri and Heishi Beihu, spring arrived over a week later than around Dogai Coring. A few *Stipa* leaves were green on 8 June 1992, but not until 16 June was there a green flush and the emergence of a few forbs. A *Draba* bloomed white on a gravel flat on 17 June, the first flower in the bleakness, and on 23 June some *Ceratoides* shrubs had tiny, pink blossoms. Heavy June snows provided ample soil moisture, especially since drainage in many places is affected by permafrost 60–70 cm below the surface. *Ceratoides* and some cushion plants also trap moisture when prevailing winds drift snow against them.

Although new growth is abundant by late June, the *Stipa* grassland still seems reluctant to enter spring. Unlike *Carex*, whose roots may penetrate soil for 40 cm or more, *Stipa* is shallow rooted, and its main period of growth and nutrient storage begins only after the onset of summer precipitation sometime in late June or July. Since the arrival time of this precipitation varies from year to year, the main growth season of *Stipa* is also somewhat unpredictable. A few *Kobresia* and *Carex* have seed heads in late June, but the main reproductive growth of graminoids is in July and early August. By late August and in September waves of feathery *Stipa* awns glint in the sun. Plant growth ceases during September, and as forbs shrivel and grass leaves turn yellow, the steppes turn back to their wintry dun and gray color.

Biomass

With much of the ground bare and most vegetation less than 15 cm tall, the Chang Tang provides little aboveground standing biomass. In the southern Chang Tang, outside the reserve, Cincotta et al. (1991) found that only 28% of their study area supported a net primary productivity greater than 10 g/m^2. To provide an indication of forage availability on alpine steppe we clipped vegetation within 1 cm of the ground on 46 plots 1 m^2 in size (fig. 12.4). Most plots were done along vegetation transects, and various plant communities were included.

The standing crop (live and dead) on plots was highly variable (table 12.4). *Stipa* was the dominant grass, often associated with *Kobresia robusta* and *K. prainii*. Graminoids represented 77% of the biomass, the remainder consisting of forbs and a little *Ajania*, *Potentilla*, and *Ceratoides* shrub. The average for forbs in the Aru Basin is unusually high because several large clumps of *Biebersteinia* and *Oxytropis falcata*—species little eaten by ungulates—were on several plots. If these forbs are eliminated from the compu-

12.4. Kay Schaller clips a plot for measuring vegetation biomass in the Aru Basin. The object in the background is a 10-point frame for sampling vegetation. (July 1992)

Table 12.4 Mean biomass (dry weight) in grams of plant types on 1 m² clip plots in alpine steppe

	Various areas July (n = 23)	Aru Basin Early Aug. (n = 13)	Various areas Late Aug.–Sept. (n = 10)
Grasses and *Kobresia*	4.5	8.0	10.2
Carex moorcroftii	1.5	2.9	0.5
Forbs	1.7	14.8	1.7
Dwarf shrubs	0.8	2.0	0.1
Total ± S.D.	8.5 ± 7.9	27.7 ± 25.0	12.5 ± 5.8

tation, mean biomass drops from 14.8 to 2.7 g/m^2, and the total biomass for the Aru Basin drops from 27.7 to 15.6 g/m^2. The standing crop of most areas probably lies between 8 and 16 g/m^2, or 80–160 kg/ha, dry weight.

Table 12.4 is divided into three periods from July to September. Dry weight of vegetation, particularly of grasses, increases as plants mature and moisture content decreases. Wet weight of the July samples averaged 53% higher than dry weight, whereas the late-summer samples averaged only 36% higher.

Wet weight of the standing crop, including a significant portion of dry graminoid material from the previous year, averaged 19 g/m^2 outside the Aru Basin and 47 g/m^2 inside the Aru Basin if *Biebersteinia* and *Oxytropis falcata* are retained in the sample. This biomass is extremely low when compared to the East African savannahs. Boutton, Tieszen, and Imbamba (1988) reported 368–466 g/m^2 in Kenya's Masai Mara Reserve and 326–499 g/m^2 in Nairobi National Park. On short-grass plains of the Serengeti, in an area heavily grazed by wildebeest, zebra, and gazelle, the biomass was about 84 g/m^2 (McNaughton 1979). Our samples were drawn from lightly grazed areas, some by livestock in addition to wild ungulates, but even so the Chang Tang biomass in most areas was only a quarter that of the Serengeti.

Availability

The standing crop is small but all of it is theoretically available to ungulates. However, *Arenaria*, *Androsace*, and *Thylacospermum* and several other cushion plants grow so compactly that their leaves and flowers are almost impervious to herbivory. Certain species are aromatic, and their secondary compounds could restrict palatability. Among these are *Biebersteinia* and sage, though the latter is extensively cropped at times. Some legumes may be toxic, *Oxytropis falcata* among them. According to pastoralists, *O. stracheyana* may cause death in sheep and goats. However, in October 1993, chirus selected for that species at one site, and kiangs at another site pawed grooves 30–40 cm long in the soil to expose the base of this legume and then bit it off, leaving the taproot. However, plant defenses may be effective in some situations but not others depending on nutrient content, passage rate, and other factors (Belovsky and Schmitz 1994).

Much of the Chang Tang is so lightly grazed that dead and senescent graminoid leaves and stems remain standing through the winter into the following summer. New growth is hidden in tufts of this old material, making it difficult for grazers to select for the green, high-quality items. This is especially true of *Carex moorcroftii*, whose old leaves are rigid and sharp-pointed. On 30 June, one *Carex* patch had a tinge of green but in a sample of 520 leaves, 73% were old. At the height of the growing season more than half of the graminoid leaves are old (table 12.5). The presence of so much senescent material gives much of the steppe a drab appearance even in summer. Exceptions are alpine meadow, whose green turf of *Kobresia* is maintained by yaks and others as a grazing lawn often no more than 1 cm tall, and pastures that are used extensively by livestock in the southern part of the reserve.

Table 12.5 Percentage of new, green graminoid leaves and of graminoid leaves cropped by herbivores on alpine steppe during July–August

	Total leaves sampled	% of leaves green	% of total leaves cropped
Aru basin	1488	58.8	7.4
Stipa steppe (various areas)	1639	44.9	6.1
Hills north of Margog Caka	144	25.7	6.9
Tian Shui Valley	92	43.5	7.6
Hills south of Tian Shui	68	26.5	0

Crude protein content

Crude protein content, as measured by the standard Kjeldahl N procedure, provides a rough measure of forage quality. Protein is a limiting nutrient, and maintenance requirements for ungulates are known, with, for example, wildebeest needing 5–6% (Sinclair 1975) and North American deer 6–8% (Koerth et al. 1984). We sampled crude protein of selected plants at all seasons (table 12.6). *Stipa*, an important forage species, has an annual nutrient cycle no doubt typical of graminoids (fig. 12.5). If 6% crude protein is taken as average maintenance level, then animals are below that from October to May. Only from June to September can grazers obtain protein and other nutrients beyond those needed for maintenance and use it for purposes such as lactation, deposition of fat, and the rut. Forbs and dwarf shrubs had, like the graminoids, protein levels up to around 20% in summer. By October, forbs were dry and brittle, but they retained more protein (up to 12%), on the average, than did graminoids (up to 7%), making them important winter feed for selective browsers such as gazelles. *Ceratoides* was probably also important in winter, especially in desert steppe.

Crude protein levels give only a rough indication of forage quality. About 10% of the ingested protein remains physiologically unavailable to an animal (Owen-Smith and Cooper 1989). However, actual intake is no doubt higher than our analyses of samples indicate because animals are selective in what plants they eat. During July, senescent *Carex* leaves had 3.3% protein, whereas green *Carex* leaves from the same site had 17.3%.

Mineral content

As McNaughton (1990) has shown in the Serengeti, grazers tend to concentrate at sites which best meet their mineral requirements. Fertile soils are associated with high-quality forage, especially soils high in copper (Cu), sodium (Na), zinc (Zn), calcium (Ca), phosphorus (P), and magnesium (Mg). The last three are important for lactating females and young. I col-

Table 12.6 Crude protein contents (in percentages) of selected species at various seasons in the Chang Tang Reserve

	June	July–Aug.	Oct.	Nov.–Dec.
Graminoids*				
Kobresia robusta	17.0	12.7	2.6	—
Kobresia prainii	—	22.4	—	—
Kobresia sp.	—	22.4	4.3	4.5
Elymus sibiricus	—	16.0	—	—
Poa litwinowiana	—	18.5, 18.5	—	—
Poa sp.	—	—	—	4.5
Deyeuxia tibetica	—	—	5.7	3.1
Trisetum spicatum	—	—	—	3.7
Agropyron cristatum	—	—	—	4.1
Littledalea tibetica	—	—	—	3.3
Festuca rubra	—	—	—	6.9
Carex moorcroftii	4.9, 18.5	3.3, 17.3	—	—
	4.0, 3.3	12.3, 17.3		
Forbs				
Oxytropis pauciflora	12.9	17.6	12.6	—
Oxytropis glacialis	—	12.9, 20.8	—	8.7
Oxytropis stracheyana	—	—	8.3	11.5
Oxytropis chiliophylla	—	—	12.4	—
Oxytropis falcata	—	—	—	11.4
Astragalus heydei	—	12.8	—	—
Biebersteinia odorata	—	17.2	—	—
Leontopodium pusillum	—	8.7	5.0	—
Potentilla bifurca	—	13.8	9.5	—
Sibbaldia tetrandra	—	—	3.2	—
Dwarf shrubs				
Ajania fruticolosa	—	9.8	—	—
Ceratoides compacta	12.6, 19.1	19.8	7.6	—

*For *Stipa* see fig. 12.5.

lected 15 samples from the top 10 cm of soil at 15 representative sites (table 12.7). The soils were all alkaline with a pH of 7.1–8.9 and they consisted mainly of fine sand, silt, and clay (66%), coarse sand (10%), and the rest gravel. Mineral contents, especially of iron (Fe), Na, and Mg, were highly variable even within one area, suggesting much patchiness in forage quality. The Aru Basin was low to deficient in Ca, Na, Mg, aluminum (Al), and manganese (Mn) but high in potassium (K) when compared to other areas. Glacial runoff may have leached the soils. The desert steppe had, on the whole, higher mineral concentrations of Ca, Cu, Na, Mg, Fe, Al, and Mn; it had a lower concentration of K than the Aru Basin. In the central Chang Tang, between the Tian Shui River and Margog Caka, the samples showed variable levels, though deficient in Cu, Zn, and K.

I collected two *Stipa* and two *Carex* samples in desert steppe during

Table 12.7 Mean concentrations of elements in soil (in ppm) in three areas of the Chang Tang

	Aru Basin *Stipa* steppe (n = 5)	Toze Kangri Desert steppe (n = 5)	Central Chang Tang *Stipa* steppe (n = 5)
Ca	5,584 (1,705–11,398)	23,479 (19,094–30,023)	19,789 (9,604–28,498)
P	8.4 (4.7–15.3)	5.6 (1.5–13.2)	8.8 (5.6–11.8)
Cu	0.3 (0.3–0.4)	1.0 (0.9–1.1)	0.3 (0.2–0.3)
Na	7.4 (4.3–13.3)	161.5 (58.8–530.3)	48.9 (16.1–153.5)
Zn	0.4 (0.3–0.5)	0.3 (0.2–0.4)	0.1 (0.03–0.4)
Mg	82.7 (57.1–106.3)	511.2 (269.5–777.0)	174.9 (72.7–335.1)
K	101.0 (54.0–125.0)	66.0 (37.0–91.0)	57.0 (36.0–88.0)
Fe	2.2 (0.7–4.8)	7.1 (4.5–10.1)	3.4 (1.1–6.3)
Al	4.2 (1.9–7.5)	28.4 (24.1–31.2)	11.7 (10.9–13.7)
Mn	18.0 (11.0–23.1)	53.6 (34.7–81.4)	27.0 (22.2–32.4)
pH	7.7 (7.4–8.0)	8.6 (7.9–8.9)	7.9 (7.1–8.3)

June to compare their mineral content with that of soil from the same area. Ca levels in the plants were lower than in the soils, and P, Na, Mg, K, Fe, Al, and Mn were much higher (table 12.8). McNaughton (1988) gave mineral levels of forage on patches where Serengeti ungulates had concentrated. Mineral levels on those sites were consistently higher than on control sites. The mineral contents of our samples were as high or higher than those favored by the Serengeti animals, even though the leaves were senescent. However, K and P were low in our grass and soil samples. Calcium and phosphorus are major constituents of bone, and an inadequate balance between the two can lead to impaired fertility and growth

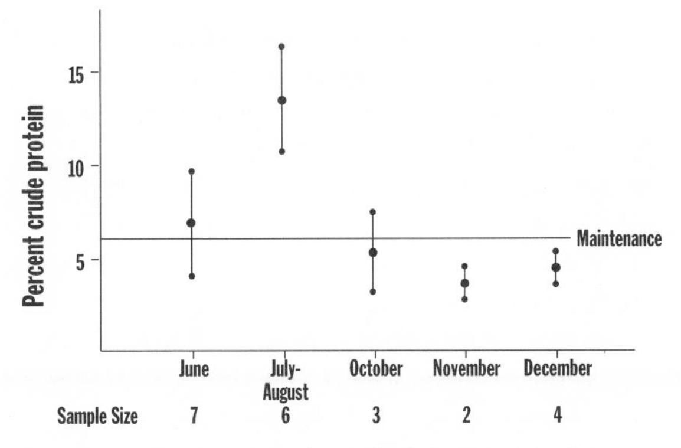

12.5. Mean percentage crude protein (with standard deviation) in *Stipa* spp. grass during various months.

Table 12.8 Concentration of elements (in ppm) in two samples of *Stipa* and two of *Carex* from desert steppe, Toze Kangri, June 1992

	Stipa sp.		*Carex moorcroftii*	
	Sample 1	Sample 2	Sample 1	Sample 2
Ca	14,500	12,700	5,980	7,830
P	304	344	137	2,720
Cu	5.3	11.6	2.2	7.7
Na	843	2,630	1,820	5,080
Zn	11.7	19.2	5.3	36.6
Mg	1,520	2,680	2,610	4,080
K	2,000	1,900	2,640	4,810
Fe	2,220	1,670	487	1,440
Al	1,680	1,200	426	920
Mn	117	74	118	137

(Murray 1995). Phosphorus may be deficient in at least parts of the Chang Tang. Seeds are generally high in P, and the summer diet of several ungulates included them.

Food habits

We examined sites where animals had recently fed to ascertain what plants had been taken. But this method can provide only fragmentary and sometimes unreliable information on food habits. Several herbivores, including pikas and hares, may have fed at the spot, and certain forbs may be totally eaten, leaving no trace. To obtain quantitative data, fecal samples for microhistological analysis were collected during various seasons. Each sample consisted of a mix of one pellet or a piece of dropping from each of 10 fresh fecal piles. One, or occasionally three, slides were made from each sample, and 20 microscopic fields were checked on each slide by the Composition Analysis Laboratory (Fort Collins, Colorado). The percent relative frequency was defined as the number of fields in which a plant species was identified divided by the number of fields with identifiable plant material of any species multiplied by 100 (Gill et al. 1983).

There are several biases in using fecal analysis as a method of quantifying food habits because of differential digestion of plant species and physiological differences between herbivores (Plumptre 1995). Graminoids are analyzed with an acceptable level of accuracy, but leguminous forbs tend to be underrepresented (Gill et al. 1983). Biases are most likely in the diet of mixed feeders, especially in the relative proportions of graminoids and forbs. However, fecal analysis is useful in determining general proportions of food types and delineating seasonal changes.

Table 12.9 Mean percent relative density of plant fragments in feces of chirus in the Chang Tang Reserve at various seasons

	Toze Kangri area 6/92	Dogai Coring area 6/94	Tian Shui area 7/91	Aru Basin 7–8/90, 7–8/92	Shuanghu–Yibug Caka		
					9/90	10/93	12/91
Sample size	9	5	6	17	9	11	9
Graminoids							
Stipa sp.	17.6%	21.4%	3.7%	4.3%	10.9%	47.3%	46.6%
Poa sp.	—	—	8.3	5.9	7.3	1.1	2.6
Elymus sibiricus	—	—	0.5	0.7	1.1	0.9	0.4
Kobresia sp.	3.8	33.1	3.9	7.3	24.4	7.1	1.1
Carex moorcroftii	14.7	16.5	22.8	13.5	0.5	10.9	13.0
Other graminoids	—	—	—	0.5	—	—	—
Forbs							
Leontopodium sp.	0.2	—	10.7	11.9	3.2	7.8	8.4
Saussurea sp.	—	—	0.7	0.1	0.4	—	—
Thermopsis inflata	—	—	—	0.2	—	—	0.6
Potentilla bifurca	0.2	—	31.1	7.7	7.8	4.2	9.4
Heracleum sp.	—	—	0.5	0.2	—	0.4	—
Astragalus heydei	—	—	4.2	2.7	1.4	0.4	0.3
Oxytropis and *Astragalus* sp.	—	1.6	6.5	28.7	16.9	5.8	9.5
Legume pod	—	—	—	1.5	1.0	—	—
Seeds	—	—	1.1	0.9	2.9	—	0.4
Other forbs	—	—	1.4	1.8	8.5	3.9	6.5
Dwarf shrubs							
Ceratoides compacta	63.5	6.2	4.4	11.1	13.7	0.2	1.2
Ajania fruticolosa	—	21.2	—	—	—	—	—

Chiru

Chirus ate graminoids, forbs, and shrubs throughout the year but the proportion of each food type varied with the season (table 12.9, figs. 12.6 and 12.7). *Stipa* is widespread, yet it formed an important part of the diet (47%) only in winter. Other grasses were consumed in small but significant amounts as well. In June 1994, a herd of male chirus foraged on a patch of green *Poa*, but the species did not appear in our fecal samples from that area. *Kobresia* was grazed on both alpine meadow and alpine steppe; the highest percentage of this sedge (33%) was noted in the June 1994 feces when animals selected for *K. robusta* on hillsides. *Carex moorcroftii* features prominently (about 13–23%) in the chirus' diet, even in areas where *Stipa* is common and succulent forbs are available. When *Carex* is senescent, chirus may paw around the base of the plant to expose and eat the soft, white corms, behavior also observed for *Stipa*. At least a third of the chirus'

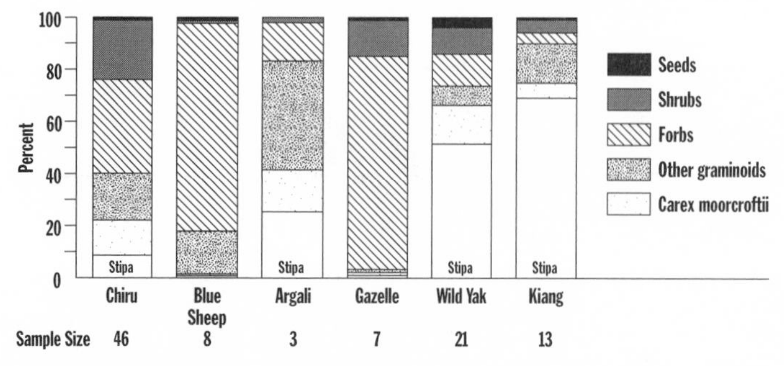

12.6. Summer (June–September) food habits of wild ungulates in the Chang Tang Reserve, based on fecal analysis.

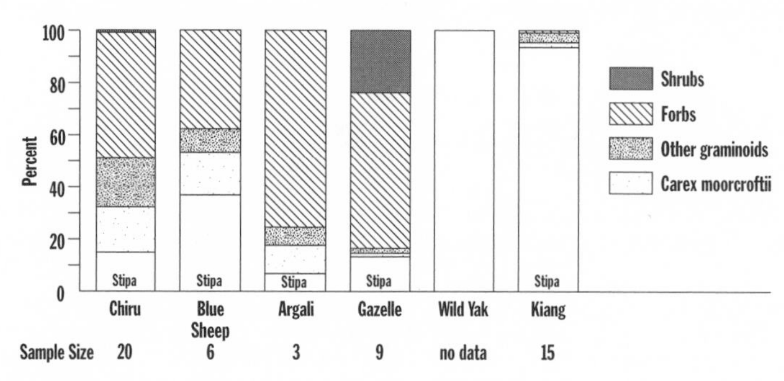

12.7. Winter (October–December) food habits of wild ungulates in the Chang Tang Reserve, based on fecal analysis.

food consists of graminoids, and in winter the proportion may reach two-thirds.

Forbs were much sought by the animals, especially in summer. Among legumes they occasionally concentrated on *Oxytropis glacialis* and the cushion plant *O. pauciflora*. When *O. pauciflora* had seed pods, chirus sometimes moved from plant to plant to gather these, almost denuding an area; such seed pods had a high crude protein content (21%). Important forbs also included *Leontopodium* and *Potentilla bifurca*. Fecal analysis did not distinguish between *Potentilla bifurca* and *Potentilla* shrub. In some areas chirus probably ate both. Other forb species were eaten only on occasion. One herd near Toze Kangri focused on *Draba*, and a herd in the Tian Shui area plucked a variety of forbs—*Ranunculus longicaulis*, *Youngia gracilipes*, *Heteropappus* sp., *Delphinium* sp., *Saussurea gnapholodes*, and *S. hookeri*—all species which did not appear in the fecal samples of that area.

Table 12.10 Mean percent relative density of plant fragments in feces of argalis in the Chang Tang Reserve at various seasons

	Yako Basin	Garco	
	6/94	9/90	10/93
Sample size	2	1	3
Graminoids			
Stipa sp.	36.8	4.0	7.1
Poa sp.	—	8.2	4.5
Kobresia sp.	26.2	62.4	1.4
Carex moorcroftii	24.9	—	10.6
Forbs			
Leontopodium sp.	—	—	27.7
*Potentilla bifurca**	7.0	12.7	22.3
Oxytropis and *Astragalus*	1.7	12.7	25.3
Other forbs	—	—	1.1
Dwarf shrub			
Ajania fruticolosa	3.5	—	—

*May include some *Potentilla* shrub.

The shrub *Ceratoides compacta* was an important chiru food in desert steppe (63.5%), and sage (*Ajania*) was prominent in the diet (21%) around Dogai Coring. However, shrubs usually constituted less than 14% of the diet on alpine steppe.

Seven rumens examined between May and December were full of graminoids, especially *Stipa*, with one also containing legume pods and another leaves of *Potentilla* shrub.

A few fecal samples from other parts of the chirus' range showed contents similar to those from the Chang Tang Reserve. One June sample from Xinjiang, from the desert steppe just west of the Arjin Shan Reserve, consisted of 52.9% *Stipa*, 24.8% *Festuca rubra*, 3.6% *Poa*, 7.3% legume, and 11.3% *Ceratoides*. A November sample from the upper Tuotuo River in Qinghai had 14.3% *Stipa*, 4.5% *Kobresia*, 31.4% *Carex*, 31.4% *Leontopodium*, 9.0% legume, and 9.3% *Potentilla*. Other Qinghai samples revealed similar diets (Schaller, Ren, and Qiu 1991; Harris and Miller 1995).

Argali

The argali fecal samples showed great variation in contents (table 12.10). Those from the Yako Basin contained mainly graminoids (88%) in June, whereas those from Garco contained much *Kobresia* (62%), *Potentilla*, and legumes in September but, in the following month, few graminoids (24%) and many forbs. In Qinghai's Yeniugou, argalis foraged on *Kobresia* (47%),

Table 12.11 Mean percent relative density of plant fragments in feces of blue sheep in the Chang Tang Reserve at various seasons

	Jangngaida Shan 7/91	Aru Basin 8/90	Noorma Co 9/90	Garco and Jangngaida Shan 10/93
Sample size	3	3	2	6
GRAMINOIDS				
Stipa sp.	2.4%	0.4%	—	36.8%
Poa sp.	3.0	13.0	11.8%	2.8
Elymus sibiricus	1.0	1.7	—	2.0
Festuca sp.	0.6	—	—	—
Kobresia sp.	1.5	3.6	16.4	3.6
Carex moorcroftii	2.0	—	—	16.6
FORBS				
Leontopodium sp.	2.8	34.9	4.4	11.2
Thermopsis inflata	0.6	—	—	—
*Potentilla bifurca**	60.9	24.3	43.6	12.6
Rhodiola sp.	—	—	2.1	—
Astragalus heydei	12.3	0.8	—	2.2
Oxytropis and *Astragalus*	7.9	20.5	21.7	12.2
Seed	0.6	0.8	—	—
Other forbs	0.6	—	—	—
DWARF SHRUBS				
Myricaria prostrata	0.9	—	—	—
Hippophae thibetana	2.9	—	—	—

*May include some *Potentilla* shrub.

Poa (13%), *Leontopodium*, legumes, and other forbs in summer (Harris and Miller 1995).

Blue sheep

Blue sheep, like argalis, showed much seasonal change in diet (table 12.11). Graminoids ranged from a low of 10.5% in July to a high of 62% in October. *Leontopodium*, legumes, and *Potentilla* (probably both forb and shrub) were also important foods. In the Yeniugou, argali foraged mainly on graminoids (92%) in summer (Harris and Miller 1995), and in Nepal grass was also a major food in winter, supplemented with shrubs and forbs (Schaller 1977b).

Tibetan gazelle

Tibetan gazelles differed from the other ungulates in that graminoids were not a major part of the diet, contributing at most 16%. Forbs, especially *Oxytropis pauciflora*, *O. chiliophylla*, and other legumes, as well as *Potentilla bifurca*, were staples. Sage was also a locally important food plant (table

Table 12.12 Mean percent relative density of plant fragments in feces of Tibetan gazelles in the Chang Tang Reserve at various seasons

	Dogai Coring 6/94	Aru Basin 7–8/90, 7–8/92	Shuanghu–Yibug Caka		
			9/90	10/93	12/91
Sample size	1	4	2	7	2
GRAMINOIDS					
Stipa sp.	3.1%	0.7%	0.8%	13.8%	14.8%
Poa sp.	—	0.2	—	—	—
Kobresia sp.	—	0.4	0.8	0.6	1.6
Carex moorcroftii	—	0.2	—	0.7	—
Other graminoids	—	—	—	0.2	—
FORBS					
Leontopodium sp.	—	—	—	5.7	—
Saussurea sp.	—	0.6	—	0.5	—
*Potentilla bifurca**	3.1	40.9	94.2	36.1	70.6
Astragalus heydei	3.1	12.2	0.8	1.9	6.0
Oxytropis and *Astragalus*	1.5	39.6	3.4	9.2	1.9
Legume pod	—	1.6	—	—	—
Other forbs	—	0.9	—	1.7	3.5
DWARF SHRUBS					
Ajania fruticolosa	89.1	1.8	—	27.8	—
Ceratoides compacta	—	0.9	—	2.5	1.6

*May include some *Potentilla* shrub.

12.12). Legumes (80%), *Leontopodium*, and other forbs were the main forage species in the Yeniugou, graminoids supplying only 10% (Harris and Miller 1995).

Wild yak

Yaks consumed mainly graminoids, especially *Stipa*, which contributed at least 30% in each area (table 12.13). *Carex moorcroftii* was prominent in the June samples (28–32%). In that month the animals around Toze Kangri foraged on senescent leaves and those around Dogai Coring partly on green ones. Our samples contained little *Kobresia* and this may reflect a bias in collecting feces. Bulls, alone and in twos and threes, often grazed on strips of alpine meadow, adeptly harvesting the short *Kobresia* stubble. Domestic yaks plucked sedges 2–3 cm tall by wedging them between tongue and dental pad (Cincotta et al. 1991). Hedin noted that "the tongue was thickly covered with horny barbs, directed backwards towards the throat. With these the yak plucks up grass, lichens, mosses, using its tongue more than its teeth and horny upper jaw in grazing" (1898). One November, I observed domestic yaks foraging on dry, brittle *Kobresia* only 1 cm tall: the animals licked the sedge off without biting or pulling. Forbs were

Table 12.13 Mean percent relative density of plant fragments in feces of wild yaks in the Chang Tang Reserve at various seasons

	Toze Kangri 6/92	Dogai Coring 6/94	Aru Basin 7–8/90, 7–8/92	Jangngaida Shan 7/91	Garco 9/90
Sample size	3	6	8	2	2
Graminoids					
Stipa sp.	67.5%	54.1%	45.5%	30.0%	67.6%
Poa sp.	—	3.1	3.6	—	1.8
Elymus sibirica	1.3	—	2.5	—	—
Pennisetum flaccidum	—	—	0.5	—	—
Kobresia sp.	—	8.1	2.5	3.2	1.8
Carex moorcroftii	27.5	32.2	0.8	10.9	—
Forbs					
Leontopodium sp.	—	0.6	11.1	—	—
Heracleum sp.	—	—	0.6	—	—
Potentilla bifurca	—	0.4	14.6	—	—
Astragalus heydei	—	—	0.7	—	—
Oxytropis and *Astragalus*	—	—	3.2	—	—
Legume pod	—	—	1.8	—	—
Other forbs	—	—	0.5	—	—
Seeds	—	—	1.2	—	28.8
Dwarf shrubs					
Ceratoides compacta	3.7	1.2	10.6	55.9	—
Ajania fruticolosa	—	0.4	0.2	—	—

important as yak forage in summer, contributing 33% to the diet in the Aru Basin. One herd in the Tian Shui Valley was crowded on a river flat cropping *Draba*. Only one sample, from the Jangngaida Shan, contained much shrub, that of *Ceratoides* (56%). The September sample from Garco had many seeds (29%), presumably from graminoids.

Elsewhere, a June sample from Xinjiang contained 98.6% graminoids, of which 81.3% was *Carex moorcroftii*, 11.7% *Stipa*, and the rest *Festuca rubra* and *Littledalea tibetica; Ceratoides* was 1.4%. In the Yeniugou, Harris and Miller (1995) found that yaks in summer consumed mainly graminoids (85.6%), principally *Kobresia*.

Kiang

Two-thirds or more of the kiangs' diet consisted of *Stipa* supplemented with *Kobresia*, *Carex*, *Elymus*, and other graminoids. The animals ate some legumes and *Ceratoides* too but such items contributed relatively little (table 12.14). However, *Ceratoides* may be important to the few kiangs that live in desert steppe. In Qinghai, kiang samples contained 99% graminoids, mainly *Stipa*, in summer (Harris and Miller 1995) and 98% in winter

Table 12.14 Mean percent relative density of plant fragments in feces of kiangs in the Chang Tang Reserve at various seasons

	Dogai Coring 6/94	Aru Basin 7–8/90, 7–8/92	Alpine steppe (not Aru) 7–8/90, 7–8/91	Shuanghu–Yibug Caka 10/90, 10/93	Shuanghu–Yibug Caka 12/91
Sample size	1	5	7	10	5
Graminoids					
Stipa sp.	78.9%	75.7%	62.8%	90.1%	99.7%
Poa sp.	—	4.5	2.9	1.7	—
Elymus sibiricus	—	5.0	2.3	3.9	—
Kobresia sp.	—	4.4	12.0	2.0	0.3
Carex moorcroftii	21.1	5.2	4.2	1.7	—
Other graminoids	—	—	0.1	—	—
Forbs					
Saussurea sp.	—	—	0.4	—	—
Thermopsis inflata	—	—	0.4	—	—
Potentilla bifurca	—	—	0.4	0.3	—
Astragalus heydei	—	—	2.5	—	—
Oxytropis and *Astragalus*	—	0.5	3.7	0.3	—
Legume pod	—	0.4	0.4	—	—
Seeds	—	0.4	—	—	—
Dwarf shrubs					
Hippophae thibetana	—	—	0.4	—	—
Ceratoides compacta	—	3.9	7.4	—	—

(Schaller, Ren, and Qiu 1991). One June sample from Xinxiang had 61.6% *Stipa*, 24.9% *Carex moorcroftii*, 6.8% *Festuca rubra*, and 6.8% *Ceratoides.*

White-lipped deer

White-lipped deer occur sparsely at the eastern edge of the Chang Tang, mainly in the Yeniugou, where they share the same habitat as the six ungulate species that occur in the Chang Tang Reserve. They tended to remain high in the hills, where they fed mainly on *Kobresia* (91%) and other graminoids (Harris and Miller 1995). Farther east in Qinghai, the deer had a more mixed diet. Based on an examination of two rumens, Zheng, Wu, and Han (1989) listed 24 food plants, among them a number of graminoid genera *(Koeleria, Stipa, Festuca, Elymus, Kobresia, Carex, Poa, Scirpus, Roegneria)* and a variety of forbs *(Ranunculus, Trollius, Astragalus, Taraxacum, Allium, Plantago)*. Based on fecal analysis, Takatsuki et al. (1988) determined that white-lipped deer near Gyaring Hu in Qinghai and in western Sichuan consumed mainly graminoids (46–85%) but also forbs and shrubs.

Table 12.15 Mean percent relative density of plant fragments in feces of domestic sheep and goats in the Chang Tang Reserve at various seasons

	South of Shuanghu 6/94	Aru Basin 7–8/90, 7–8/92	Apine steppe (not Aru) 7–8/90, 7–8/91	Shuanghu–Yibug Caka		
				9/90	10/93	12/91
Sample size	2	7	8	3	10	3
GRAMINOIDS						
Stipa sp.	42.1%	38.7%	40.6%	20.2%	87.2%	87.6%
Poa sp.	1.6	1.0	2.2	—	—	—
Elymus sibirica	—	1.1	1.2	—	—	—
Kobresia sp.	4.9	1.4	9.6	23.0	0.7	0.9
Carex moorcroftii	12.9	3.4	2.8	—	5.2	1.0
Other graminoids	—	—	0.4	—	1.1	—
FORBS						
Leontopodium sp.	—	27.5	1.2	4.0	0.2	0.9
Saussurea sp.	—	—	—	1.4	—	—
*Potentilla bifurca**	35.3	17.9	22.5	36.9	2.5	7.7
Astragalus heydei	—	—	3.8	—	0.2	—
Oxytropis and *Astragalus*	3.2	3.7	0.4	1.3	0.3	—
Legume pod	—	0.2	1.0	—	—	—
Seeds	—	—	0.5	—	—	—
Other forbs	—	0.2	1.4	7.7	1.2	1.9
DWARF SHRUBS						
Ceratoides compacta	—	4.5	8.1	4.2	0.9	—
Ajania fruticolosa	—	—	—	1.3	0.3	—
Hippophae thibetana	—	—	3.2	—	0.2	—

*May include some *Potentilla* shrub.

Livestock

Four domestic ungulate species have been imposed on the residents in the reserve: a total of about 1.5 million head consisting of 60% sheep, 30% goats, 8% yaks, and 2% horses (Miller and Schaller 1996). Sheep and goats are usually herded in mixed flocks, and their feces were analyzed together, even though their food habits differ somewhat. Cincotta et al. (1991) observed feeding animals and found that sheep selected more graminoids than did goats (62% versus 46%) and browsed less on shrubs (18.5% versus 31.5%). About two-thirds of our combined samples consisted of graminoids, principally *Stipa*, and the rest consisted of *Leontopodium*, legumes, and a few other forbs, as well as of available shrubs (table 12.15, fig. 12.8). In the Yeniugou, Harris and Miller (1995) found that sheep and goats also foraged mainly on graminoids (87%).

Domestic yaks were primarily grazers (91%; table 12.16), as also found by Cincotta et al. (1991). Horses subsisted almost wholly on grass, mainly *Stipa* (fig. 12.8).

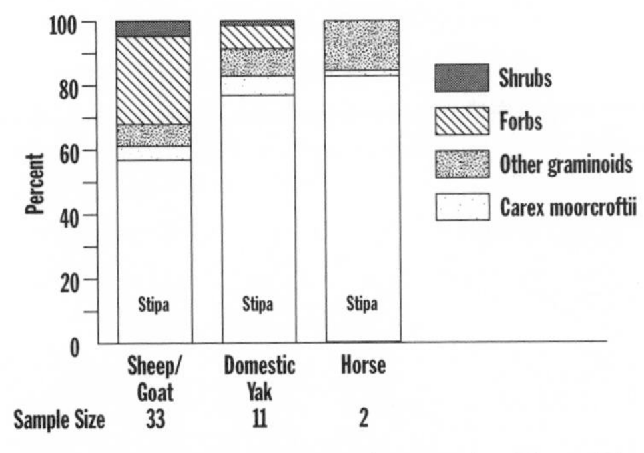

12.8. Mean annual food habits of domestic ungulates in the Chang Tang Reserve, based on fecal analysis.

Table 12.16 Mean percent relative density of plant fragments in feces of domestic yaks in the Chang Tang Reserve at various seasons

	Aru Basin 7/92	Alpine steppe (not Aru) 7–9/90–93	Shuanghu–Yibug Caka	
			10/93	12/91
Sample size	3	4	2	2
GRAMINOIDS				
Stipa sp.	50.0%	81.8%	89.7%	94.1%
Poa sp.	3.9	5.6	—	—
Elymus sibiricus	4.7	—	—	—
Kobresia sp.	3.2	10.3	1.7	1.0
Carex moorcroftii	16.9	2.1	4.4	1.8
FORBS				
Leontopodium sp.	9.8	0.7	—	—
Potentilla bifurca	11.5	0.7	—	—
Oxytropis and *Astragalus*	—	—	4.4	—
Other forbs	—	—	—	2.1
DWARF SHRUBS				
Ceratoides compacta	—	0.6	—	1.0
Hippophae thibetica	—	0.7	—	—

The domestic species appeared to forage on a smaller variety of forbs than the wild ones, but this possibly reflects availability in the vicinity of herders' camps. Sheep and goats had the largest proportions of forbs and shrubs in their diet, similar to the medium-sized argalis, blue sheep, and chirus. Domestic yaks were dietary analogues to wild yaks, and horses to kiangs.

Comparisons of summer and winter diets

As figures 12.6 and 12.7 indicate, several ungulate species show a conspicuous difference in the amount of forbs and *Stipa* they consume in summer and winter. Taking only the legumes among the forbs as example, Mann-Whitney U tests showed that chirus ($P = 0.027$) and domestic sheep and goats ($P = 0.011$) ate a significantly higher percentage of legumes in summer than in winter, and that argalis ($P = 0.05$) and domestic yaks ($P = 0.05$) ate significantly more in winter than in summer. Blue sheep and gazelles showed no significant differences, and no winter data for wild yaks could be collected. *Stipa* was eaten in significantly greater amounts ($P = 0.003–0.001$) during winter than summer by chirus, blue sheep, gazelles, domestic sheep and goats, domestic yaks, and kiangs; the argali sample was too small for a conclusion (fig. 12.9). Chirus consumed significantly more shrubs in summer than in winter.

Comparisons between species

Six wild ungulate species coexist in the Chang Tang, and a seventh, the white-lipped deer, occurs at the eastern margin. These species avoid serious competition and have done so historically, or they would not have persisted together, and the question is how they divide the limited resources. Species can separate spatially, including seasonal migrations, by selecting for topography, soil types, and other landscape features, and by eating certain food types, plant species, and plant parts. Most plant species have so wide a distribution and the altitudinal range from plains to the upper limit of vegetation is so narrow that all are available to animals, requiring only short movements upslope or downslope.

All ungulates ascend high into hills but there is a difference between those that are habitually in such topography and those that select it intermittently to forage and for other reasons. Blue sheep were usually within a few hundred meters of precipitous terrain, to which they retreated in times of danger, and this severely restricted their distribution (see Fox et al. 1988; Oli 1996). We usually encountered argalis in steeply rolling hills and on mountainsides, though they ventured into nearby gentle terrain too. Blue sheep and argalis occupied at times the same terrain, especially at midday, when both often rested high on ridges and scree slopes. Wild yaks, in particular the nursery herds, were usually in or near hills as well, but whether this reflected only habitat preference or also an attempt to avoid human persecution was not clear. Solitary bulls and bull herds were widespread, on plains as well as in hills; they often were in moist valleys and draws, where they foraged on small patches of green growth which could not support a nursery herd for long.

In contrast to these three species, the gazelles, chirus, and kiangs were basically plains animals. However, each also ascended hills, often, it seemed, in search of high-quality forage. Before the summer rains, when the plains were exceptionally dry, the hills offered forage of greater variety and nutrient content. Female Tibetan gazelles often moved into hills during June and there gave birth, as noted also by Harris and Miller (1995). The broken terrain provided not only better forage but also better cover in which young can hide. Male gazelles often remained on the plains or

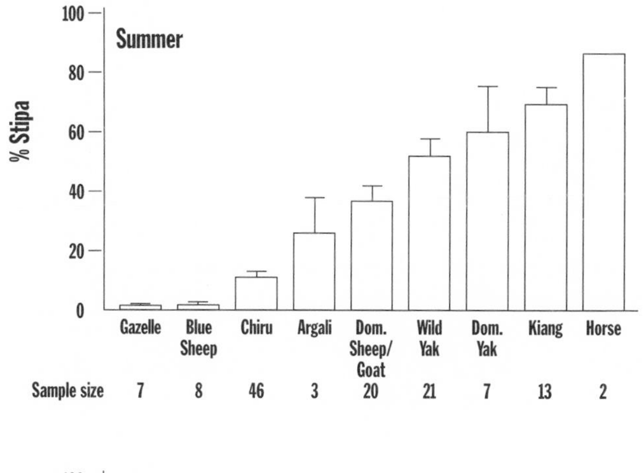

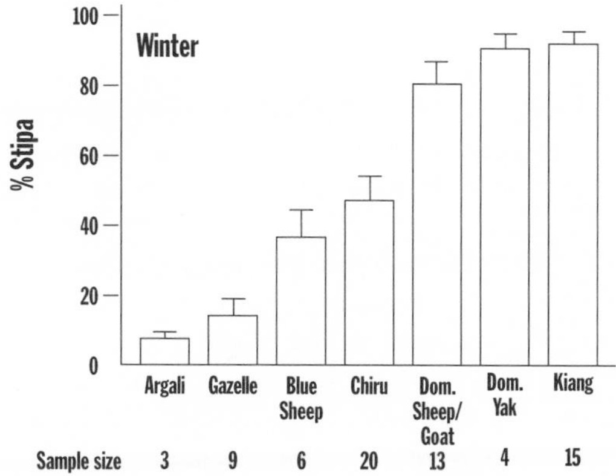

12.9. Mean percentages of *Stipa* grass in the summer and winter diets of ungulates in the Chang Tang Reserve. The bars represent the standard error. (Adapted from Miller and Schaller 1998.)

around the base of hills, but some were on the high crests above the females. Chirus segregated by sex for much of the summer, most females traveling through all types of terrain to and from the calving grounds. The males were scattered widely at that season, mostly in rolling hills. They moved higher as the season progressed until some were in rather barren uplands above 5100 m. Large male aggregations occurred in certain areas such as the Aru Basin. Herds clustered on alkali flats, on patches of *Aufeis*, and even on snowfields above the vegetation zone, apparently to escape warble and nasal botflies. Those that had gathered on alkali flats during the midday hours moved to lower slopes in late afternoon to forage. In the Yeniugou, where chirus were nonmigratory, males shifted high into hills in summer, whereas females were in the broad basins (Harris and Miller 1995). In late autumn both sexes congregated in basins and other gentle topography and there they later rutted. Small kiang herds were scattered in all types of terrain but especially among rolling hills during summer, and by October most had congregated on broad grassy plains for the winter. All ungulates seemed to avoid soils which during summer became waterlogged and muddy. Thus a complex mosaic of factors determined where and for what reasons an ungulate species selected particular terrain.

Plains and hills had plant types (graminoid, forb, shrub) in different proportions and this affected use by ungulates. *Stipa* covered much of the plains in alpine steppe. This grass and other graminoids represented over three-fourths of the vegetation composition, an ideal situation for grazers. On hills, graminoids contributed proportionately less to the composition, and forbs and shrubs more (figs. 12.1 and 12.2). Of the Chang Tang ungulates, the kiang was the most dependent on grass, especially on *Stipa*, the dominant grass in terms of composition and biomass; *Carex moorcroftii*, the second-most common graminoid, was eaten but not favored unless it had much green growth. This food preference affected kiang distribution in that the animals were rare in desert steppe and other areas with little grass, even if *Carex* was abundant. Wild yaks were also dependent on graminoids, but unlike kiangs, they readily ate senescent *Carex*, as well as various forbs. Being large, these two ungulates were able to take advantage of graminoids, an abundant food supply, but they also had large food requirements that required them to move continually to sites with ample forage.

Medium-sized ungulates were generally mixed feeders obtaining a nutritional balance by selecting a variety of food types, the amount of each depending on locality and season. Blue sheep, argalis, and chirus foraged on graminoids, forbs, and shrubs. All three species are members of the subfamily Caprinae. It is tempting to speculate that they have adapted to different habitats—steep terrain, hills, plains—to minimize competition for resources. The white-lipped deer, another mixed feeder, was only marginally associated with the other three species.

The small Tibetan gazelle was a browser, although it also ate a little grass, especially in winter, when forbs were scarce. High-quality browse was a scattered and uncommon resource. In desert steppe and in arid parts of alpine steppe it apparently became a limiting resource because the gazelles were generally not found there. As Hedin noted, "the region was perfectly barren except for a few small patches of hard, sharp yellow blades, one or two inches high; this was called grass . . . hard as whalebone, and pierced like a needle through even thick clothes. Yet in that inhospitable country this was the only fodder available" (1903).

One measure of the ability with which ungulates can select small food items, whether green grass leaves among dead ones or parts of a forb, is the width of the incisor tooth row. I measured a sample: male Tibetan gazelles, 1.9–2.3 cm; female blue sheep, 3.2–3.4 cm; female chirus, 3.2–3.6, and males, 4.0 cm; female and male kiangs, 6.7–7.3 cm; a female wild yak, 7.7 cm, and males, 9.0–11.0 cm. No measurements for Tibetan argalis were obtained but a Gobi argali female had a width of 3.1 cm and males had widths of 4.6–4.7 cm. The broad incisors of kiangs and yaks are obviously well suited for mowing grass.

Kiangs in particular have to consume much bulk because they digest food poorly, as shown by their feces, which contain many grass stalks and other visible plant parts. Plant particle size in feces of ruminants is correlated with body weight (E. Dierenfeld, pers. comm.). For example, average fecal particle size of Tibetan gazelles was 121 microns, chirus 137, and wild yaks 181. Kiangs are smaller than yaks but their particle size was 400 microns.

Selection for or avoidance of certain plant species, rather than plant types, was difficult to ascertain. With plant distribution patchy and our vegetation transects few in number, statistics that compare availability and amounts in feces may provide spurious conclusions. Furthermore, selection is influenced by the nutrient content of the plant and it may vary greatly within a short distance. At times a herd selected a particular plant for unknown reasons. Kiangs in one area dug up and ate *Oxytropis stracheyana*, and chirus in the Aru Basin fed on senescent *Carex* at one site even though ample green forage was available nearby.

All ungulates probably consumed every available graminoid species to some extent. Kiangs clearly selected for *Stipa*, and gazelles selected against it as well as against most other graminoids. Among forbs such genera as *Rhodiola*, *Draba*, *Thermopsis*, *Adonis*, and *Sibbaldia* were local in distribution and their presence in feces merely indicated that they are palatable. *Youn-*

gia, *Ranunculus*, and some others were also eaten but were not detected in the feces. Among the shrubs, only *Ceratoides* was widely available but was most common at high elevations with unsettled soils, places where most ungulates spent little time.

However, there were a few small and widely available forbs that can be used as indicators of preference because foraging animals were likely to encounter them often. *Saussurea stoliczkai* was common in *Stipa* grassland and less so on hills, where other *Saussurea* species occurred. *Saussurea* generally constituted 0.2–1.7% (and in one instance 3.1%) of the vegetation in transects in which this genus was recorded. Fecal samples of gazelles, chirus, and kiangs showed densities of 0.1–0.7%, indicating no selection for or against the plant by these ungulates. *Leontopodium pusillum* was a common, mat-forming plant throughout the alpine steppe. Most transects had *L. pusillum* at 0.7–8.3% of the composition and once at 14.9%. In the Aru Basin, the percentage of *L. pusillum* was 3.4–7.0% of the composition but 34.9% of the fecal density in blue sheep, 11.9% in chirus, and 11.1% in wild yaks. In summer, the blue sheep seemed to select for that species. Argalis near Garco showed 27.7% *Leontopodium* in their feces, again a seeming preference. Other ungulates had fecal percentages of 0.2–11.2, depending on area, eating the plant without marked selectivity.

Potentilla bifurca is an inconspicuous, widespread forb, and *Potentilla* (probably *P. parvifolia*) is a common shrub about 15–25 cm tall. Fecal analysis did not distinguish between the species. In the Aru Basin, where the shrub was scarce (it did not appear in any transect), *P. bifurca* was 0.3–4.3% of the composition. The density of *Potentilla* fragments in feces of chirus was 7.7%; wild yaks, 14.6%; blue sheep, 24.3%; and gazelles, 40.9%. This forb was obviously much favored in summer. Outside the Aru Basin, 0.6–7.6% of the plant composition was *P. bifurca* and 0.8–14.1% was *Potentilla* shrub. High densities of *Potentilla* were found in several samples, up to 22.3% in argalis, 31.1% in chirus, 60.9% in blue sheep, and 94.2% in gazelles. The gazelle samples were from an area where *Potentilla* shrub was scarce. The importance of the shrub to wild ungulates is unknown, but livestock commonly browsed on it (Cincotta et al. 1991).

Legumes are among the dominant forbs in alpine steppe and all ungulates eat them. Except for *Astragalus heydei*, fecal analysis did not distinguish species. *A. heydei* occurred sparsely in the Aru Basin, 0.3–0.4%, yet fecal density in chirus was 2.7% and in gazelles 12.2%, suggesting selection at least by the latter. Taking all legumes together, they usually composed 0.9–5.8% of the vegetation, with as much as 18–20% on a few hills. In the Aru Basin, chirus had 31.4% legumes in their feces, and gazelles 51.8%, the highest for any samples, a strong indication of selection. Indeed, chirus were observed as they sought *Oxytropis glacialis* and *O. pauciflora*, moving from plant to plant.

How do these ungulates partition their resources? Different studies have derived different conclusions based on particular herbivore communities. In the Virunga Volcanoes of Rwanda there was little dietary overlap between elephant, buffalo, mountain gorilla, bushbuck, and black-fronted duiker (Plumptre 1995). The six large ungulate species in Uganda's Ruwenzori National Park overlapped least in their diets during the dry season, when nutritious forage was most scarce (Field 1972). In the Indian Himalaya, Green (1986) found that the four ungulate species were separated largely by habitat and food habits: muskdeer and serows being browsers; gorals, grazers; and sambars, mixed feeders. In the Serengeti, spatial and temporal use of habitat and emphasis on certain plant types helped reduce competition, but selection for plant parts was of major significance (Jarman and Sinclair 1979). However, in these areas there is vertical stratification of the vegetation from grass to trees.

Even a broad comparison of these studies with our work in the Chang Tang indicates that resources are used somewhat differently. The wild ungulates overlapped greatly, in fact almost completely, in the plant species they ate. Five graminoid genera provided most of the forage and all ungulates consumed these, gazelles doing so sparingly. Similarly, of the palatable forbs and shrubs, all were probably cropped to some extent by the various ungulates. The animals differed considerably in the proportion of plant types they consumed, not in the species they chose. Selection for plant parts could not be studied. Since species and plant types were similar on hills and plains, animals tended to select not for habitat per se but for topography and nutrient content.

Food was obviously abundant on alpine steppe. At the end of the growing season, *Stipa* seed stalks rippled in the breeze, indicating light grazing pressure. By then the nutrients were stored underground, and graminoids could tolerate heavy grazing. The following summer much of this old growth remained standing, and less than 10% of the leaves had been cropped (table 12.5). Forbs also remained easily available, especially in hills. I hypothesize that resource partitioning was based mainly on two factors: spatial differences in using terrain and body size, with all that it implies nutritionally. Species of different size can coexist on the same food supply (Sinclair 1983), but if species are of roughly the same size, as are blue sheep, argalis, and chirus, they then differ to some extent spatially.

Is there actual competition for resources? An answer has to remain elusive in part because wildlife populations are now well below what they were a hundred years ago. That food can be severely limiting is shown by the sparse populations of all species in desert steppe. As East (1984) has noted for African savannahs, a positive correlation exists between herbivore biomass and rainfall. There may, in addition, be competition in local situations. The heavy persistent grazing by yaks on some *Kobresia* swards probably makes these less inviting to other species. Conversely, yaks and kiangs probably facilitate grazing on *Stipa* and *Carex* when they crop the senescent material, making green growth accessible to more selective grazers such as chirus. Competition is most likely in winter, when high-quality forage is in shortest supply. Such competition probably led to an ungulate community with differences in body size and use of terrain. What we investigate today is the product of evolutionary history, the coexistence of species in an ungulate community within which any competitive exclusion occurred long ago.

However, four domestic ungulates were added to the ecosystem. They eat the same plants in roughly similar proportions as their native counterparts: domestic yaks as wild yaks, horses as kiangs, sheep and goats as the wild mixed feeders. In the Chang Tang, livestock numbers have not as yet increased to a level where forage shortage has led to obvious competition. Once this occurs, it is unlikely that a natural evolutionary process will be allowed to eliminate those that are less well able to adapt.

Water requirements

Drinking water may be critical to ungulates in arid environments in that it can restrict their movements to the vicinity of streams, springs, and other such sources. Fresh water is seasonally scarce in the Chang Tang. Most lakes and streams are brackish or saline. Traveling cross-country in desert steppe, we sometimes found no drinking water for 100 km or more. Occasionally a puddle with seepage water occurred in a dry streambed, and kiangs and yaks visited such sites to drink. Blue sheep sometimes descended to a rivulet, and chirus may drink at a stream, though they usually waded across without halting. The ungulates made no conspicuous treks to water. It was my impression that they obtained most of their moisture from eating vegetation and snow, but that they drank occasionally. In winter all potable water is frozen, and chirus have then been observed to lick the surface of ice (Bonvalot 1892). Domestic sheep and goats need to drink at least once every two days, according to the nomads.

Two species used water for purposes other than drinking. Chirus sometimes crowded knee-deep into a pond, apparently to escape the attentions of oestrid flies. Wild yaks, with their thick black coats, seemed to cool themselves at times by standing belly-deep in water. Black hair absorbs

Table 12.17 Summer food habits of Tibetan ungulates compared with their ecological counterparts in Mongolia, based on mean percent relative density of plant fragments in feces

	Sample size	Graminoid	Forb	Shrub	Seed
Tibetan gazelle	7	1.7	83.1	14.3	0.9
Goitered gazelle	9	11.8	38.0	50.2	—
Chiru	46	40.4	35.6	22.8	1.2
Mongolian gazelle	10	23.2	40.0	36.8	—
Tibetan argali	3	83.5	14.3	2.2	—
Gobi argali	8	16.2	21.2	62.5	—
Kiang	13	89.9	4.1	5.6	0.4
Kulan	7	78.6	7.7	13.7	—

Note: The Mongolian samples were collected in Aug. 1994.

considerably more heat from solar radiation than does light-colored hair (Grenot 1991).

Comparisons with Mongolian ungulates

The Chang Tang's ungulates have their ecological equivalents in the semideserts and steppes of the Gobi in eastern Mongolia, except for wild yaks, which vanished from that fauna after the Pleistocene (Guthrie 1990). There are ibex instead of blue sheep, goitered instead of Tibetan gazelles, migratory Mongolian gazelles instead of chirus, Gobi instead of Tibetan argalis, and kulans instead of kiangs. The animals are similar in size, they occupy similar terrain, and the question obtrudes if they evolved similar food habits as well. A main difference in habitats, other than elevation, is that the semidesert has a much greater variety of shrubs than the Chang Tang has. In August 1994 I collected fecal samples of all species but ibex.

Goitered gazelles, like Tibetan gazelles, ate little grass (table 12.17), and instead browsed much on shrubs (50%), particularly *Artemisia* sp., *Anabasis brevifolia*, *Ceratoides* sp., and *Caryopteris mongolica*, and on forbs (38%) such as *Agrophyllium arenarium*, *Kochia sieversiana*, *Tribulus terrestris*, and legumes. The preferred habitat of these gazelles is semidesert, and indeed they occur in similar conditions in parts of the Tibetan Plateau, where they are sympatric with Tibetan gazelles.

Mongolian gazelles inhabit steppes which resemble those of the Chang Tang, yet, unlike chirus, they consumed only modest amounts of grass (23%) even when it was green; *Cleistogenes songarica* and *Stipa* sp. were the principal grasses in the diet. One forb, *Astragalus junatovii*, was a main forage species (39%) and the sages *Ajania* and *Artemisia* were also important (37%). The gazelles had a limited diet, even though many plant species were available, in contrast to the eclectic foraging of chirus. Five winter samples from Mongolian gazelles showed 3% graminoids and 95% sage, again a quite different selection of plant types than that by chirus. The fact that both species migrate in huge aggregations is not reflected in the number of available species they select. In this context it is of note that Mongolian, goitered, and Tibetan gazelles have first incisors that are disproportionately wide, designed for plucking discrete food items, whereas chirus have incisors more equal in size, suited for less selective foraging.

Sage, an aromatic shrub, was a major food item of Mongolian gazelles, and it was also eaten locally by Tibetan gazelles and chirus. Saigas in Kazakhstan browse much on sage, especially in winter (Bannikov et al. 1967), as do pronghorns on the American plains (Einarsen 1948).

Gobi argalis grazed little (16%), mainly on *Cleistogenes* and *Stipa*, and they browsed on a few forbs with emphasis on *Astragalus junatovii* (13%). They fed principally on shrubs such as *Carypteris mongolica* (43%) and sage (17%), and also on *Anabasis brevifolia*, *Ceratoides* sp., *Amygdalis mongolica*, *Caragana leucophloea*, and *Zygophyllum xanthoxylon*. Tibetan argalis, by contrast, had few shrubs available and they grazed much and browsed little.

The kulan, like the kiang, was mainly a grazer, its forage consisting of *Aristida neumanni* (25%), *Stipa* sp. (20%), *Cleistogenes* (16%), and *Eragrostis minor* (14%), among others. Forbs and shrubs were insignificant components of the diet, except for *Ceratoides* (11%).

In sum, shrubs provide the Mongolian ungulates with an important, varied, and probably nutritious food source which the Chang Tang ungulates almost lack. Mongolian gazelles and chirus seem to differ in the variety of plants and probably the amount of grass they eat.

13

Phylogeny of Tibetan Steppe Bovids

Morphological and Molecular Comparisons

WITH GEORGE AMATO

Genus *Pantholops* Hodgson . . . The classification of this genus presents considerable difficulty.

G. E. Pilgrim (1939)

The most beautiful thing we can experience is the mysterious. It is the source of all true art and science.

Albert Einstein (1879–1955)

THE GOAL OF our research in the Chang Tang was conservation of the ecosystem, particularly of its unique large-mammal community. Our initial focus was the natural history of ungulates, as shown in previous chapters. Molecular biology has become an integral part of conservation, basic to understanding evolutionary processes, and it has been applied to several important problems.

One such problem is to measure the amount of inbreeding in small populations. Inbreeding reduces genetic variability, which in turn may decrease viability of individuals and adaptability of populations (Soulé and Wilcox 1980). Population genetics was not an urgent component of our project, except for concern over the small populations of argalis and wild camels, and no research was done.

Molecular biology can also help to clarify systematics at the species and subspecies level. Based on morphological criteria, some of dubious validity, various taxonomic units have been proposed. Are there more than one subspecies of argali and kiang on the Tibetan Plateau? Are Tibetan and Przewalski's gazelles conspecific or not? Deciding what is an evolutionarily significant degree of genetic variation below the species level, rather than a minor variation between populations, remains a problem that has yet to be resolved by molecular or morphological methods (Amato 1994). However, the traditional concept of subspecies retains importance in conservation. Subspecies and species are sometimes listed uncritically in compendia of endangered organisms, with no effort made to establish conservation priorities. If the taxonomy is at fault, much money and effort may be expended on saving a trivial variety at the expense of a highly distinctive organism. Most phenotypic variation at the subspecific level is small, based on minor differences in skeletal measurements, coat color, horn shape, or other character. Such variation is often encoded in polymorphic nuclear genes, which may be under considerable environmental selection pressures. Other characters may result from phenotypic effects in certain environments. Our work involved mitochondrial ribosomal DNA (mt-rDNA), a conserved genome that is less rapidly affected by evolutionary processes (Ballard and Kreitman 1995). The DNA sequences we selected for analysis failed to resolve taxonomic issues below the species level, except to note a minor degree of genetic distinctiveness between wild and domestic camels. Other more quickly evolving mitochondrial genes or nuclear markers would be appropriate for population genetics research.

One issue particularly interested us: the phylogenetic position of Tibetan Plateau bovids, their relationships, and their adaptive radiations. The chiru's morphology points to two subfamilies, the Antilopinae and the Caprinae, and some have presumed the closest relative to be the saiga. What is the chiru's ancestry? Also, does the gazelle genus *Procapra* warrant separation from *Gazella*, and should the aberrant Mongolian gazelle remain in the same genus, *Procapra*, with the Tibetan gazelle? Is the blue sheep more closely allied to sheep or goats? Answers to such questions are not mere exercises in curiosity and scientific tidiness but form a part of any conservation priorities that may be established. Should certain species have priority in conservation?

With more and more species approaching oblivion, conservation will increasingly have to consider priorities. Many factors, from aesthetics to economics, determine at present whether an effort is made on behalf of a species. Because saving genetic diversity is a basic conservation goal, the genetic distinctiveness of a species is a parameter that might also be considered when dealing with endangered organisms. One could argue that the most genetically distinct species within a taxon, such as the family Bovidae, the ones containing the most evolutionary novelty, and those without closely related forms should be given a measure of priority in conservation over those that have radiated into a cluster of genetically similar species. If this idea has merit, it demands phylogenetic trees and cladograms that accurately reflect relationships (Vane-Wright et al. 1991). As noted earlier, this project involved species whose systematic position remained ambiguous. By discussing these in the context of the whole bovid family, we hope to provide insight into the degree of their genetic uniqueness.

A species can be defined as one or more populations of individuals that exhibit reproductive cohesion, sharing diagnostic heritable characters in their karyotypes, genetics, morphology, and behavior. Systematics has until recent years been based mainly on anatomical characters, but the advent of genetic analyses, particularly of the polymerase chain reaction (PCR) technology, has made it possible to sequence particular genetic sites and provide additional characters for diagnosing species. Certain morphological traits have evolved several times in bovids to create convergence that obscures true relationships (Gentry 1992). Molecular data may for various reasons also be flawed (Novacek, Wyss, and McKenna 1988). But the combination of morphological and molecular data does add a new level of confidence to the results.

The family Bovidae is difficult to classify in part because it appears to represent a rapid, early radiation into many forms with only one obvious link—bony horn cores covered with horn sheaths (Gatesy et al. 1992). Based on morphological characters, Simpson (1945) classified the family into a number of subfamilies and tribes, and these were later amended, most recently by Gentry (1992) and Gatesy et al. (1992, 1997). A number of genera within the tribes remain monotypic, and this reflects some uncertainty in relationships. For example, Gentry (1992) placed the chiru into an indeterminate tribe. The following subfamilies and tribes are here provisionally accepted:

Subfamily Bovinae	
Tribe Boselaphini	Nilgai, four-horned antelope
Tribe Tragelaphini	Nyala, bushbuck, eland, and others
Tribe Bovini	Yak, bison, buffalo, wild cattle
Subfamily Cephalophinae	
Tribe Cephalophini	Duikers
Subfamily Hippotraginae	
Tribe Hippotragini	Roan, sable, oryx, and others
Tribe Alcelaphini	Wildebeest, hartebeest, topi, and others
Subfamily Reduncinae	
Tribe Reduncini	Reedbuck, kob, lechwe, and others
Tribe Peleini	Rhebok
Subfamily Aepycerotinae	
Tribe Aepycerotini	Impala
Subfamily Antilopinae	
Tribe Neotragini	Dik-dik, steenbok, oribi, and others
Tribe Antilopini	Gazelle, saiga, blackbuck, dibatag
Subfamily Caprinae	
Tribe Rupicaprini	Goral, serow, chamois, mountain goat
Tribe Ovibovini	Takin, musk ox
Tribe Caprini	Sheep, goat, blue sheep, aoudad
Tribe Pantholopini	Chiru

This chapter discusses mainly morphological and mt-rDNA differences between Tibetan Plateau bovids and related bovids to assess phylogenetic relationships. (Chapter 14 concentrates on behavioral comparisons.) Morphological data are from published sources. The molecular data are based on samples collected during this project (table 13.1) and on other material obtained by George Amato, Elisabeth Vrba, and John Gatesy (see Gatesy et al. 1997).

Methods

We reconstructed the phylogenetic relationships of selected bovid taxa by the cladistic method, the identification of shared features (Nelson and Platnick 1981). This method depends on two assumptions. One is that groups are linked by shared, derived characters. These characters are determined by outgroup analysis, which compares characters for taxa that are related but are outside the group of interest. The second assumption is that the phylogenetic tree should be constructed by using the minimum number of steps required to unite the taxa with these derived characters.

Several factors complicate cladistic analysis. One problem is that the use of DNA sequences as characters provides many data for analysis. To compare a particular base or site between individuals requires that the characters are at the same site, that they are homologous. To assess homology between species is difficult because DNA insertions and deletions accumulate over evolutionary time as taxa diverge. Such insertions and deletions result in gaps when an attempt is made to align the sequences. Powerful mathematical algorithms have been developed to aid in aligning multiple sequences that may have a number of equally likely alignments. A second problem in reconstructing phylogenies with cladistics is the weighting of specific characters, of assessing their importance. A knowledge of certain DNA characters, derived from previous molecular work, may enable a researcher to give additional importance to some characters. It is also possible to explore the data for robustness by observing changes in patterns when different weighting assumptions are used. In our previous work (Gatesy et al. 1997), we identified useful weighting schemes for mt-DNA sequences in bovids by comparing 14 different analyses. Yet another problem in constructing cladograms is that when more than a few taxa are examined, the number of possible trees that need to be considered before the simplest or most parsimonious tree can be identified is enormous. Complex algorithms are available to manage this problem, and these provide the best estimates by doing heuristic, rather than exhaustive, searches for possible trees.

We employed the following techniques to construct cladograms. Skin

Table 13.1 Collecting locality and number of samples of central Asian bovids used in molecular analyses

Species	Locality	Sample size
Wild camel	Great Gobi National Park, Mongolia	2
Domestic camel	Eastern Gobi, Mongolia	4
Wild yak	Chang Tang Reserve, Tibet	2
Domestic yak	Lhasa, Tibet	4
Tibetan gazelle	Chang Tang Reserve, Tibet	1
Mongolian gazelle	Eastern steppe, Mongolia	1
Goitered gazelle	Eastern Gobi, Mongolia	1
Saiga	Sharghyn Gobi, Mongolia	1
Chiru	Chang Tang Reserve, Tibet	3
Tibetan argali	Chang Tang Reserve, Tibet	2
Gobi argali	Eastern Gobi, Mongolia	1
Domestic goat	Altay Mountains, Mongolia (1); zoo (1); Genbank (1)	3
Asiatic ibex	Altay Mountains, Mongolia (2); zoo (1)	3
Blue sheep	Chang Tang Reserve, Tibet	2

and muscle samples were collected from animals found dead in the field (table 13.1) and from zoological gardens and museums; published DNA sequences (Genbank) were also included. Total genomic DNA was isolated from blood samples by using standard phenol/chloroform procedures (Caccone, Amato, and Powell 1987), and from skin and other tissues by using a chelating resin (Walsh, Metzger, and Higuchi 1991). Fragments of two ribosomal subunits, the 12S and 16S mitochondrial genes, were PCR amplified and sequenced for about 639–892 bases, or nucleotide positions, with universal vertebrate primers. Domestic goats and ibex from widely different parts of their range showed no variation in the mt-rDNA sequences, demonstrating the usefulness of these genes in phylogenetic studies. PCR products were sequenced in both directions on an ABI Automated Sequencer. To align sequences objectively requires the elimination of some ambiguous nucleotide positions and the weighting of others. We used the multiple-sequence program MALIGN 2.1 (Wheeler and Gladstein 1994) for sequence alignment and exploration of various weighting strategies (Gatesy et al. 1997). Maximum-parsimony cladograms were constructed based on searches with the PAUP 3.1.1 program (Swofford 1993; Gatesy, DeSalle, and Wheeler 1993; Gatesy et al. 1997). These complicated procedures of replicating, ordering, weighting, and combining molecular character sets, produced figure 13.2, and with the addition of Gentry's (1992) morphological data (fig. 13.1) they produced figure 13.3. In addition, a subset of the bovid taxa was analyzed separately for several central Asian species of Antilopinae (using Maxwell's duiker as an outgroup) and of Caprinae (using the gemsbok as an outgroup) to provide a strict-consensus tree of the most probable relationships of these species

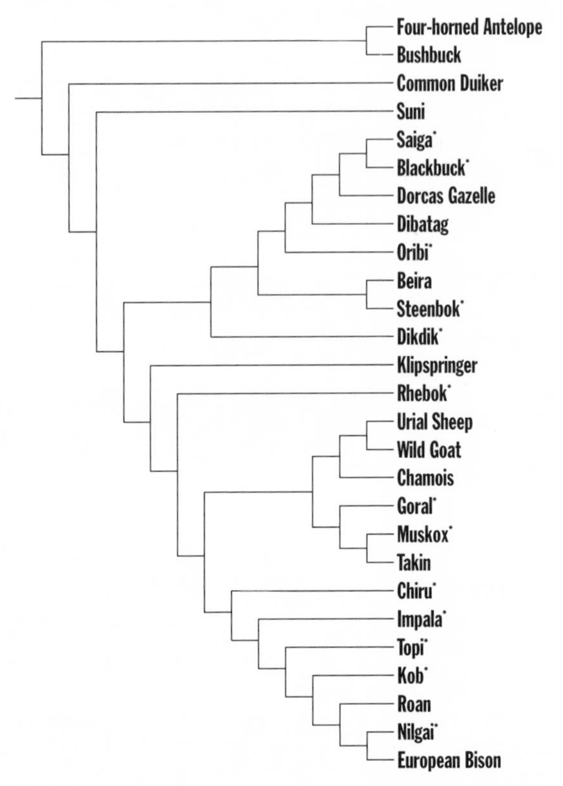

13.1. A cladogram based on skeletal measurements of 27 species of Bovidae (adapted from Gentry 1992). The asterisks indicate species that are also shown in figs. 13.2 and 13.3.

(fig. 13.4). These analyses followed the same alignment parameters and weighting scheme as used by Gatesy et al. (1997).

The 12S and 16S ribosomal mitochondrial genes reflect genetic differences in evolutionary lineages at the appropriate level for our studies in that they tend to be constant within a species but variable even between closely related species (Amato et al. 1995). They are, therefore, ideally suited for indicating significant levels of genetic divergence between species, but they are less sensitive in revealing differences at the subspecific level, as revealed by the analyses below.

Yaks and Bactrian camels, wild and domestic

The scientific nomenclature for domestic animals and their wild progenitors remains unsettled. Sometimes the same subspecific or specific names

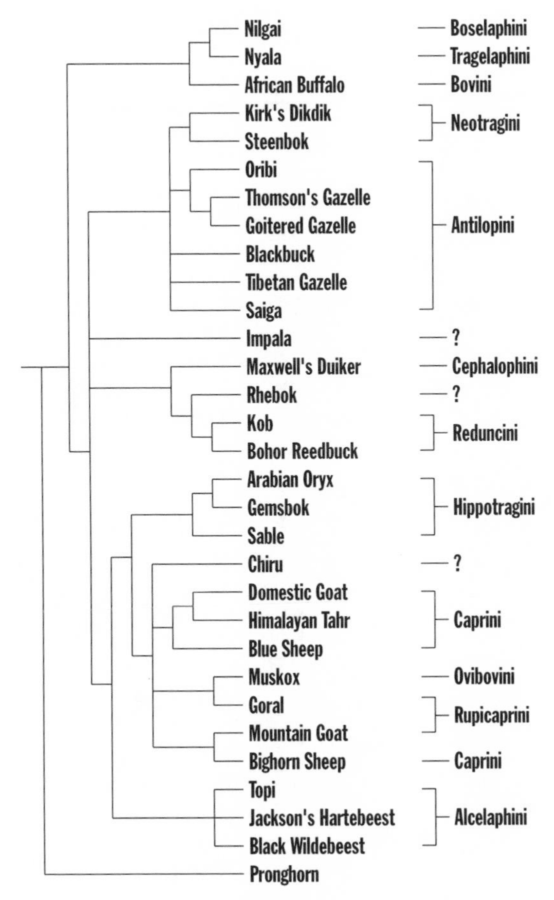

13.2. A cladogram based on mt-rDNA analysis of 31 species of Bovidae. (Adapted from Gatesy et al. 1997, whose cladogram included 57 taxa.)

are used for both; at other times different names are applied, or only a vernacular name is given to domestic animals. Whatever the preference, domestic and wild forms tend to differ in appearance (see chapters 7 and 9).

Domestic and wild yaks had identical mitochondrial haplotypes in the gene fragments that were tested. The samples from the domestic animals were obtained in Lhasa, far from any wild population, and it seems unlikely that the animals have had contact with wild ones for many decades.

Domestic and wild camels varied by three base substitutions (0.005%) in the 16S fragment. Too few animals were analyzed to treat any conclusion with confidence. However, the animals were genetically differentiated at a level large enough that they could qualify as subspecies.

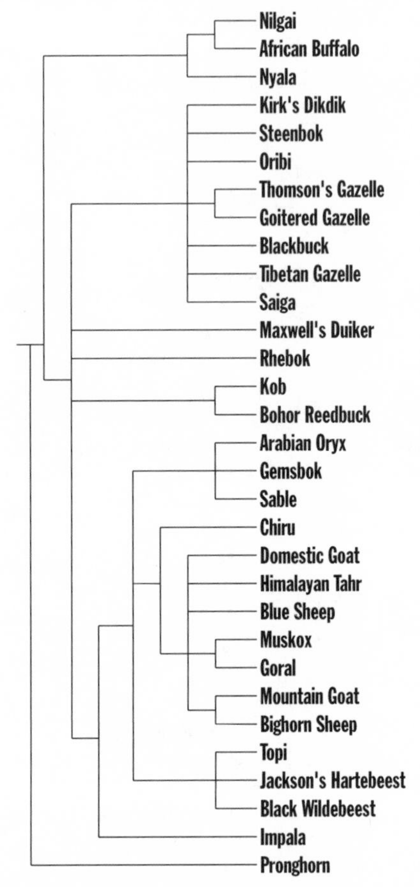

13.3. A cladogram based on a combination of Gentry's (1992) skeletal characters and the mt-rDNA data. (Adapted from Gatesy et al. 1997.)

The difference in the results between yaks and camels may represent different histories in levels of contact between wild and captive populations.

Argalis

The taxonomy of argalis has been a contentious issue (see chapter 4). We tested two subspecies whose morphological differences are minor, the Tibetan argali (*O. a. hodgsoni*) and the Gobi argali (*O. a. darwini*). The two had identical haplotypes in the gene fragments. These data argue against

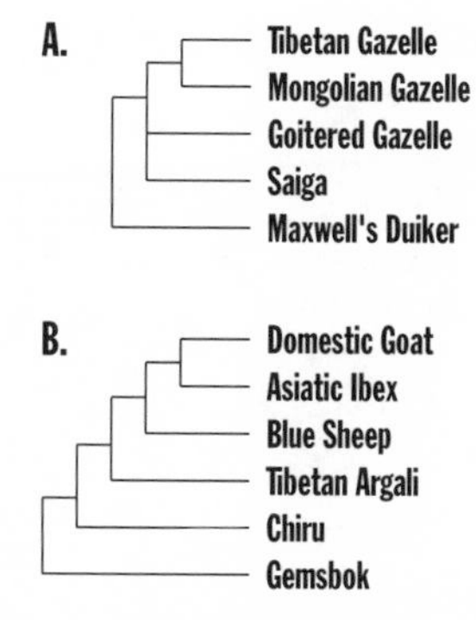

13.4. Strict-consensus trees based on mt-rDNA analyses using the data of Gatesy et al. (1997) and partitioned analysis of several species of (A) Antilopini and (B) Caprini.

an evolutionarily significant amount of time of separation between these populations.

Chiru and saiga

Pantholops and *Saiga* have been taxonomically bound to each other, with Simpson (1945) placing both in the tribe Saigini, subfamily Caprinae, and von Roy (1958) placing them in the tribe Antilopini, subfamily Antilopinae (compare figs. 3.6 and 13.5). However, the two species had long refused to be categorized with the sheep and goats, with the antelopes, or with any other bovid group. Sterndale considered them antelopes but "connected by cranial affinities with the sheep" (1884). Pocock (1910) separated the two, each species into its own subfamily, and Bannikov et al. (1967) gave the saiga a subfamily status and placed the chiru with the Caprinae. Two questions need to be answered: Are chirus and saigas related, and how should they be classified?

Pilgrim (1939) listed the skull characters by which chirus and saigas resemble each other, including a small bulla, broad face, large nasal aperture, and backward position of the fronto-parietal suture. The females of both species lack horns. The two species have air sacs in an enlarged muzzle, but their structure differs: the saiga's air sacs are created by a greatly enlarged nasal passage, and the chiru's consist of a small bulge on each side of the nostril. Air sacs evolved independently several times, being also found, for example, in Speke's gazelles. In chirus, "the sinuses within the frontals, while like the Caprini, link it also with *Saiga* and *Antidorcas*," and several palatal and tooth characters also are similar in chirus and saigas (Gentry 1992).

13.5. A Mongolian saiga antelope male.

Both saigas and chirus have morphological characters that link each to the Antilopini and Caprini. The saiga's complicated interfrontal suture and high position for the lateral tubercle on the radius are similar to those in the Antilopini, whereas the wide anterior tuberosities of the basioccipital and several other characters resemble those of Caprini (Gentry 1992). The chiru has a mix of traits too, sharing, for example, many limb bone features with the Antilopini and long horn cores with the Caprini, as well as with impalas and sables. "The medio-lateral compression of the horn cores and the linked presence of keels are like Caprini other than *Ammotragus*, but like only *Gazella* and *Litocranius* among Antilopini. The wide orbital rims are like *Saiga* and several Caprinae" (Gentry 1992). Saigas and chirus differ considerably in their skin glands (table 14.2).

This morass of conflicting characters emphasizes that a sharp line between the Antilopinae and Caprinae is difficult to draw (Simpson 1945). Pilgrim concluded that saigas and chirus "are much closer to the Caprinae than to the Gazellinae" (1939). Gentry (1992) moved the saiga to the Antilopini and the chiru to the Caprinae, although his cladogram (fig. 13.1) actually shows the chiru as having its closest affinities to the Hippotraginae and impala.

The analysis of mt-rDNA clearly separates saigas and chirus (fig. 13.2). In agreement with Gentry's (1992) morphological analysis, the saiga is

Table 13.2 Percent sequence divergence for 463 bases of unambiguously aligned 12S and 16S mitochondrial ribosomal gene segments

Antilopini				
	Saiga	Tibetan gazelle	Mongolian gazelle	Goitered gazelle
Saiga	—	4.53	4.96	4.10
Tibetan gazelle		—	4.75	4.96
Mongolian gazelle			—	4.96
Goitered gazelle				—

Caprini					
	Chiru	Tibetan argali	Blue sheep	Asiatic ibex	Domestic goat
Chiru	—	5.18	5.61	6.47	6.04
Tibetan argali		—	4.96	4.53	4.53
Blue sheep			—	4.31	4.31
Asiatic ibex				—	3.02
Domestic goat					—

associated with the Antilopinae, whereas the chiru clusters with the Caprinae. The percent sequence divergence between saiga and chiru is 10.03. Features linking the two species are, therefore, either primitive or the result of convergence. The saiga is further discussed below with the gazelles.

The ambiguous position of the chiru in Gentry's (1992) results was resolved when Gatesy et al. (1997) combined morphological and molecular characters (fig. 13.3). The cladogram shows the chiru as a sister group or basal member of the Caprinae. The chiru lacks some of the morphological specializations of the subfamily, and in certain features, such as the limb structure, resembles the Antilopinae. Unlike other Caprinae, chiru females lack horns (table 14.2), and body size and horn length of adult males do not increase with age. But such divergent characters may reflect the chiru's position as a basal member of the subfamily.

In Gatesy et al.'s (1997) analysis, the chiru is equally distant (but unresolved) from the other Caprinae. Our separate analysis revealed the chiru as a sister taxon of the Caprini (fig. 13.4). As table 13.2 shows, the percent sequence divergence links the chiru with that tribe.

Blue sheep

Schaller wrote that blue sheep "show a bewildering combination of sheep- and goat-like traits. They lack beards and calluses on the knees, they have no strong body odor, and the females have small, almost nonfunctional horns, all characters typical of sheep. On the other hand they resemble goats in their flat broad tail with a bare ventral surface, the conspicuous markings on the forelegs, and the large dew claws. The horns are shaped like those of Dagestan tur" (1977b). As Lydekker pointed out, "the structure and colour of the horns are the same as in goats" (1898), and the shape of the basioccipital bone also resembles that of goats. He concluded that the blue sheep is intermediate between sheep and goats, in contrast to Pocock (1910) who viewed it as an aberrant goat; other investigators placed it closest to sheep (see Wang and Hoffmann 1987). Karyotypic and chromosomal analyses indicated that ancestral caprine stock led first to Capra ($2n = 60$) and blue sheep ($2n = 54$) and then to other genera of Caprini (Bunch, Nadler, and Simmons 1978; Bunch and Nadler 1980).

Molecular analysis shows that ibex and domestic goat are sister taxa and that the tahrs are also closely related to them (fig. 13.2 and 13.4). The blue sheep is just outside this species cluster, but more closely related to the goats than to the argali and other *Ovis* (fig. 13.4 and table 13.2).

Gazelles

Five gazelle species, including the saiga, occur in central Asia. Three of these species are in the genus *Procapra* and one in *Gazella*. The skull of *Procapra* can be distinguished from *Gazella* by its lack of preorbital depressions and by its long nasal bones that are pointed at the end (Groves 1967). Glands in *Procapra* (excluding Mongolian gazelles) are either absent or rudimentary, in contrast to goitered gazelles and other *Gazella*, which have prominent preorbital, pedal, inguinal, and carpal glands (table 14.2). *Procapra* is said to have a postcornual gland behind each horn (Pocock 1918).

Among *Procapra*, Przewalski's gazelles were considered to be a subspecies of Tibetan gazelles by Allen (1940), but as Groves (1967) has shown, they are valid species with distinctive horn shape and skull characters. (The mt-rDNA of Przewalski's gazelles has not been analyzed as yet.)

The Mongolian gazelle is an ambiguous *Procapra*, so much so that it has been at times placed into its own genus, *Prodorcas* (Pocock 1910; Allen 1940). It resembles the other two *Procapra* species in being a buff-colored gazelle without facial and lateral stripes, and in having a large, white rump patch surrounding a short (10–13 cm), black-tipped tail (fig. 13.6). The horns are stout, uniquely shaped, and about 20–25 cm long (fig. 6.2). Preorbital and carpal glands are rudimentary or absent, but, unlike other *Procapra*, the species is said to have inguinal glands, as well as a preputial gland not found in other gazelles (Pocock 1918). The laryngeal region of males has a swelling like a small dewlap. The species is larger than other *Procapra* and shows marked sexual dimorphism. Of 169 adults weighed on Mongolia's eastern steppe in November and December 1993, the mean body mass

13.6. Male Mongolian gazelles migrate across the lush grasslands of the eastern steppe in Mongolia. Their goiters are prominent. (September 1989)

of 102 males was 41.1 ± 4.5 kg (maximum 52.6 kg); the mean body mass of 67 females was 32.4 ± 2.2 kg (maximum 37.2 kg) (S. Amaglanbaatar, pers. comm.).

The mt-rDNA analyses support monophyly for the gazelles and saiga. The goitered gazelle is a sister taxon to the African Thomson's gazelle (fig. 13.2), and, similarly, the Tibetan gazelle is one to the Mongolian gazelle (fig. 13.4, table 13.2). Soma et al. found that the Mongolian gazelle and saiga have an identical karyotype and postulated "a very close taxonomic affinity" between them (1979). In our strict-consensus tree, the phylogenetic relationships of the three central Asian gazelle lineages (*Gazella*, *Procapra*, *Saiga*) are not resolved and remain a trichotomy.

Discussion

The logic of deriving phylogenetic relationships based on cladistic analysis has been well articulated by Nelson and Platnick (1981), and, as shown in this chapter, the addition of molecular data to morphological data is helping to resolve several ambiguous issues in bovid phylogeny and taxonomy. It is not known to what extent the molecular clock hypothesis applies to this group. We have included information on the percent sequence divergence in several species of central Asian Antilopini and Caprini to provide a rough indication of relative ages since the branches diverged. In addition, divergence data provide anecdotal support for justifying categories above the species level, given the subjective nature of such classifications (Avise and Aquadro 1982).

The various analyses lead us to the following conclusions:

Our work and that of Gatesy et al. (1997) strongly support the inclusion of the chiru into the Caprinae and the saiga into the Antilopinae. The tribe Saigini should, therefore, be abandoned.

The chiru is either a sister taxon to the rest of the Caprini or remains unresolved at the base of the other Caprinae lineages, possibly tracing its history back to the great bovid radiation in the Miocene when caprids and antelopes diverged in their evolutionary paths (Vrba 1987). A precursor of the chiru may have evolved on the Tibetan Plateau, the animal already a caprid, adapted not to rugged terrain but to plains like a typical antelope. While there remains no question about the chiru's affinity, its affiliation with a tribe remains problematic. If monophyly is used as a criterion, the Caprini and Rupicaprini are not natural groupings (Groves and Shields 1996; Gatesy et al. 1997). The musk ox, an Ovibovini, clusters with the goral, a Rupicaprini, and the bighorn sheep has an anomalous position with the mountain goat (fig. 13.2). Thus, tribal nomenclature of the Caprinae may not reflect phylogenetic relationships and probably requires reorganization.

Should the chiru be in a separate tribe? The percent sequence divergence is not markedly different between the chiru and other caprid species (table 13.2). However, in its morphology the chiru is highly distinctive, and the cladistic analysis of mt-rDNA suggests that it is the most basal member of the Caprinae (fig. 13.4). These characteristics suggest that the chiru belongs in a tribe of its own, the Pantholopini.

The morphological analyses show an anomaly with respect to the chiru's relationship to other bovid subfamilies. Pilgrim (1939) and Gentry (1992) emphasized the anatomical similarities linking chirus either to the Antilopinae or Caprinae. Yet the cladistic analyses of morphology (fig. 13.1), mt-rDNA (fig. 13.2), and combined results (fig. 13.3) clearly show the Caprinae closest not to the Antilopinae but to the Hippotragini and Alcelaphini. A cladogram represents an objective ordering of facts into a hierarchy, and, in spite of other evidence, the characters used in the analyses provide the best information: they indicate that the chiru's link to the gazelles is more ancient than its link to the wildebeest and sable.

The blue sheep appears intermediate between sheep and goats in various morphological characters, but molecular analyses place the species closest to the goats.

Among the gazelles, the saiga shows a close affinity to the Antilopini and for the present belongs in that tribe. However, the Neotragini and Antilopini may not represent natural groupings (Gentry 1992; Gatesy et al. 1997), and the subfamily will probably be revised in the future. *Procapra* and *Gazella* are generically distinct, based on morphological and molecular criteria. Whether the Tibetan and Mongolian gazelles both belong in *Procapra* or in separate genera remains unresolved in our analysis. The mt-rDNA data show that they are sister taxa and that the percent sequence divergence falls within the level of genetic distance observed in other related taxa. However, sequence divergence between Tibetan and Mongolian gazelles is greater (4.75%) than, for example, between goitered and Thomson's gazelles (2.20%). Yet the Mongolian gazelle is aberrant in its glands and several other physical features. The evidence leaves us ambivalent about the most suitable generic designation for the Mongolian gazelle, but a parsimonious decision is to leave it as a *Procapra* for the present.

We have considered only a few central Asian bovids in this chapter, yet these include several species, among them chiru, saiga, blue sheep, and Mongolian gazelle, whose phylogenetic relationships have been particularly obscure. Although our analyses have clarified some issues, answers to others remain elusive, including the inherently subjective problems in classification at the genus and tribal levels.

14
Phylogeny of Tibetan Steppe Bovids
Behavioral Comparisons

Hard is the Journey
Hard is the Journey
So many turnings
And now where am I?

Li Po, Chinese poet (701–762)

CLADISTIC ANALYSIS OF morphological and molecular characters has provided a general phylogenetic framework for the Bovidae, but uncertainties about the relationship of some genera and tribes persist (see chapter 13). Modes of behavior have usually not been included in evolutionary studies of bovids, yet they may be so invariable that their form and pattern can be treated almost like morphological features (Lorenz 1961). As Mayr and Ashlock noted, "Behavior is undoubtedly one of the most important sources of taxonomic characters" (1991). Perhaps behavior can be used to elucidate the systematic placement of those bovids whose affinities remain unclear or would benefit from validation. These include several species, such as the chiru and Mongolian gazelle, that were part of this study. Comparisons to assess homology must take into account the adaptations to the functional demands of the environment. The Bovidae, though diverse in size, are similar in general morphology, and many of the species have successfully adapted to fairly open environments (Eisenberg 1981). This should facilitate contrasting the results of behavioral comparisons with the results obtained from morphology and mt-rDNA analysis.

All bovid societies have rules of behavior based on a repertoire of visual, olfactory, tactile, and auditory signals whose display helps to provide predictability in interactions between individuals. The behavioral repertoire of many bovids has been described, including species that occur on the Tibetan Plateau or have relatives there (Schaller 1977b; Walther 1979; Estes 1991). Bovids exhibit a wide variety of patterns at least some of which reveal species-specific components. The following section briefly describes certain aspects of behavior of blue sheep, chirus, Tibetan and Mongolian gazelles, saigas, and yaks to show the diversity and social context of displays and other patterns of behavior. The next section compares the behavior of various bovids, representing all tribes, and leads, in the final section, to a discussion of the congruence of these behavior patterns with the results of the morphological and molecular analyses.

Behavior of species

No detailed behavioral work was done on this project except on chirus. My fragmentary observations are therefore supplemented with data from the literature. Males tend to interact with others more often and intensively than females do, especially in sexually dimorphic species (see Schaller 1977b), and I focus on their behavior. Distinctive patterns are most evident during aggression and courtship. Aggression can be either indirect, the animal using displays to intimidate, or direct, with overt threats and attacks. Visual signals are most prominent, at least to the human observer, because the role of odor is difficult for us to detect and interpret; vocalizations only supplement other signals on occasion.

Blue sheep

Is the blue sheep a sheep or a goat? Goats (*Capra*) prefer precipitous terrain, a habitat choice that is reflected in their stocky build. By contrast, sheep (*Ovis*) have two types of body build: most Eurasian species are lithe animals, adapted for speed in rolling terrain; North American species are stocky, but, unlike goats, they retreat to cliffs primarily to escape danger. Blue sheep (*Pseudois*) have a physique similar to goats and North American sheep, yet in their habitat preference they are like the latter. Schaller (1977b) discussed the behavioral affinities of blue sheep at length.

In general, the behavior of the various Caprini is very similar, but certain patterns distinguish goats from sheep. All Caprini clash horns to establish, assert, or test rank. Goats rear-bolt upright on their hindlegs before lunging down to make horn contact with an opponent, a technique well suited to precipitous terrain. Sheep may run at each other for 5 m or more, either on all fours or partly unbalanced on hindlegs, to clash. Blue sheep usually fight like goats, but on occasion like sheep, selection for a fighting style being apparently less rigid away from cliffs.

Most aggression in goats consists of overt threat and attack, and subordinate individuals avoid these by retreating. Sheep, however, tend to use a repertoire of displays to test each other's strength and to intimidate. Indeed, a threatened subordinate male may display a friendly gesture that signals low status: he rubs his face and horns against those of the dominant animal. Blue sheep also rub, but the other's rump, not face (fig. 14.1).

Displays of mild dominance in the Caprini include the low-stretch, twist, and kick, and these may be given singly or in combination. In a

14.1. One male blue sheep rubs the rump of another in a friendly, subordinate gesture. The male on the left displays by inserting his penis into the mouth. I took the photograph in Nepal.

typical sequence, a male approaches another animal from the rear in a low-stretch (body held low and muzzle forward), and, when close, he rotates his head sideways in a twist away from the other, a pattern unique to Caprini. Then he may stand by the other's rump and kick one or more times with a foreleg, delivering gentle, glancing blows or not even making contact. These displays are used during courtship, and, more rarely, during aggressive contacts between males. Sheep often kick during such aggressive encounters, whereas goats do not. Blue sheep behave like goats in this context. A rutting goat male mouths his penis and sprays urine upon himself, behavior not shown by sheep. Blue sheep commonly insert the penis into the mouth but do not spray urine (fig. 14.1).

What can be deduced about the taxonomic affinity of blue sheep based on these selected patterns? Sheep and blue sheep rub as a friendly gesture, but the part of the body they rub is divergent and the evolution of the behavior seems convergent. Blue sheep clash much like goats. They also mouth their penises like goats but without spraying urine, as if they are at an earlier and less complex behavioral stage. The evidence suggests that blue sheep are most closely related to *Capra* but that they have some *Ovis*-like traits.

Chiru

Many aspects of chiru behavior have been described in chapter 3, and this section limits itself to dominance displays of males and to courtship.

Dominance displays. Most dominance interactions are subtle, such as when one male casually approaches another, which then just as casually turns aside or begins to forage. A male's striking pelage color in winter and the long horns alone convey status, and he often emphasizes these traits, as well as his size, by holding himself erect and moving stiffly. Viewed head-on, a male is a fearsome apparition with his rapier horns and black face and legs that contrast with white chin and neck, at least to a human observer. A male commonly stands alone and erect for up to several minutes with forelegs extended forward and hindlegs back as if advertising his presence. Sometimes a male stands broadside to another, displaying his profile, or two males walk or trot parallel 2–3 m apart. One male may block the advance of another by overtaking him and standing broadside, a prevalent behavior when depriving access to a female.

Males exhibit three additional and distinctive displays. In the head-up posture, a male raises his neck and lifts his muzzle, exposing his white throat (fig. 14.2, left). The display may range from an upward flick of the muzzle to a prolonged gesture. The animal often walks with short, jerky steps. A seeming antithesis to the head-up is the head-down, in which a male arches his neck so far down that his muzzle almost touches the ground as if grazing (fig. 14.3, right). The animal walks stiffly, rump tucked in, ears laid back, and horn tips pointing forward. A male usually walks broadside past an opponent but occasionally also directly toward or parallel to him. Two males may also try to intimidate each other by both assuming this display, which, unlike the head-up, is solely directed at other males. In a neck-low display, a male holds his neck low with muzzle obliquely down. His horns tip forward, as if in readiness to jab, and his 15-cm tail is horizontal or at an upward angle, the long, white hairs on the underside conspicuous like a pennant (fig. 14.2, right). In this posture, a male trots or runs at another animal, male or female, or he moves through a herd, displaying to no one in particular. When displayed with vigor, the neck-low seems to be overtly aggressive, designed to startle and displace animals, which often dodge aside or flee. But when directing the neck-low at females, a male sometimes displays at a walk, appearing almost submissive.

Rutting males often growl or croak like muffled toads, or they emit "a deep-throated roar . . . bellowing challenges to all and sundry" (Rawling 1905), the nasal sacs no doubt acting as resonators. A roaring male usually lowers his neck and raises his muzzle slightly, mouth open (fig. 14.3, left), but he may also vocalize in a head-up or neck-low, the harsh sound adding

14.2. Displays by male chirus: head-up (left) and neck-low (right). (Art by Richard Keane.)

14.3. Displays by male chirus: bellow (left) and head-down (right). (Art by Richard Keane.)

emphasis to these displays. Roars may be directed at both males and females (fig. 14.4).

In direct aggression, a male jerks his head down and points his horns at an opponent, or he lunges or trots toward him, gestures which usually terminate an encounter. Chases with one male pursuing another for a few meters up to 1 kilometer are common, especially in competition for females, but they seldom result in physical contact; indeed the pursuer, close on the heels of his opponent, sometimes gives the impression of being careful not to overtake him or hinder his flight. Fights in the form of a horn-to-horn sparring match were seldom observed, only twice in 20 hours of observation during December 1991 (fig. 14.5). Once during November in Qinghai, two males faced each other, raised their muzzles in unison,

14.4. A male chiru roars while facing another, who turns his head aside. A third male stands by with neck lowered in a submissive gesture.

14.5. Two chiru males spar lightly.

jerked down their heads, and locked horns. They pushed and twisted, broke apart, and clashed again, behavior they repeated several times until one turned aside and was briefly chased by the other. The horns are potentially such dangerous weapons that chirus usually avoid a confrontation that could result in injury or death, as reported by Rawling (1905) and Prejevalsky (1876).

Most interactions involve only two males, but sometimes three or more participate. For example, one male gave a head-up to another, which then turned aside and assumed a head-down as he passed in front of a third male. Occasionally several males erupt in a melee of head-ups, neck-lows, chases, and other displays of rank that may persist for a minute or more.

In 20 hours of observation, I tallied 54 dyadic interactions, and of these 28% consisted of head-ups, 28% of neck-lows, 22% of chases, 18% of head-downs, and 4% of sparring matches.

Chirus usually convey submission by turning or moving away, but they also assume a posture with neck and head lowered obliquely (fig. 14.4). This display resembles the aggressive neck-low, except that the animal behaves subdued, mute, tail lowered, almost immobile.

COURTSHIP BEHAVIOR. When we reached the eastern Chang Tang on 18 November 1986, the behavior of animals indicated that the rut had begun, and we observed it again in the Chang Tang Reserve during December 1991. On 12 December the rut seemed to enter a new phase when males became more attentive and gentler toward the females. Mounting was observed soon thereafter, on 16 and 22 December. After a pre-rut of about a month, the animals appeared to have entered the main rut, which extends into January. The animals formed concentrations of a hundred or more individuals at certain locations such as on flats and gentle slopes with adequate forage (see table 3.13). A male with one or more females formed a characteristic grouping within a concentration, but solitary males, female herds, and mixed herds with two or more adult males were also there (see table 3.14).

All adult males 4 years old and older are of about the same size and have similar adornments; unlike other Caprinae, they lack a graded system of rank, such as horn length, by which an animal can at a glance assess an opponent's relative age and power. Yet each male must within a few weeks gain access to estrous females in a society containing many other adult males, most of them vigorous competitors. Bovid males usually use one of three mating systems: they establish territories from which other males are excluded but in which females are encouraged to remain; they associate with one estrous female at a time; or they collect and try to maintain a harem of females without reference to a specific site (Jarman 1983). Observations indicate that chirus have a harem system.

Solitary males chase after and try to accost lone females and female herds, using the head-up and neck-low displays accompanied by roars (fig. 14.6). Females often flee when approached by a male, and he may either pursue them for 100 m or more or stop abruptly and wait for others to wander near. If, however, the females remain near him, the male stands erect by them as they forage and rest. Sometimes one or two females break away. Pursuing them, he blocks their way and attempts to drive them back. Harems are usually small, probably because a male finds it difficult to control many females. In a total of 312 harems with one adult male tallied in December 1991, 70% contained one to four females, 25% five to eight

14.6. A male chiru approaches a female herd in a head-up display. (December 1991)

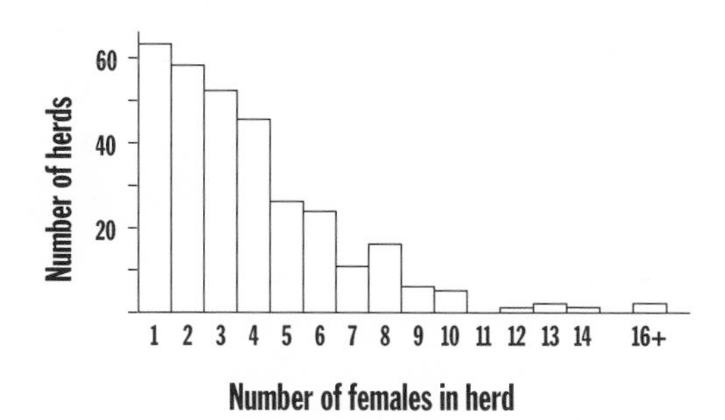

14.7. The number of female chirus in one-male harems during the rut, December 1991, based on 312 herds and 1191 animals.

females, and the rest nine or more (fig. 14.7). Most associations between a male and the females are brief, especially during the pre-rut, with females or sometimes even the male leaving abruptly. Two or more adult males seldom remain in a harem, one usually driving any others to the periphery, though a yearling male may be tolerated. Two examples illustrate the dissolution of harems:

> A male and six females forage slowly. Suddenly three females trot away together, and the male starts in pursuit but stops abruptly when he sees two males approaching about 100 m away. He runs to them, the three interacting in a medley of neck-low displays, chases, and roars until the newcomers leave. But in the meantime another male has herded two females away. On

returning to his harem, the male finds only one female, and she leaves at a walk, and when he pursues her, at speed. His harem defunct, the male trots toward several other chirus.

A male has eight females. One leaves abruptly at a run. The male chases after her a short distance before returning to the others. (A solitary male joins the escaping female and follows her to several others to form a new harem.) Fifteen minutes later, a second female flees, running fast with him at her heels, but she continues and he loses her too. Soon two males trot toward the herd. The harem male runs to meet them but they veer and then linger about 100 m from the herd. While the male is occupied with them, yet another male, as if by prearrangement, rushes into the herd displaying the neck-low and roaring. Rushing to meet this challenge, the harem male clashes horns once with the newcomer. Now the other two males enter the herd. The displays and chases scatter the females. One male follows three females and another male takes one. A single female remains and she then walks off. The herd male lingers a few minutes before also departing.

As the rut progresses, the behavior of females changes somewhat and males display new patterns of behavior. The female often tolerates the male's proximity. When fleeing, she gives the impression of inviting the male's pursuit rather than attempting an escape, and she often circles back, the two chasing each other around the harem. If she halts, the male may run at her in a neck-low posture, as if to make her flee again, or he may stand close beside or behind her. If she walks, he may follow in a head-up and use a distinctive gait with forelegs raised stiffly and high in a goose step. Moving close to her rump, he may deliver one or more straight-legged kicks that barely skim the side of her thigh, and his tail flicks back and forth. In one partial and one complete copulation observed, the male reared upright and balanced on his hindlegs with forelegs hanging down and lightly touching her for support. As the female walked slowly, he took bipedal steps and thrust several times before dropping to all fours. A courtship sequence of a male with five females went as follows:

The male stands at the periphery of the herd and roars. When the females walk slowly, he follows in head-up, then roars again. Stepping close to a female, his lower neck against her rump, he kicks the air on each side of her with a foreleg, but she dodges aside. He turns to another female, following her with a high-stepping gait and roaring. She halts. He moves close, kicks, and rears upon his hindlegs. But she walks ahead, away from him. He approaches, kicks, and again stands upright, and once more she moves ahead. On his next attempt she runs away, he after her in head-up, roaring. Returning to his remaining females he walks among them in head-up and, when they walk away, he follows in a neck-low posture.

Males of various bovid species approach females from behind in a characteristic low-stretch posture. By moving slowly and holding the body low, sometimes in an actual crouch, a male apparently reduces the probability of chasing the female away. Chirus rarely and fleetingly exhibited this posture, if at all; I was unsure whether males on occasion displayed the low-stretch or a neck-low display of low intensity. Many bovid males also sniff, nuzzle, and lick the vulva, behavior which may stimulate females to urinate. A male then sniffs the urine, either as it leaves the female or on the ground, raises his muzzle, and retracts his upper lip in a lip-curl. This apparently opens the ducts to the vomeronasal organ in the palate, where presumably the stage of the female's estrus is determined by odor (Estes 1972). No chiru male was seen to sniff or nuzzle a female, but on a few occasions a male sniffed where a female had urinated or defecated and then lip-curled. The low-stretch is often followed by urine testing in bovids, and this linked behavior is obviously rare in chirus.

In 20 hours of observation, 204 interactions between males and females were recorded, a male sometimes using more than one pattern of behavior. Of these, chasing was the most prominent (32%), followed by the neck-low display (26%), head-up (22%), goose steps (14%), a flurry of kicks (3%), lip-curl (2%), and mounts (1%). In general, chiru courtship gave the impression of being rather abrupt and brief, with displays and contacts cursory.

Unlike blue sheep, whose affinity to the Caprinae, and even a tribe and genus within that subfamily, is clear, the behavior of chirus does not point readily to any tribe, as discussed below.

Tibetan and Mongolian gazelles

The behavior of *Procapra* has not been studied, and I can present only incidental observations. Adult male Tibetan gazelles use latrine sites. On visiting a latrine, a male may paw the ground with a foreleg and then stand with hindlegs extended back and belly lowered as he urinates. After that he brings his legs forward to squat deeply and defecate at the same spot, in a linked urination-defecation sequence. Males are often alone and sedentary at certain sites which they are reluctant to leave when disturbed. Such patterns in the Antilopinae are indicative of territorial behavior (Walther, Mungall, and Grau 1983). Unlike *Gazella*, Tibetan gazelles lack preorbital glands and do not obviously mark the vegetation with secretions, but they do horn grass tufts and low shrubs. During aggressive encounters two males may face each other, jerk down their heads, and leap forward to clash horns, and then push and twist with forelegs spread and rigid. At times they go through the sparring motions without touching horns. Tibetan gazelles may bound with a spronking gait when disturbed, the hairs on their large, white rump patches erect and flaring, behavior that is more readily elicited in young than adults. Newborns are altricial, spending much of the time during the first two weeks or more after birth crouched with neck extended along the ground. With the exception of preorbital marking, the behavior of Tibetan gazelles resembles that of various *Gazella* species (Walther 1979; Walther, Mungall, and Grau 1983).

Mongolian gazelles are migratory or make extensive local movements, often in herds numbering thousands (Lushchekina et al. 1986); one herd we observed in July 1993 probably contained at least 25,000 animals. Although males tend to segregate from the females, mixed herds are common at all seasons, unlike Tibetan gazelles. For example, one herd of 280 counted in September comprised 42 adult males, 13 yearling males, 157 females, and 68 young. The ratio of males to 100 females in this herd was 35:100, or only somewhat less than the 40:100 (n = 5100) in my total population sample, or 50:100 (n = 5133) in another sample (Suchbat et al. 1989). Animals in a traveling herd emit plaintive, high-pitched bleats, reminiscent of the call of high-flying geese. Spronking was rare and brief, consisting of two to four low bounds sometimes followed by one high leap. No latrine sites were found and no marking behavior was observed. Adult males occasionally urinated with their hindlegs extended back and they also squatted to defecate, but I did not observe the two patterns linked as in Tibetan gazelles and goitered gazelles. Newborns are precocial, following their mothers within a day, though they rest often (A. Lushchekina, pers. comm.).

I briefly observed the rut on 22 December 1989. Adult males singly and in twos, as well as small female herds, were scattered over the plain. Mixed herds typically consisted of one male with one or more females and young: of 29 mixed herds tallied, 76% had one male and the rest two to seven. When a male approached or chased a female, he jerked his muzzle up repeatedly or kept it raised, his dark-gray goiterlike growth conspicuous, and his stumpy tail was held vertically. Males followed females or trotted toward roaming ones without evidence of territorialism, their behavior resembling that of chirus.

In their lack of latrines, their use of a loud contact call and abbreviated spronks, and their precocial young, Mongolian gazelles differ considerably from Tibetan gazelles. They also seem to lack territories, in contrast to Tibetan gazelles, which probably have them. On the basis of behavior alone, one could speculate that the two species belong in different genera.

Saiga

The saiga, like the chiru and Mongolian gazelle, may migrate. Also, like these two species and unlike *Gazella* and the Tibetan gazelle, the saiga's sexes are dimorphic, with males averaging 43 kg and females 31 kg in body mass (Bannikov et al. 1967). During the rut, a male collects a harem of

5–10 or more females, which he tries to maintain by chasing any female that leaves and attacking any male that approaches. A male with his harem roams over a nonexclusive area of 3–10 km^2 (Heptner, Nasimovic, and Bannikov 1966). His display consists of raising muzzle and tail, inflating his proboscis-like nose, and emitting grunts and roars (Pohle 1974). Saiga apparently do not defecate at latrine sites. After a gestation period of about five months, a month shorter than Mongolian gazelles (Miyashita and Nagase 1981), the female gives birth to one or two altricial young, which remain crouched for about 8–10 days (Heptner et al. 1966), a shorter period than that for Tibetan gazelles and *Gazella*. It remains unclear whether saigas spronk, but they occasionally make one high "observation jump," especially at the beginning of flight (Sokolov 1974), behavior that is reminiscent of Mongolian gazelles. Saigas run in a distinctive posture with head and neck held low, resembling a scurrying rodent. When moving in a large herd, animals emit hoarse bleats (Heptner et al. 1966). The saiga, like the Mongolian gazelle, is a highly distinctive Antilopini, a species whose behavior resembles that of chirus.

Wild Yak

The family Bovidae is divided into two major clades, with the tribes Boselaphini, Tragelaphini, and Bovini in one clade and all other tribes in the second (fig. 13.2). The yak is the only representative of the former clade on the Tibetan Plateau.

Except for male Asian water buffalo, which may be territorial under certain ecological conditions (Eisenberg and Lockhart 1972), all Bovini are nonterritorial. A bull yak remains with an estrous cow, standing beside her, bodies sometimes touching, or he trails her closely. While tending her, he may lick her face or anal area, and he may step away to horn the ground or swing his horns at any bull that has ventured near.

Aggression in domestic yaks was described by Schaller (1977a), and wild ones behave similarly. The most distinctive indirect pattern is a lateral display during which an animal presents its impressive profile, especially the hump and mantle of hair. Opponents may stand either facing the same direction or head-to-tail about 3–6 m apart, remain motionless or circle slowly, always broadside, a display that may continue for five or more minutes. A direct threat consists of a lunge or charge with lowered head, horn tips toward the opponent, and sometimes a sideways and upward hook of a horn. Sparring, with animals pushing and twisting head-to-head, was occasionally observed, in one instance continuing for at least 15 minutes. An encounter between two bulls illustrates various displays:

> A yak herd with a large bull, a medium-sized bull, a subadult bull, and six cows forages over a ridge and moves into view of a solitary large bull about

Table 14.1 The occurrence of some indirect threat patterns in several Bovini

	Yak	Domestic cattle	Gaur	American bison	African buffalo
Loud grunt or bellow	+	+	+	+	–
Horn vegetation or earth	+	+	+	+	+
Rub face and neck on ground	+	+	+	+	+
Paw ground	(+)	+	(+)	+	–
Roll in mud wallow	–	–	–	–	+
Roll in dust wallow	+	–	(+)	+	–
Urinate in wallow	(+)	–	–	+	+
Grind teeth	+	–	–	–	–
Lateral display	+	+	+	+	+
Head-down display	+	+	+	+	+
Head-up display*	–	–	–	–	–

Sources: Data are based on Sinclair 1977 and Schaller 1977a.
Note: Parentheses indicate that trait is present but infrequently shown.
*Commonly shown in the Bovini by Asian water buffalo.

100 m below. The large herd bull walks toward the other, stops, lies on his side in the dust and hooks the soil with a horn, and then walks on, his tongue hanging out as he emits hoarse grunts, almost roars. Twice more he rolls in the dust, flailing legs skyward, before advancing, head low and swaying from side to side, his tasseled tail vertical and waving. The other male dust-wallows once and retreats grunting. The herd male goes to the vacated wallow and also rolls at that site.

The yak shares various aggressive patterns with other Bovini (table 14.1). Frequent rolling in dust wallows, sometimes coupled with urinating and defecating, distinguishes it from other *Bos* (Sinclair 1977) but resembles *Bison*, placing the yak in a seeming intermediate position between the two genera (see chapter 7).

Kiang

Equids and bovids are not closely related, yet they share a number of behavior patterns, as shown by kiangs. I describe these briefly to place the bovid information into a broader perspective, using another ungulate that confronts similar environmental challenges.

When threatening another individual, a stallion displays a head-up with muzzle raised high, and he sometimes approaches with a prancing gait, ears laid back, and tail horizontal. Harsh grunts or a wheezing bray may accompany the display. Indian wild asses and kulans use a similar posture (Feh, Boldsukh, and Tourenq 1994; Neumann-Denzau 1991), whereas feral horses raise the neck sharply arched with chin tucked in (Berger 1986). A chase may follow a display, the animal in the lead slashing back with its hooves if pursued too closely. Equid stallions fight by kicking, rearing up on the hindlegs, neck-wrestling, and biting, the last sometimes while circling tightly, standing head-to-tail, and finally kneeling on all fours, preventing injury to the rear legs (Berger 1981). Fighting was not observed in kiangs, but they probably use similar techniques.

When a stallion herds several mares, he trots around them and cuts off and drives back any that veer away, using a typical posture with head and neck held obliquely down, ears retracted. Feral horses and kulans behave similarly (Berger 1986; Feh, Boldsukh, and Tourenq 1994). Equids, whether territorial or not, may defecate at latrines, which are used mainly by stallions (Moehlman 1985). Such latrines of kiangs and kulans are much pawed, sometimes creating a shallow trench 2–3 m long and 1 m wide.

Behavioral comparisons

To assess systematic affinities based on behavior, I have selected 26 bovid species representing all tribes for comparison. To this list the pronghorn, family Antilocapridae, was added as a relative of the Bovidae and a species that was once included with that family. Only one or two species are listed for certain subfamilies, but the Caprinae and Antilopinae are represented by several because these subfamilies are of main concern in this discussion. Of the 27 species tabulated in tables 14.3 and 14.4, 21 are also represented in the cladograms of chapter 13 (figs. 13.2 and 13.3). Vrba and Schaller (forthcoming) used the behavioral data to construct a cladogram and compare it to results obtained from the analysis of morphology, molecular biology, and the fossil record.

Anatomical structures may have an influence on the expression of behavior, and glands show this particularly well. Pocock (1910, 1918) used the presence or absence of specific glands as a basis for assigning species to genera and tribes. For instance, he noted that glands located by the false hoof or dewclaws are found only in the Tragelaphini and Boselaphini. However, glands are extremely labile in evolution, as is evident from the great variation in glands within tribes such as the Antilopini (table 14.2). The chiru shows no tribal affinity with respect to glands. Some species have glandular areas that appear nonfunctional. Pocock (1910) refers to "aborted pedal glands" in *Capra*, blue sheep, and tahrs. I suspect that such glands are disappearing, that their function has been lost before the anatomy has been fully modified. A number of species have uniquely placed glands, among them the metatarsal gland in impala, a gland below the ear in oribis and reedbucks, and three glands in pronghorns, one below the ear, a second dorsally above the tail, and a third on the posterior side of the hock. The effect of most glands on the behavior of the individual and of those that perceive the scent remains unknown.

Table 14.2 Comparisons of glands and display organs

	Functional gland			
	Preorbital	Pedal*	Inguinal	Post-cornual†
BOVIDAE				
Caprinae				
Chiru	–	–	+	–
Goral	+?	++	–	–
Japanese serow	+	++	–	–
Mountain goat	–	+	–	+
Takin	–	–	–	–
Musk ox	+	–	–	–
Himalyan tahr	–	–	–	–
Ibex	–	–	–	–
Argali	+	++	+	–
Blue sheep	–	–	–	–
Cephalophinae				
Maxwell's duiker	+	++	–	–
Antilopinae				
Oribi	+	++	+	–
Mongolian gazelle	–	–	–	+
Tibetan gazelle	–	++?	–	–
Goitered gazelle	+	++	+	–
Blackbuck	+	++	+	–
Saiga	+?	++	+	–
Hippotraginae				
Kob	+	–	+	–
Sable	+	++	–	–
Rhebok	–	++	–	–
Wildebeest	+	+	–	–
Hartebeest	+	+	–	–
Aepycerotinae				
Impala	–	–	+	–
Bovinae				
Nilgai	+	++	–	–
Nyala	–	–	+	–
Yak	–	–	–	–
ANTILOCAPRIDAE				
Pronghorn	–	++	–	–

Note: A + indicates that trait is present and a +? indicates that the gland may not be functional. Not all glands have been tabulated (see text).

*A + indicates that pedal glands are found only on forelegs, and a ++ indicates that they are on all four legs.

†Pocock (1918) reported postcornual glands in Tibetan gazelles, but I observed no evidence of them in several specimens.

‡A ± indicates that some have horns, some do not.

Functional gland				Female has horns‡	Long ruff, mane, or mantle
Caudal	Preputial	Carpal	Metatarsal		
–	–	–	–	–	–
–	–	–	–	+	–
–	+	–	–	+	+
–	–	–	–	+	+
–	–	–	–	+	+
–	+	–	–	+	+
+	–	–	–	+	+
+	–	–	–	+	–
–	–	–	–	+	–
–	–	–	–	+	–
–	–	–	–	±	–
–	–	+	–	–	–
–	+	–	–	–	–
–	–	–	–	–	–
–	–	+	–	±	–
–	–	+	–	–	–
–	–	+	–	–	–
–	–	–	–	–	–
–	–	–	–	+	+
–	+	–	–	–	–
–	–	–	–	+	+
–	–	–	–	+	–
–	–	–	+	–	–
–	–	–	–	–	+
–	–	–	–	–	+
–	–	–	–	+	+
+	–	–	–	+	–

Bovids exhibit a large number of behavior patterns, but only certain ones can provide useful insights into evolutionary relationships. As in molecular analyses, comparisons must be weighted to reduce incongruities. Several factors affect the suitability of a pattern as an indicator of systematics:

1. Certain behaviors are so widespread, and presumably primitive, that they lack special affinities. These include the head-up, head-low, and lateral displays, which are found in all tribes. Some species may, however, lack a particular pattern, and its absence may help to provide significant insight. Wildebeests, takins, and yaks do not exhibit a head-up, and small, sexually monomorphic species usually lack a lateral display. In some species the displays are accentuated by prominent ruffs, manes, or mantles, but such hairy adornments are erratically distributed among several tribes, though confined to relatively large species (table 14.2). Behaviors that show little variation in my sample of taxa have not been included.

2. A few patterns appear to be ancient and have been lost, or at least are seldom shown, in most species. The head-to-tail fight, in which animals attack each other's flank and hindquarters while standing side by side, and the neck fight, in which one animal places its neck over another's and presses down, are of this type. Such behavior is often shown by species with small, pointed horns (such as nilgais and mountain goats, and also by equids and camelids that lack horns), which make head-to-head combat awkward. On rare occasions, blue sheep (Wilson 1984), aoudads, and young chamois also have a neck fight, behavior that seems vestigial in these species.

3. Behavior may have evolved several times independently, a convergence that is partly the result of environmental pressures. For instance, oryx dig shallow pits beneath shrubs apparently to avoid sun, chirus paw pits to escape detection by parasitic flies, and mountain goats excavate pits as part of the rutting ritual.

4. A number of displays are species specific, making them of limited value for comparisons. Rump rubbing in blue sheep is of this type.

5. The behavior of a species may vary depending on population density, resource availability, and other parameters. Pronghorn males either are territorial or merely defend a harem of females (Kitchen and O'Gara 1982), and gorals also seem to show such behavioral flexibility (S. Lovari, pers. comm.). Male goitered gazelles usually deposit feces at latrines, but in one population they apparently did not (Habibi, Thouless, and Lindsay 1993).

6. A widespread pattern in a genus or tribe may have been lost in one member. Such an absence cannot be used to evaluate phylogenetic relationships. For example, *Damaliscus* and *Alcelaphus*, both Alcelaphini, are the only known bovids that do not urine-test females and lip-curl (Estes 1991). The blackbuck does not kick with a foreleg during courtship, a common gesture in other Antilopini.

Ideally analysis should be based on discrete and unambiguous patterns that are not under strong environmental influence and are consistent within a tribe. But few fit these criteria. Furthermore, the evolutionary history of the bovids must be considered. Bovids appeared in the Middle Miocene and radiated rapidly. By the Upper Miocene, 5–7 million years ago, most of today's bovid tribes are evident in the fossil record (Vrba 1987, 1995; Gentry 1992). However, some of the early lineages are extinct, making it difficult to trace behaviors to an ancestral group. Given these strictures, I selected 13 behavior patterns for comparison (table 14.3).

Territorial behavior, or the defense of a plot of land, is widespread among bovids. Although influenced by environmental conditions, it tends to follow subfamily divisions. The Bovinae and most Caprinae, including the chiru, are not territorial. The Hippotraginae, Cephalophinae, and impala are territorial, as are most Antilopinae, the saiga and Mongolian gazelle apparently being exceptions.

Territories are marked with olfactory and visual signals, usually just by the male but in the Neotragini by both sexes. Such signals include glandular secretions and the animal itself. Repeated defecation at particular sites creates latrines, but these are not always indicative of territorial behavior. Nilgais make prominent latrines without being territorial (Sheffield, Fall, and Brown 1983). The feces of Himalayan gorals accumulate at certain rest sites, but this species is probably not territorial (Lovari and Apollonio 1994). The territorial Alcelaphini, impala, serows (Lovari and Locati 1994), and Neotragini have latrines, and so do the Antilopini, except the saiga and Mongolian gazelle. Conversely, many territorial species do not exhibit the behavior, among them the Cephalophinae and Hippotragini.

Blue sheep, aoudads, kobs, and rarely *Ovis* and saigas use an extended penis as a visual signal during aggressive encounters, a pattern that seems to have evolved independently in several lineages. Several Caprini—tahrs, blue sheep, aoudads, *Capra*—and the goitered gazelle (Blank 1992) insert the penis into the mouth. As a next level of complexity in the penile signal, an animal enhances its odor by urinating on itself, behavior that is found in various Caprinae, such as chamois, takins, tahrs, and *Capra*, and in the lechwe, tribe Reduncini (Walther, Mungall, and Grau 1983).

Linked urination-defecation (an animal urinates and defecates in sequence at the same site) is typical of the Neotragini, in which both sexes perform the pattern, and of most Antilopini, impala, and pronghorns, in which only males exhibit it. Among the two aberrant Antilopini, the Mongolian gazelle displays the individual components of the display but a linkage was not observed, and the saiga's behavior has not been reported. A chiru male often assumes the urination stance, usually without urinating, apparently as an advertisement display. On two occasions, a male then squatted low and defecated, the linkage in the patterns obviously rare.

Certain forms of indirect and direct aggression are found in most bovids

Table 14.3 Comparisons of behavior patterns

	Territorial	Defecate at latrine	Mouth penis	Spray urine on pelage	Linked urination-defecation	Head-to-tail fight
BOVIDAE						
Caprinae						
Chiru	–	–	–	–	(+)	–
Goral	±	+	–	–	–	
Japanese serow	+	+		–	–	
Mountain goat	–	–	–	–	–	+
Takin		–		+	–	
Musk ox	–	–	–	–	–	
Himalayan tahr	–	–	+	+	–	+
Ibex	–	–	+	+	–	–
Argali	–	–	–	–	–	–
Blue sheep	–	–	+	–	–	–
Cephalophinae						
Maxwell's duiker	+	–	–	–	–	
Antilopinae						
Oribi	+	+	–	–	–	
Mongolian gazelle	–	–	–	–	+	
Tibetan gazelle		+	–	–	+	–
Goitered gazelle	+	+	(+)	–	+	–
Blackbuck	+	+	–	–	+	–
Saiga	–	–	–	–		
Hippotraginae						
Kob	+	–	–	–	–	–
Sable	+	–	–	–	–	–
Rhebok	+	–	–	–	–	
Wildebeest	+	+	–	–	–	–
Hartebeest	+	+	–	–	–	–
Aepycerotinae						
Impala	+	+	–	–	+	–
Bovinae						
Nilgai	–	+	–	–	–	+
Nyala	–	–	–	–	–	
Yak	–	–	–	–	–	–
ANTILOCAPRIDAE						
Pronghorn	±	–	–	–	+	–

Note: A + indicates that trait is present; a (+) that it is rare; a blank that there is insufficient information; and a ± that species may or may not be territorial depending on situation.

Neck fight	Fight on carpal joint	Low-stretch	Foreleg kick	Mount upright	New-born precocial	Spronking gait
–	–	(+)	+	+	+	(+)
–	–	+	+	–	+?	–
–	–	+	+		+	–
+	–	+	+	–	+	–
	–				+	–
	–		+	–	+	–
+	(+)	+	+	+	+	–
	–	+	+	+	+	–
	–	+	+	+	+	–
(+)	–	+	+	+	+	+?
	–	+	+	+	–	–
	–	+	+	+	–	+
–	–				+	(+)
–	–				–	+
–	–	+	(+)	+	–	+
–	–	–	–	+	–	+
	–	+			–	+?
–	–	+	+	+	–	–
+	+	+	+	+	–	–
	–	–	+	+	–	+
–	+	+	–	–	+	–
–	+	+	–	–	+	+
–	–	+	–	+	+	+
+	+	+	–	–	–	–
–	–	+	–	–	–	–
–	–	–	–	–	+	–
–	(+)	–	–	+	–	+

and include postures that accentuate an animal's physical attributes, lunges, chases, and locking horns. A few patterns are limited to certain genera or tribes (table 14.3), among them head-to-tail fights, neck fights, and horn or neck combat while opponents kneel on their forelegs, as shown by all Alcelaphini and nilgais and rarely by sable and some Tragelaphini (Walther 1979).

The low-stretch and foreleg kick are widely used singly and in sequence by bovid males during courtship, and among the Caprini also during aggressive encounters. To these displays the Caprini have added the twist, in which the male rotates his horns away from the other animal after approaching in low-stretch. Tongue-flicking and grunting often accompany the displays. The low-stretch is found in most tribes. In chirus it occurs in such rudimentary form that it usually is not recognizable as a distinct pattern; the low-stretch is absent in rheboks, blackbucks, the Bovini, and pronghorns. It presumably is being lost or has been lost in evolution from the repertoire of these species. The kick follows a tribal pattern: it is absent from the three Bovinae tribes, Alcelaphini, impala, and pronghorns.

Bovids have two basic mounting postures. In one the male extends his head and neck along the back of the female while resting his body heavily on her. In the other the male stands with head and neck erect either with his chest resting on the female and clasping her or barely touching her except to hold her lightly and intermittently with his forelegs to retain balance. These mounting postures have tribal affinities. Rupicaprini, Ovibovini, Alcelaphini (except blesbok), and the three Bovinae tribes mount with head and neck over the back of the female, and the Hippotragini, Reduncini, Cephalophini, Caprini, rhebok, and impala mount with head and neck upright but leaning into and clasping the female. The Neotragini and Antilopini balance upright, barely touching the female; the former tend to have forelegs folded and the latter dangle them. The pronghorn posture resembles that of the Antilopini most closely. The two chiru mountings observed were most similar to the posture of the Antilopini.

Newborn bovids may struggle to their feet and follow their mother within hours after birth or remain hidden for a few days. Young that follow their mothers within seven days after birth are here considered precocial, or followers. The young of certain species may remain hidden for one to four weeks or longer, and these are altricial, or hiders. Precocial young are found in all Caprinae tribes, including the gregarious chiru and solitary serow (Kishimoto 1989), Alcelaphini, Bovini, and impala. One Antilopini, the Mongolian gazelle, is also a follower. The remaining bovids are altricial, the behavior being primarily an antipredator strategy (Leuthold 1977).

The high-bounding spronking, or stotting, gait is prominent in the Antilopini, Alcelaphini (except wildebeest), rhebok, impala, oribi, and pronghorn. Several Caprinae, including chirus, urial sheep, and blue sheep, may give a series of low, four-legged bounds when fleeing but it is not clear if this behavior is homologous to spronking. However, Mongolian gazelles exhibit similar bounds, and, given the prevalence of spronking in the Antilopini, the low bounds in this species probably represent a modification of the more usual pattern. Mongolian gazelles and saigas may also display one high leap, which possibly represents an abbreviated spronk.

Discussion

Behavior patterns often show affinity to certain tribes. Can they be used to clarify the systematic position of species, and do the results correlate with morphological and molecular data?

Various behavior patterns of blue sheep place them into the Caprini, most closely related to goats, a conclusion that agrees with morphology. The molecular data show blue sheep as an early branch of the goat lineage, the other branches leading to *Capra* and *Hemitragus* (fig. 13.2). The three types of analyses independently reach the same conclusion.

The Tibetan gazelle (*Procapra*) and other gazelles (*Gazella*) do not differ in those aspects of behavior that have been observed, except as they relate to the presence of certain glands. The two genera are, however, anatomically distinct, and molecular analysis shows that each represents an independent lineage of considerable antiquity. Behavior cannot with certainty distinguish *Procapra* and *Gazella* but it can be used to question the inclusion of the Mongolian gazelle in the former genus. Unlike the Tibetan gazelle, the Mongolian gazelle seems to lack territorial behavior, it does not defecate at latrines, and its young are precocial, to mention three traits. It can be argued that its migratory habits have affected behavior, but other migrants, such as wildebeests, springboks, and white-eared kobs, maintain seasonal territories. Unlike the Przewalski's gazelle (Jiang et al. 1994) and Tibetan gazelle, the Mongolian gazelle shows much sexual dimorphism. In its glands and several other anatomical features it does not align with *Procapra*. Based on this evidence, the inclusion of the Mongolian gazelle with *Procapra*, rather than in its own genus *Prodorcas* as suggested by Pocock (1918), remains a matter of debate.

The saiga is even more anomalous than the Mongolian gazelle. During its taxonomic history it has been designated as a *Capra*, *Antilope*, and *Gazella*, and, as noted in chapter 13, it was once lumped with the chiru as a distinct Caprinae tribe. Behaviorally the saiga resembles the chiru, as well as the Mongolian gazelle, in the absence of territorial behavior and latrines. It is similar to gazelles in its altricial young, a rapid courtship circling while standing head-to-tail (Pohle 1974), and some other patterns. On the basis of behavior, the saiga cannot be unambiguously placed into a certain subfamily. The morphological traits show affinity to both Antilopinae and Caprinae, though most point to the former. Molecular analysis has resolved the matter by showing that the saiga is an Antilopinae, specifically a taxon associated with the Antilopini.

Chiru behavior lacks diagnostic specializations: several subfamilies could claim affinity, but, as with the case of the saiga, the Caprinae and Antilopinae offer the most similarities. Like most Caprinae, the chiru is not territorial, it does not defecate at latrines, and its young are precocial. Its head-down display is extreme, the neck arching so far down that the muzzle almost touches the ground, an exaggerated display found in several other Caprinae, including ibex and takins. Three chiru patterns—linked urination-defecation, spronks, and upright mounts while barely touching the female—resemble those of Antilopini. However, the first two of these are rare and appear to be vestigial. A courting chiru male may also follow a female with high-stepping, rigid forelegs in a display reminiscent of Grant's gazelles and other *Gazella.* The chiru also shares patterns, such as linked urination-defecation, with the impala, a species of ancient lineage. Rutting chiru and impala males chase other males or just race around a herd in a characteristic display with tail raised to reveal a flag of long, white hairs as they emit roars. In sum, behavior does not link the chiru unambiguously to a subfamily although it points to a relationship with the Caprinae and Antilopinae. The evolution of these two subfamilies remains poorly understood, but their lineages parallel each other (Gentry 1992; Vrba and Schaller forthcoming). The morphological analysis described earlier also produced uncertain results, suggesting either an ancient relationship between the two subfamilies or a remarkable amount of convergence. The mt-rDNA work has provided crucial insights into the evolutionary position of the chiru: the species is a Caprinae, either basal to the subfamily or a sister group to the others.

However, as noted in chapter 13, morphological and molecular analyses show that the Caprinae are most closely allied to the Hippotragini and Alcelaphini, not the Antilopinae, even though various anatomical features and several behavior patterns suggest a relationship between chirus and gazelles. Nevertheless, similarities between the Antilopinae and Caprinae may not be fortuitous. The two subfamilies have a common ancestor, and the chiru perhaps retained some behavioral and anatomical characters from a time in the Miocene when subfamilies radiated; other characters may be convergent. Whatever further research reveals, the molecular data have performed best in pointing to the chiru's systematic relationships, followed by morphology and lastly by behavior. This is not a reflection on the effectiveness of each analysis but on the rigor and precision with which it was done. Each kind of analysis made a useful contribution, the three complementing each other.

The results further show that the chiru is genetically, morphologically, and behaviorally so distinctive that its loss would be far greater than that of a species with a number of close and similar relatives. Isolated on the Tibetan Plateau, the chiru evolved there and adapted to the rigors of the Chang Tang, needing only space to roam, scant forage, and special places to give birth and spend the winter. In the coming century, when saving biological diversity will be humankind's most important challenge, special taxa, such as the chiru, must remain a focus for conservation.

15

Nomads, Livestock, and Wildlife

Conservation of the Chang Tang Reserve

We abuse land because we regard it as a commodity belonging to us. When we see land as a community to which we belong, we may begin to use it with love and respect.

Aldo Leopold (1949)

THE WILD YAK, chiru, and other species have existed on the Tibetan Plateau since at least the early Pleistocene, long before humankind arrived there. But sometime within the past millennia, bands of hunters began to roam the Chang Tang and left stone artifacts as evidence of their passing. Still later, certainly during the past thousand years and perhaps much earlier, the first nomadic pastoralists moved into the cold uplands. The nomads lived there with their livestock and the wildlife when the first Westerners penetrated the region in the late 1800s and left a written record of their passing. The travelers marveled at the huge herds of wild yaks, the migrating chirus in their vast throngs, and the inquisitive herds of kiangs. And they marveled that nature could support such multitudes on such meager pastures. Wildlife in many parts of the Tibetan Plateau was depleted during this century, but that in the Chang Tang endured until the political upheavals of 1959 and the subsequent two decades. Nomads were resettled in the north, where few had lived before, roads were built into remote places, and rigid government decrees affected every household. And the wildlife was slaughtered in large numbers.

When I began work in the Chang Tang of Qinghai in 1985 and Tibet in 1988, the region was in a major period of transition. The large wild herbivores had been reduced to remnants, and livestock production and development had become a priority for the pastoral areas. I was eager to study the various species for the first time and provide them with a written history. Most chapters in this book represent a record of this work. But no biologist can for long observe yaks, chirus, and other species without a sense of anxiety and foreboding, without a feeling that their existence may be brief unless efforts are made to protect them. One of my tasks was to be an advocate for conservation. The government of the Tibet Autonomous Region was remarkably receptive to the idea of a conservation pro-

15.1. Several nomad families with their flocks settled in the Aru Basin in 1991. (July 1992)

gram in the Chang Tang and, among various actions, it established the Chang Tang Reserve.

Conservation problems are mainly social and economic, not scientific. A reserve often creates hostility because restrictions limit traditional activities of local people, whether hunting or livestock grazing. To establish a reserve is relatively easy, but to maintain and manage it is a complex and difficult endeavor. Innovative solutions must be found to conservation conflicts, solutions based on local interests, cooperation, skills, and traditions. The government now has the challenge of managing this relatively pristine ecosystem in such a way that the nomads can continue their pattern of life and the wildlife can roam the uplands as in the past. No plans exist as yet for accomplishing this goal, and to develop them will require a long-term effort.

The focus of my work was on the wildlife. However, any account of the Chang Tang must include the nomads, who will ultimately determine the future of the reserve by their livestock practices and attitudes toward wildlife (fig. 15.1). To provide a brief overview of human presence in the reserve and to comment on conservation issues are the main purposes of this chapter. It is worth emphasizing that we all, not just the nomads, are responsible for the fate of the reserve. To subsist, the nomads raise sheep for wool, which is much prized in the carpet and other industries for its high-quality fibers, and they hunt chirus for their extraordinarily fine and light wool, which is used to weave expensive scarves. The fate of the Chang Tang also rests with traders, manufacturers, and shop owners. And finally it is in the hands of those wealthy enough to buy carpets, scarves, and other products, who affect the lives of people and creatures so remote that they have never given them a thought.

Prehistory

In the summer of 1991, I found several sites in the reserve with stone artifacts and on subsequent trips located more. Some sites had only a few artifacts, but in others, no doubt encampments, the ground was littered with flakes and discarded implements. A total of 25 sites were found in the eastern part of the reserve, from the southern border north to the base of the Rola Kangri massif at 35°15′ N (fig. 15.2). The sites at the Tianshui He, Yako Basin, and a hot spring along the northern flank of the Jangngai Range were particularly rich. Twenty-one sites were located by a spring or on the bank of a freshwater stream. Three sites were on ancient beach lines (two of them in dry lake basins), and one was on a broad mountain pass. All artifacts were surface finds; no excavations were attempted. A sample of artifacts was collected and some were deposited with the Tibet Historical Relic Administration and some given for analysis to J. Olsen at the Department of Anthropology, University of Arizona (Brantingham, Olsen, and Schaller in press).

The distribution of artifacts raises three points of interest. First, I found sites only in the eastern half of the reserve, although I searched for them in the Aru Basin and on the desert steppe to the north. Few people may have penetrated that western area because of extensive glaciers and lakes, and erosion may have covered sites. However, a few artifacts and petroglyphs have been found in the west (Hou 1991). Second, the abundance of artifacts in the eastern part, as well as many sites farther south, indicates long-term and possibly permanent occupancy of the region. Nomadic hunters were the sole inhabitants at first, but pastoralists may also have been responsible for some sites during the past two millennia or more. It is not known when livestock first reached the Chang Tang. Third, many sites are located in terrain that is currently uninhabited and has been unoccupied, except for occasional hunting bands and gold seekers, for over a hundred years in part because good pastures are scarce and the area is remote from population centers. As noted in chapter 2, the Holocene climate there was generally warmer and more humid before 4000 B.P. than it is today, with stands of trees in some locations, and this would have made life easier for hunting bands during certain periods.

A total of 220 stone artifacts were analyzed by Brantingham, Olsen, and Schaller (in press). The artifacts were technologically diverse and included flakes (36%), flake tools (30%), blades (11%), utilized flakes (10%), microblade cores (8%), and a few core tools, microblades, and others (fig.

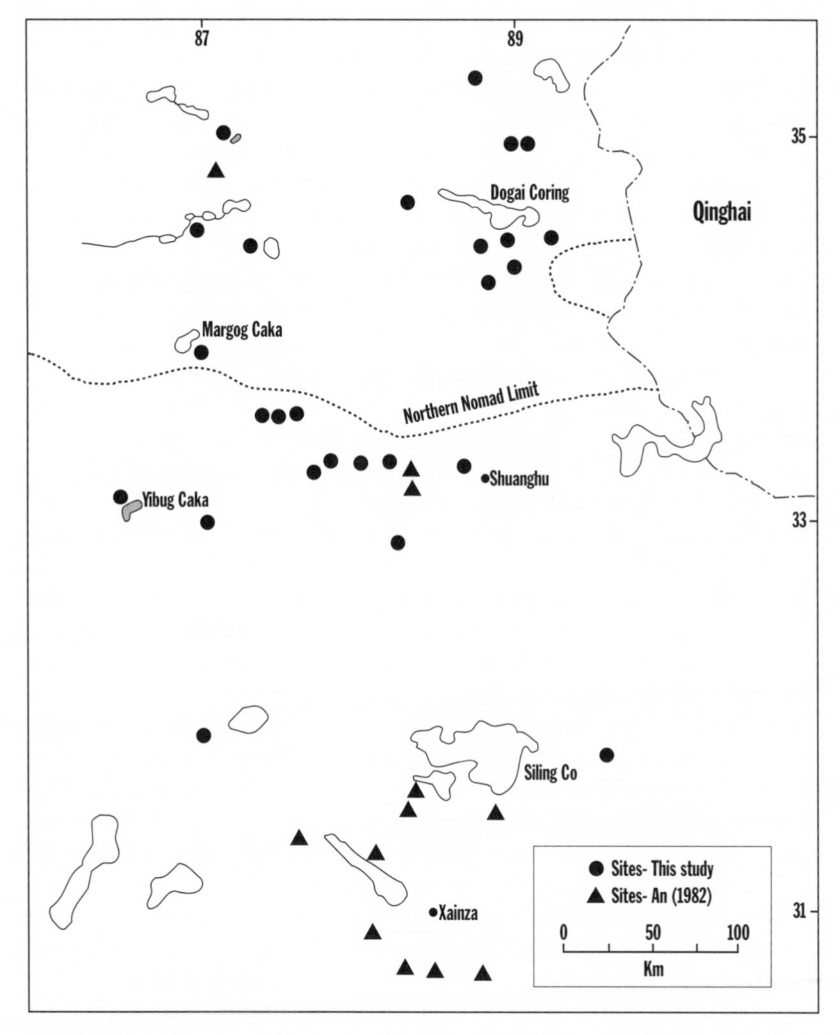

15.2. Location of stone artifact sites in the eastern Chang Tang of Tibet, based on finds during this project and reported by An (1982).

15.3). Three technological types were evident: generalized cores, large blades, and microliths. Among the 74 tools in the sample, unstandardized retouched flakes were most frequent (26%), followed by side scrapers and end scrapers at 22% each, and a small assortment of retouched blades, double-edged scrapers, knives, points, and a biface. No burins, gravers, or drills were found.

Several raw materials were used in the manufacture of the artifacts, among them chert, chalcedony, obsidian, basalt, mudstone, quartzite, and several fine-grained metamorphic rocks. Unstandardized cores and flakes consisted mainly of igneous material, and microlithic cores and tools such as thumbnail scrapers were predominantly of fine-grained cherts and chalcedonies.

The sample of tools contained a relatively high number of large blades and blade fragments, especially from the Tianshui He and Yako Basin sites, and most of these were made of fine-grained, vitreous volcanic rock. Many of the blades had been retouched with notches and scraper edges to serve as multipurpose tools. The microliths consist primarily of cores in various stages of reduction. "Waste flakes" from the microblade cores had often been retouched and used.

Because the artifacts were collected on the surface, it is difficult to age them. However, as Brantingham, Olsen, and Schaller (in press) noted, relative degrees of weathering provide a general indication of the length of time an artifact has been exposed. In general, the unstandardized flakes show the most weathering, the blades an intermediary amount, and the microblade cores the least, suggesting marked age differences between these technological types.

The Chang Tang artifacts reveal interesting similarities with the Upper Paleolithic technology of north China. The blades resemble those from the Shuidonggou site in Ningxia and the Shiyu site in Shanxi, where radiocarbon dates indicate an age of 20,000–30,000 years (Li 1993). Archeological material from a site in the Qaidam Basin is thought to be of similar age, based on radiocarbon dates of shells (see Brantingham, Olsen, and Schaller in press). These dates may also apply to some of the Chang Tang material. The microlithic technology in China probably arose much later, at the end of the Pleistocene, perhaps 10,000–12,000 years ago (Gai 1985).

These preliminary results show that nomadic hunters may have occupied the high elevations of the Chang Tang for at least the past 30,000 years, since before the last glacial maximum. This early date coincides with the Middle to Upper Paleolithic elsewhere in Eurasia and attests to the adaptability of these early humans. However, the timing of the occupation of the Chang Tang needs to be tested and confirmed with stratigraphic excavations.

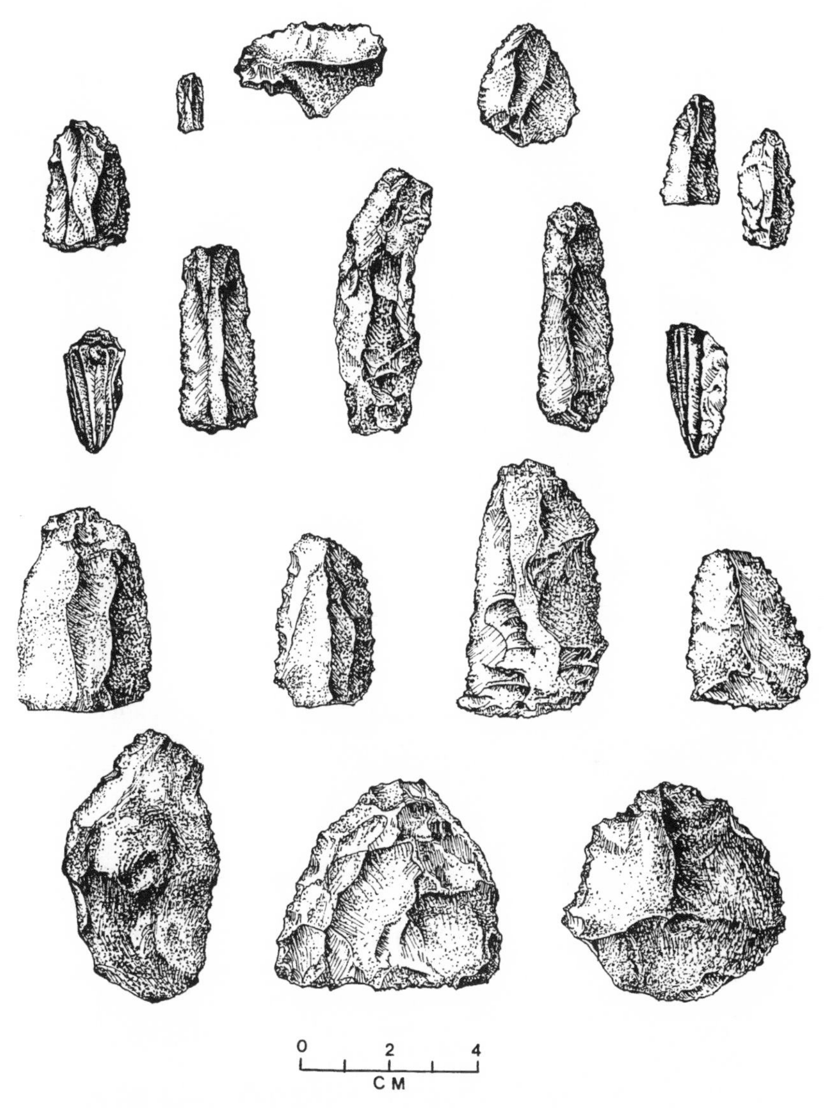

15.3. A sample of stone artifacts from the Chang Tang Reserve. Second row, extreme left and right: wedge-shaped microblade cores. Top row and three central artifacts of second row: small retouched prismatic blades. Third row: large retouched prismatic blades. Bottom row, two artifacts on left: side scrapers on flakes; on right: circular scraper on a flake. (Art by Sharon Wirt.)

Recent history

Until recently, the harsh climate and high elevations of the Chang Tang discouraged permanent settlements, although people had often traversed the region. A trade route existed from Rutog north to the oases along the southern edge of the Taklimakan Desert. A Mongolian general, Tsering Dondub, led his army from Khotan south over the Kunlun Shan and at an angle across the Chang Tang to Nam Co (Tengri Nur), from where he invaded and looted Lhasa on 2 December 1717 in a dispute over the succession of the Dalai Lama. Later that century, Emperor Chien Lung of the Qing dynasty opened a trade route from Nam Co north to Yutian (Keriya), but it was so difficult that it was abandoned during the 1800s (Hedin 1922). A major trade route followed the eastern edge of the Chang Tang from the Qaidam Basin over the Burhan Budai Shan across an uninhabited "no-man's-land," as Rockhill (1894) described it, through the Tanggula Pass to Nagqu. This route was used by Huc and Gabet ([1850] 1987), Przewalski (Prschewalski 1884), and others, and today the Lhasa-Golmud highway partly traces it.

Nomads have long taken livestock to the limit of good grazing, and hunters have penetrated uninhabited terrain, but their numbers were few and seasonal. Starting in Xinjiang, Hedin traveled south and saw no people until northeast of Linggo Co near Purog Kangri, where he met "three yak hunters; two or three heads and some hoofs were lying about" (1903). Taking a route similar to Hedin's, Bonvalot (1892) first encountered evidence of nomads in the southern foothills of Purog Kangri not far from the present-day Shuanghu. After leaving Xinjiang, Littledale (1896) first came across people near Garco. On his trip through southwest Qinghai into Tibet, Rockhill (1894) first saw people near Zige Tangco (Namru Co), close to the current reserve border at about 91° E. When Grenard ([1903] 1974) came south from Xinjiang, his caravan met no humans for two months, but then at Ringmo near Yibug Caka he found nomads. Crossing the Chang Tang from the northwest, Hedin (1909) had not seen any Tibetans for 79 days when he found a camp at Gomo Co. These nomads had come from Gerze, which lies over 200 km southwest. Hedin (1903) also met nomads in a valley east of Lemchung Co, southeast of the Aru Basin. At that time the Aru region had "an evil reputation for brigandage" and "five chukpas or robbers were caught here two years ago by Tibetan officials and beheaded" (Rawling 1905). Bower (1894) and Hedin (1903) also expressed their concern about bandits, and Deasy's (1901) camp southeast of the Aru Basin was attacked and looted of baggage and pack animals. Thus a century ago, the northern limit of pastoral activity was

15.4. Several villages exist now in areas which a quarter century ago were visited only seasonally by nomads. This village is located in the Amu Range, one of two belonging to the Garco commune. In the foreground is a *chorten*, or stupa. (October 1993)

much as it is today—the Aru Basin, Gomo Co, Yibug Caka, Garco, Purog Kangri—except that most nomads were seasonal.

The situation changed in the 1950s and 1960s when roads were built into the Chang Tang and administrative posts such as Shuanghu were established. In 1976 the government encouraged nomads to move north from the heavily settled areas around Xianza. Several communes did so, and after establishing and abandoning various settlements, largely because of a shortage of freshwater, the present communities of Rongma, Tsasang, and Garco (two villages) came into existence (fig. 15.4). In 1993, these three administrative units contained 271 households, totaling 1633 persons (table 15.1). The population has grown considerably since its arrival a quarter century earlier, in part due to a high reproductive rate (table 2.5) and in part to continuing immigration. For example, one family had fled Tibet in 1958 for northern Xinjiang, but in 1987 it returned without livestock, settled in Rongma because of available grazing land, and acquired a small herd. Garco increased from 36 to 64 families between 1976 and 1993. Attempts were also made by the government to settle marginal areas north of the traditional sites. In 1978, several households were moved into

Table 15.1 Number of pastoralists and livestock in those parts of Shuanghu and Nyima Counties that lie within the Chang Tang Reserve, listed by administrative unit *(xiang)*

	Families	People	Sheep	Goats	Yaks	Horses	Total livestock	Total SEU*
SHUANGHU COUNTY								
Xiti	244	1,380	80,156	26,201	4,355	1,198	111,910	139,600
Doma	209	1,212	56,916	18,391	4,026	1,111	80,444	106,887
Tsasang	109	706	37,163	17,191	1,834	358	56,546	66,173
Garco	79	439	28,067	8,524	2,197	93	38,881	49,416
Baling	177	938	46,965	13,964	2,844	675	64,448	82,113
Mema	191	955	38,681	11,920	3,455	434	54,490	73,568
Nouchuozhe (Tsolo)	340	2,194	91,860	48,171	5,834	1,376	147,242	180,675
NYIMA COUNTY								
Oju	271	1,542	60,212	40,572	3,533	693	105,010	122,538
Rongma	83	488	24,681	14,440	1,059	174	40,354	45,019
Hurdo	295	1,498	58,150	38,565	4,253	580	101,548	122,113
Total (both counties)	1,998	11,352	522,851	237,939	33,390	6,692	800,873	988,102

Source: Data from county records, 1993 (see Miller and Schaller 1996).
*SEU = sheep equivalent unit: 1 sheep = 1.0, 1 goat = 0.86, 1 yak = 6.0, 1 horse = 9.0.

the Yako Basin, about 175 km from the nearest nomads, but because of the isolation and shortage of potable water the people left the following year. A similar venture in the Tianshui Valley also failed. Yet the government continues to discuss plans for settling the uninhabited north even though it is marginal for livestock production.

The southwestern corner of Qinghai adjoining the Chang Tang Reserve was uninhabited until the construction of the Lhasa-Golmud highway during the 1950s. With road access and government assistance, many families settled the region. Now 328 households, with 1738 persons, are just east of the reserve in the Sewu and Jiri areas, and many more are near the highway. In the early 1990s some of these nomads began to graze their livestock seasonally in the reserve between Purog Kangri and Dogai Coring, a formerly uninhabited area.

Within the past 50 years, and especially the past 25 years, good rangelands of the Chang Tang that had few or no nomads have been permanently settled, and there is pressure to develop marginal pastures as well. The Chang Tang Reserve extends over parts of five counties, including Shuanghu, which is officially a Special Administrative Area, slated for development. In 1993, over 4100 households comprising 22,000 people used the reserve (tables 15.1 and 15.2). A few nomads venture into the reserve only seasonally; for example, households from Shenchen travel to the Aru Basin in summer to graze their livestock. But most reside in the reserve

Table 15.2 Number of pastoralists and livestock in administrative units (*xiang*) of Gerze, Rutog, and Amdo Counties that use the Chang Tang Reserve

	Families	People	Total livestock	Total SEU*
GERZE COUNTY				
Sumdo	127	709	45,600	—
Changdong	122	622	32,300	—
Donge	177	904	44,000	—
Luku	128	631	37,300	—
Dingu	136	703	31,200	—
Ebtsa	155	777	38,000	—
Shenchen	131	640	31,900	—
Jatso	130	658	41,800	—
Total	1,106	5,644	302,100	417,821
RUTOG COUNTY (incomplete)				
Dongru	78	337	14,550	19,536
Sumxi	51	219	9,490	13,706
AMDO COUNTY (incomplete)				
Zhaqu	167	872	47,219	68,501
Qiangma	315	1,634	82,628	127,139
Gangnyi	252	1,039	68,917	119,967
Marong	148	761	34,158	98,250

*SEU = sheep equivalent unit. See table 15.1.

permanently. These numbers do not include officials with families, construction workers, traders, and others whose occupation is not pastoralism.

The lure of gold has long enticed prospectors into the Chang Tang. Rawling (1905) came across an old gold seeker's camp east of the Aru Basin and an active camp about 80 km south of Lumajangdong Co. Prospectors invaded the Arjin Shan Reserve in Xinjiang and the eastern part of the Chang Tang in Qinghai during the early 1990s and killed much wildlife (Wong 1993; D. Miller, pers. comm.). In June 1994, we saw many gold seekers from Qinghai on their way to Nyima along the southern border of the Chang Tang Reserve, where gold had been found. A major potential threat to the reserve is the discovery of oil. Two exploration teams were in the reserve during 1994. Oil in commercial quantities has been discovered just east of Siling Co, where several drilling sites are active along the reserve border. Oil development would cause long-lasting damage with its roads, clutter of buildings, pipelines, and refuse.

Hunting

Western perception generally credits Tibetans with such reverence for life that they do not kill other creatures, a perception reinforced by descriptions such as this one by Migot:

> The sun was warm and, as we talked, everyone let down his *shuba* and embarked on a louse-hunt in its well-stocked coverts. It was a very humane form of sport, since Buddhism forbids its adherents to take the life of any living creature, however humble; whenever, therefore, anyone found a louse he placed it tenderly on the floor beside him. (1957)

A different view was expressed by Ekvall:

> There are many areas in Tibet, even far up in the *aBrog* [high pasturage], that are designated as animal sanctuaries where hunting is prohibited by decree, either because of Buddhist scruples or because it is considered poaching on the preserves of the mountain gods, some of whom are "gods of the hunt." Nevertheless, in wide areas, still teeming with game, the nomads are enthusiastic hunters. They carry firearms at all times; the demands of pasturing take them close to the haunts of the herbivores of the higher country; they hunt beasts of prey in order to protect their herds, and they are accustomed to taking life for meat and skins. It is most natural to turn to the herds of wild yak, orongo antelope [chiru], wild sheep, and gazelle when meat is, or may be, scarce, and to hunt onager [kiang], wolf, fox, bear, and lynx for pelts, musk deer for its pods, and stag for antlers in-the-velvet, which bring high prices. (1964)

In the agricultural areas of southern Tibet, where people were under the strong influence of Tibetan Buddhism, a segment of the population treated living beings according to decrees such as this one issued by the Dalai Lama in 1901: "Nobody, noble or humble, should do violence to them or harm them." However, nomads in the Chang Tang traditionally hunted to subsist and to obtain animal products for barter, as Ekvall (1964) noted. Life was so difficult that hunting did not reflect on religious sentiment, which was and still is strong, but on the need to survive. Hedin described a household at bare subsistence level:

> The two tents contained nine inmates—two grown-up men, two women, three girls, and two boys. The older man was named Purung Kungga, and he owned 150 sheep and 4 dogs, but no other animals. . . . In one of the tents lay the hides and meat of nine antelopes. The people lived almost exclusively on the game they caught in their snares. (1909)

To survive, a family of that size would need the meat of 30–40 sheep annually, and they would trade wool for tea and barley, which when roasted and ground becomes *tsamba*, a Tibetan staple. But a cull of so many animals would cut into herd capital or at least prevent herd recovery, leaving no reserves for the inevitable seasons of high lamb or adult mortality. The family's only alternative was to hunt.

Hunters often shot animals with muzzle-loaders. Since these guns were not effective at distances over 40 m, a hunter sometimes dug a shallow, circular depression near a wildlife trail and waited in it for the close approach of an animal. Dogs may be used to drive blue sheep to a cliff and keep them at bay while hunters shoot them at leisure (Goldstein and Beall 1990). The most efficient method of catching all but yaks and kiangs was with a leg-hold trap, as described by Bower (1894) and Rockhill (1894) a century ago. The trap consists of a ring about 16–18 cm in diameter, made of chiru horn which has been soaked until pliable. The ring is covered with yak yarn. Up to a dozen sharpened prongs of horn or wood converge down and inward. The trap is concealed over a meter-deep hole, and anchored with a rope to a buried horn or other object. When an animal steps through the ring the prongs prevent it from withdrawing the leg. Such traps are often placed on trails and, in winter, on the chiru's mating grounds (fig. 15.5).

Scattered throughout this book are representative quotations of subsistence hunting which I gleaned from the accounts of early travelers. Wildlife is hunted by the same methods today but with some important differences. Some nomads have access to modern rifles through purchase, on loan from the government ostensibly to protect livestock from wolves, or on illegal rental from officials who expect wildlife products in payment. Traders come in trucks to communities and individual households, making it easy for nomads to send chiru wool, wild yak meat, and other items to market. From the 1960s onward, killing became more and more commercial. There is an ethical distinction between killing to subsist and killing to plunder. Roads also made wildlife accessible to opportunistic hunters such as officials, truck drivers, and soldiers. As a result, wildlife numbers plummeted, a trend accelerated by an ever-increasing human population.

Although several species—snow leopard, kiang, chiru, wild yak—are fully protected by law in China, such laws had no impact in the Chang Tang, and indeed many officials seemed ignorant of them. In 1991, with the passing of the Tibet Wildlife Act, the situation began to improve. Within a short period, the Tibet Forest Bureau had made every official and household aware of the laws and began to enforce them. When I visited the county town of Baingoin in winter 1991, I was told that a local official had, in cooperation with the military, organized the killing of at least 1000 chirus annually but that this year's harvest was only about 200. The official was later arrested. Nomads and truck drivers were also apprehended for illegal hunting. Such incidents had a marked impact on the public, but not everyone heeded the laws, as a few personal experiences illustrate:

> 20 December 1991. A Party secretary from Nagqu arrives in Rongma on tour. In the back of his Toyota Land Cruiser is a freshly shot gazelle.
>
> 21 December 1991. We come across a camp with two hunters from Gerze in the hills west of Rongma. They had arrived 10 days ago and in

15.5. A nomad carries a muzzle-loader and three leg-hold traps. A trap is hidden on a trail, and when an animal steps into it, the sharp prongs hold the leg. (December 1988)

15.6. A pile of frozen chiru carcasses and several heads are in the camp of some poachers near Yibug Caka. (December 1991)

> that time killed 22 chirus [fig. 15.6]. [The meat was for their own consumption and the horns and hides would be sold.]
>
> 4 June 1992. A truck from near the Aru Basin arrives at a small trading post in Domar. In the back of the truck is a fresh bear hide and three male chirus, the latter shot en route to trade for supplies.
>
> 31 May 1994. We find a camp high on the slopes of Purog Kangri on the migration route of chirus. Three brothers from the Shuanghu area are in camp. Their kill so far consists of one yak bull and 9 chirus, of which three were pregnant. One of the brothers is a bankrupt trader who hoped to return to business by selling chiru hides. [They were later fined the equivalent of $1750, to be paid in sheep wool, and two rifles were confiscated.]

Wildlife is killed for meat, hides, and medicinal products. Tibetans generally do not eat carnivores, marmots, or hares, but the pelts of most species are useful trade items. Among ungulates, only the kiang is not hunted in some areas, but in others its meat is consumed. Many animal parts have medicinal value, and local people sell such products or barter them, as, for example, to the Tibetan Medical College in Lhasa, where they receive treatment in return. Mixed with other ingredients, blood from the heart of wild yaks reputedly helps to cure blood disease, chiru horn kills bacteria and stops diarrhea, kiang meat is good for digestive and kidney problems, and snow leopard bones reduce mental discomfort. Certain products bring a high price on the international market. Bear gall is said to cure ulcers and reduce fevers, musk deer glands are used to treat everything from asthma to pneumonia and typhoid, and over 10 deer parts, including antlers in velvet, penises, and tails, have curative powers over a wide spectrum of maladies. The market for animal products is so lucrative that hunters from other areas of China and even neighboring countries such as India and Bhutan enter Tibet.

I have devoted space to a description of hunting because it has been and continues to be the most urgent conservation issue in the Chang Tang Reserve. Of particular concern is the illegal trade in chiru wool, to which I have made passing reference but which requires a more detailed discussion.

The chiru wool trade

Chiru wool, known in the trade as "Shahtoosh" (King of Wool), is one of the finest animal fibers known (10–12 microns). This wool has for centuries been traded from Tibet to Kashmir (C. Jest, pers. comm.), where weavers produced scarves and shawls, often mixed with goat cashmere. The scarves were warm yet so light that they could be pulled through a finger ring, a sign of excellence. In northern India a mother would start saving to buy a scarf for her daughter's dowry as soon as she was born (A. Kumar, pers. comm.). In recent decades a market for Shahtoosh began to develop in Western countries and in Japan. On passing through Gerze in 1988, I watched traders from eastern Tibet in front of their tent plucking wool from chiru hides, and in a nearby shop were hides and sacks of wool (fig. 15.7). This wool, I later learned, would be smuggled into India by several routes: from the Tibetan border towns of Nyalam and Burang into Nepal; over the Lipulake Pass, which lies at the site where the borders of Nepal, China, and India meet (Talwar and Chundawat 1995); and west out of Tibet to Ladakh.

I have no idea how many chirus were slaughtered during the 1980s and 1990s except to note that tens of thousands of animals must have been killed. A chiru provides about 150 g of wool. In the winter of 1992 an estimated 2000 kg of wool reached India, and consignments totaling 600 kg were seized (and released) in India during 1993 and 1994 (Bagla 1995). This amount alone represented at least 17,000 chirus. Traders from Tibet use Shahtoosh to barter for tiger bones, which are used as ingredients in traditional medicines, a two-way traffic in endangered species. In Tibet, a truck driver was arrested with 300 chiru hides. Five households moved permanently to the Aru Basin in 1991 to hunt chirus. Profits were so great that they jointly bought a truck. The day before we visited these people in 1992, the truck had left with 50 chiru hides to be bartered in town for gasoline and other items. When we halted at a tent outside the Aru Basin

15.7. Nomads sell chiru hides to traders, who here pluck the wool for smuggling to India and Italy, where it is woven into scarves. (August 1988)

that year, the herdsman offered to provide us with any number of chiru hides for the equivalent of $28 each. In Xinjiang, Wong wrote that a "routine check of passing vehicles intercepted four trucks out of a 17-truck convoy and discovered 674 antelope skins. . . . At Huatugou, they tracked down one group with 360 skins and at Ulan, another group with 300 skins" (1993).

In 1993, a herdsman received the equivalent of about $60/kg for chiru wool, compared to $9.5/kg for goat cashmere and $1.0/kg for sheep wool. By the time Shahtoosh reached India it traded for $1250–$1500/kg. And in Western markets the scarves and shawls sold for $2000–$8500 each, depending on size.

Even though this international trade is illegal, in contravention of the Convention on International Trade in Endangered Species (CITES) since 1979, Shahtoosh became a style symbol in the West, "the ultimate emblem of New Age snobbery," as one newspaper article phrased it. Officials and buyers were usually unaware of how the wool was obtained, and traders made no effort to enlighten them. For example, the New York store Bergdorf Goodman advertised this "royal and rare" fabric in 1995:

> The source of the wool is the Mountain Ibex goat of Tibet. After the arduous Himalayan winter is over, the Ibex sheds its down undercoat by scratching itself against low trees and bushes. . . . A difficult process then commences as local shepherds, called Boudhs, from the region of Changthang, Tibet, climb into the mountains during the three spring months to search for and collect this matted hair.

Most supposed facts in this quotation are wrong, but the idea that a product was obtained sustainably by poor, hardworking, indigenous people made Shahtoosh an ecologically and politically correct luxury item.

During the early 1990s the Tibet Forest Bureau began a major effort to suppress the chiru wool trade. For instance, by 1993 a total of 1127 hides had been confiscated in Lhasa alone; the fine for killing a chiru rose to the equivalent of $118 (and for a wild yak, $588; a kiang, $59; and an argali, $35); and 10 chiru poachers were arrested in only a two-month period in 1993. Tibet and the central government provided funds for a guard force, and checkpoints were established at Shuanghu, Nyima, and several other sites. Antipoaching patrols twice penetrated the remote Koko Xili area of Qinghai during the winter of 1995–1996 and there encountered several teams of motorized Chinese hunters with high-powered rifles. Over 1600 chiru hides were confiscated. Concerned about the level of slaughter, China's State Council issued a directive in 1996 that chirus must receive better protection.

Pressured by two nongovernmental organizations (TRAFFIC-India and Wildlife Protection Society of India), the government of India began to enforce its laws by confiscating 172 Shahtoosh shawls in Delhi in December 1995. CITES in Italy and France took action against the trade. Several major fashion houses in Europe and North America removed scarves and shawls from their shelves. In February 1997, police in London seized 200 Shahtoosh shawls worth an estimated $0.5 million. These actions benefited the chirus, but the trade has continued, fueled by the wealth of some nations. Before the chiru slaughter had reached Qinghai and Xinjiang, I wrote that "the chiru as a species is not threatened with extinction, the bleakness of its habitat having so far protected it" (Schaller, Ren, and Qiu 1991). No longer. Remoteness provides little protection when an animal becomes a valuable commodity. The chiru as a species is now highly vulnerable, its future uncertain.

Local attitudes toward wildlife

The core of Buddhism is based on compassion and reverence for life. Therefore religion predisposes Tibetans against killing wildlife, except predators of livestock. Certain areas were at one time considered animal sanctuaries (Ekvall 1964), and monasteries protected wildlife in their vicinity (Harris 1991). Traditions tended to be submerged during the turmoil of the 1960s and 1970s, but they have reasserted themselves. In their study of nomads at Pala, south of Tangra Yumco, Goldstein and Beall noted that hunting is taking on the stigma of former times. They quote a herdsman: "We are Buddhists, and as such should not kill other creatures. Most nomads in Pala, therefore, are like myself and nowadays do not hunt. I myself did so to maintain my family during the Cultural Revolution when my wealth was confiscated and I was excluded from the commune. But these days I do not need game to survive and will not take the life of another sentient creature" (1990).

Cultures are always in flux. Nomad society is not the same as in 1959, and it will continue to change as outside market forces, population increases, "modernization," and emphasis on maximum livestock production rather than on subsistence affect the lives of people. In the reserve, the attitude of nomads toward wildlife was ambivalent. Tolerance for any species that affected livestock was low. Nomads near Nyima wanted kiangs shot because they grazed on winter pastures that households had saved for their sheep and goats. As one herdsman asked us: "Why are there government limits on the amount of livestock I can own when there are several hundred kiangs in the area? Kiangs are useless but eat much grass." Herdsmen at Garco shot wild yak bulls because they broke fences, drove off domestic yak cows during the rut, and were reputed to attack people. Some nomads expressed positive sentiments. "It is good to protect wildlife. But we would like to hunt *zhi* [chirus] because we are poor." In contrast to nomads, who were forbidden to kill wildlife, officials sometimes hunted for fun and profit. One chiru hunter in the Aru Basin told me: "If the officials obey the law and stop hunting, we will too." And one nomad said simply, "I like to see wild animals around."

It is widely believed that 100 animals will appear from the mysterious north, a place where few nomads have been, for every animal that is shot. Instead of correlating excessive killing with the decrease in animal numbers, some people blame the decline on the prohibition of hunting!

Nomads have a special antipathy toward wolves. Since the year 1642, the Dalai Lama had annually issued a decree "to prevent the killing of all animals, except hyenas and wolves," to quote a statement made by him in 1944 (*Tibet* 1992). An edict against wolves is still in effect in the reserve, and they are killed indiscriminately. Wolves prey on livestock, but their toll, though significant, is relatively modest. In 1993, the Rongma administrative unit *(xiang)* lost an estimated 0.7–1.0% of its livestock to wolves, and other *xiang*s reported similar or lower predation rates.

Monks, officials, biologists, and other concerned individuals need to begin a dialogue about conservation with nomads and provide them with spiritual reinforcement and facts. My attempt to spread the conservation message was limited to talking about the need to protect wildlife and to giving nomads a small card for the family shrine. One side of the card has a drawing of the eleventh-century Tibetan philosopher and poet Milarepa seated among peaceful animals, including a gazelle, and of a kneeling hunter who has laid down his arms, a sword, bow, and quiver of arrows. The other side has one of Buddha's sayings in Tibetan script:

> All beings tremble at punishment,
> To all, life is dear.
> Comparing others to oneself,
> One should neither kill nor cause to kill.

Livestock and wildlife

Nomads in some areas already view wildlife as being in competition for forage with their livestock, and such friction is certain to increase. Potential conflict cannot be resolved on the basis of emotion and preconception but must be addressed with detailed information and with imaginative solutions. Data on such topics as methods of livestock management, the total number of domestic and wild ungulates in an area, and the food habits of each species are particularly relevant.

Livestock management

Nomadic pastoralism developed during the same period as agriculture, probably less than 10,000 years ago (Goldstein and Beall 1990). Archeological excavations at the Karou site in eastern Tibet, which is about 5000 years old, revealed an agricultural community. Bones at the site were of wild animals, with the possible exception of domestic pigs (*Karou* 1985). A site in the Qaidam Basin indicated that domestic sheep may have been there about 4000 years ago (Miller 1994). Chinese sources from 1400 B.C. mention the Qiang, a tribe related to Tibetans, in the eastern part of the plateau (Stein 1972), where their descendants today practice both agriculture and pastoralism. Such a dual lifestyle is characteristic of Tibetans living below an elevation of 4300 m, and it seems likely that pure nomadism at high elevation evolved from it.

By the ninth century it was noted that Qinghai was "good for pasturing flocks" (Stein 1972). Penetration of the uplands was probably made possible, or at least easier, by domestication of yaks. A nomad family needs at least six yaks to transport household goods, and these animals also provide so many essential products that yaks almost define this nomadic culture.

By the eleventh century, nomads *('brog-pa)* had an identity different from other Tibetans *(brod-pa)* in local records (Stein 1972). Thus livestock and wildlife on the plateau have, in effect, coexisted for at least a millennium. But during this century wildlife has markedly declined as livestock increased. In Tibet alone there were 11 million sheep, 8 million goats, and 4 million yaks in the mid-1990s.

The life of nomads conjures up a romantic vision of people moving free as the wind toward distant horizons in search of good pastures. Tibetan nomads have had no such freedom in recent centuries (Goldstein and Beall 1990). Their movements have been limited to certain areas by the power of monasteries and aristocratic families to whom the nomads owed allegiance, paid annual taxes in butter and wool, and provided free labor. Bound by heredity to an estate, a family was not free to leave, but it owned its own livestock. The power of the landlords extended far into the Chang Tang, and only a few families escaped it, some by becoming bandits. A household was bound to certain pastures, which were adjusted and reallocated every three years according to the amount of livestock. (One yak was counted as the equivalent of 6 sheep and 7 goats.) There was little or no communal grazing land. Each household had a home base that it occupied much of the year. Livestock was shifted and pastures rotated according to local conditions, and often a satellite camp was established a few hour's walk from the base. To save forage near the base, livestock was usually moved to a distant area for several months between August and December (Goldstein and Beall 1990). Such movements were generally 15–50 km, but in the northern Chang Tang they sometimes extended 100 km or more.

After China took control of Tibet in 1959, the nomads continued their lives much as before except that the triennial pasture reallocation ceased. With the arrival of the Cultural Revolution in the late 1960s, private ownership of livestock was abolished, religious practices were barred, communes were established, and wealthy nomads were ostracized and persecuted as "class enemies." The new system was rigidly enforced by local Tibetan leaders. Forced sales of livestock and high taxes soon resulted in widespread poverty (Goldstein and Beall 1989). The nomads had lost control over their lives as never before. The situation changed abruptly in 1981, when communes were abolished and religious observances were again permitted. Every person received an equal amount of livestock. Taxes were eliminated and had not been reinstated by 1997. The lives of nomads soon improved. Within a decade there were again successful and wealthy nomads and poor ones, even though all had started equally (fig. 15.8). Poor nomads had mismanaged their livestock or sold it, and they now worked for the well-to-do on menial tasks such as building corrals and slaughtering livestock. Household wealth, as measured by livestock numbers, has direct relevance to wildlife conservation: nomads with little or no livestock are most likely to supplement their diet and income by killing wildlife. For example, of the 109 families in Tsasang, the poorest 17 families did the most hunting.

The communes were converted into local administrative units called *xiangs*, and these were based on the traditional household production system. Garco was an exception in that in 1981 it voted to remain a commune. Its families did as well or better financially than those in *xiangs*. In 1993 the average household in Garco had 492 head of livestock (625 sheep equivalent units), whereas the other households in Shuanghu and Nyima Counties within the reserve had an average of only 402 head (498 SEU).

Each *xiang* is coordinated by two Tibetan leaders (a Party Secretary and an administrative head), who supervise activities with varying degrees of rigor. Every household is officially limited to a certain number of animals per person. Depending on range condition, this number varies from about 40 to 70. If numbers exceed the quota, a forced reduction can be ordered without reference to the actual condition of the rangelands. This policy creates much resentment. Nomads had traditionally increased their flocks to provide a surplus for the hard times that were sure to follow heavy snows, disease, and other inevitable disasters. The current restrictions on numbers provide little margin of safety and they are viewed as a means by

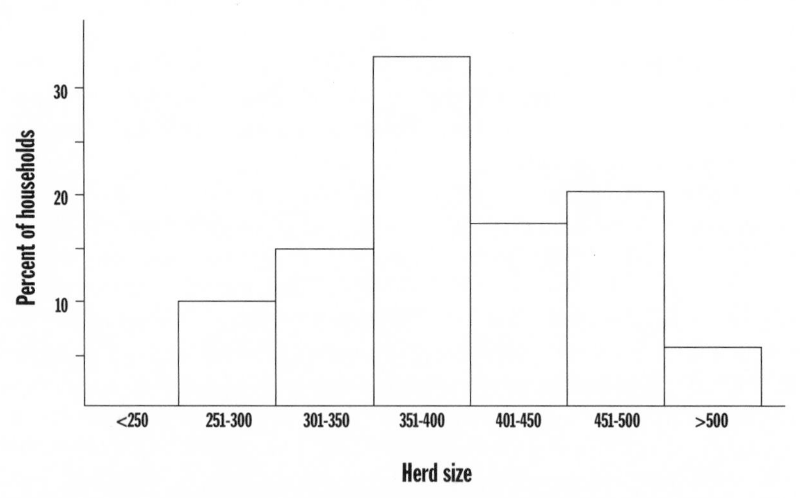

15.8. The total number of livestock per nomad household in Nyima and Shuanghu Counties within the Chang Tang Reserve. Households without livestock are not shown.

which the government keeps people poor. And nomads with little livestock are more likely to kill wildlife, as noted above.

Ecological comparisons between wild and domestic ungulates

Six species of wild and four species of domestic ungulates share the rangelands of the reserve. Species in this diverse assemblage range from 15 kg gazelles to 600 kg wild yaks. Whatever the specific requirements of a species, all depend on an ample resource base. Of the three major vegetation zones (see fig. 2.7), the *Carex-Ceratoides* zone, or desert steppe, is little used by ungulates, wild or domestic. Good grazing grounds are patchy in the *Stipa-Carex* zone. Consequently, few nomads inhabit it, and wild ungulates occur in moderate numbers only locally. The best rangelands are in the *Stipa* zone and there wild and domestic ungulates are most abundant. Supposed competition between livestock and certain species of wild ungulates has already become an issue there. Because argalis are rare and gazelles small, they are of little concern to nomads. Wild yaks have been almost exterminated in the *Stipa* zone, making any problem local. But nomads consider kiangs, and to a lesser extent chirus and blue sheep, competitors of livestock. Is this perception valid?

NUMBERS AND BIOMASS. Table 15.1 lists the livestock of *xiangs* in Shuanghu and Nyima Counties living within the reserve or grazing there seasonally. We obtained total livestock numbers for all eight *xiangs* in Gerze County that use the reserve but not the numbers of sheep, goats, yaks, and horses for three of these *xiangs*; we also lack data for a few *xiangs* in Rutog County, in the western tip of the reserve, and in Amdo County, at the eastern end (table 15.2). In the early 1990s, the total number of domestic animals in the reserve for at least part of the year exceeded 1.36 million and probably reached 1.4 million. Of these, 64% were sheep, 29% goats, 6% yaks, and 1% horses. The ecological density of livestock was 4.6–4.7 animals/km^2 in the reserve (334,000 km^2 minus the 11.6% composed of lakes, glaciers, and barren peaks). Domestic yaks thrive best on alpine meadow, which is scarce in the reserve except at the eastern tip in Amdo County. There a higher percentage of the livestock population consisted of yaks and a smaller percentage of goats than elsewhere (fig. 15.9). But numbers alone provide only a limited measure of the actual impact of animals on the rangelands: a yak is not the equivalent of a sheep. The various species can be converted into sheep equivalent units (SEU), based on feed consumption, in which 1 sheep equals 1.0, 1 goat equals 0.86, 1 yak equals 6.0, and 1 horse equals 9.0 (tables 15.1 and 15.2). Total SEUs in the reserve were 1.85 million, or 6.3/km^2. Biomass offers yet another means of comparing the contribution of each species to an area, and it is obtained by multiplying numbers by estimated weight of each species. Ideally the weight of each age class and sex should be calculated. I lack such information for livestock and instead used as an average the approximate minimum female weight provided by Epstein (1969), Bonnemaire (1976), and Cincotta et al. (1991). This overestimates weights of young animals and underestimates weights of males in sexually dimorphic species, but one presumably compensates for the other. Minimum total biomass in the reserve was 59.68 million kg, or 202 kg/km^2, of which sheep made up 43.8%, goats 16.6%, yaks 35.1%, and horses 4.5%. By all three measures the sheep was the dominant domestic animal in the reserve (see table 15.3).

In previous chapters, I attempted to calculate the number of each wild ungulate species based on transects, sample counts, and a large measure of guesswork. It is with great reluctance that I present specific figures, because they will be quoted as fact even though they convey only an order of magnitude. Furthermore, populations have not remained static in the years since surveys were made. Yet conservation requires the use of even tentative data upon which to consider action. According to my calculations, the reserve contained about 103,000 wild ungulates, or 0.35/km^2, of which the chiru was the most abundant (37%), followed by kiang (24%), gazelle (20%), and then blue sheep, wild yak, and argali (table 15.3). Sheep equivalent units, based on relative size, were assigned to each species: gazelle, 0.5; chiru, 1.2; blue sheep, 1.4; argali, 2.5; wild yak, 7.0; and kiang, 10.0. SEUs totaled about 373,000, or 1.3/km^2. By this measure, the kiang was the most important species (67%), followed by wild yak (14%) and then chiru (13%). Biomass of species was computed on the basis of the weights of the sexes and age classes (tables 15.3 and 15.4). Total biomass was 9.7–11.5 million kg, or 32.8–38.9 kg/km^2 of available habitat. The kiang had the highest biomass (58%), yak next (25%), and chiru, which was the most abundant species, third (10%).

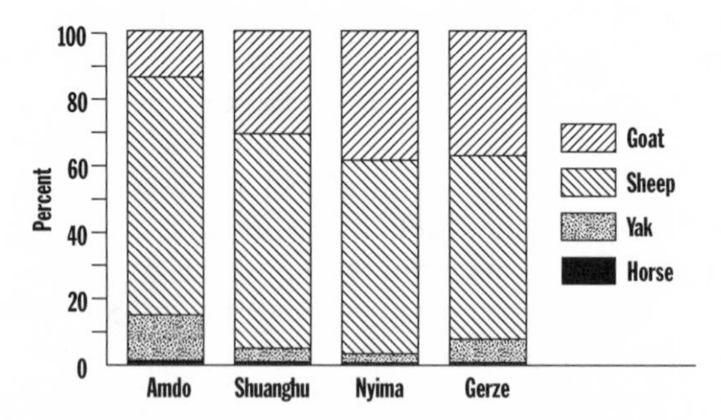

15.9. Percentage of livestock by species in several counties within the Chang Tang Reserve.

Table 15.3 Estimated number and biomass of wild and domestic ungulates in the Chang Tang Reserve

	Estimated total number	Number/km²	Biomass (kg/km²)	SEU*/ km²
WILD				
Argali	400–500	trace	0.09–0.12	trace
Blue sheep	10,000	0.03	1.36	0.05
Chiru	40,000	0.14	3.58	0.16
Gazelle	18,517–22,428	0.06–0.08	0.83–1.00	0.03–0.04
Yak	7,000–7,500	0.02–0.03	8.83–9.46	0.17–0.18
Kiang	21,743–28,006	0.07–0.09	18.15–23.38	0.74–0.95
Total	97,660–108,434	0.32–0.37	32.84–38.90	1.15–1.38
DOMESTIC				
Sheep	871,690	2.95	88.57	2.95
Goat	395,686	1.34	33.50	1.15
Yak	80,580	0.27	70.95	1.64
Horse	11,979	0.04	9.13	0.37
Total	1,359,935	4.60	202.15	6.11

Note: See tables 15.4 and 15.5 for population compositions and weights used in computations. The number/km² is based on ecological density: the 334,000 km² of reserve minus 11.6% (38,744 km²) for lakes, glaciers, and barren peaks, or a total of 295,256 km².
*SEU = sheep equivalent units. See table 15.1.

Table 15.4 Approximate population composition (in percentages) of wild ungulate species used in computing biomass

	Male	Female	Yearling*	Young
Argali	30	50	—	20
Blue sheep	33.5	44.5	—	22
Chiru	23	47	12	18
Gazelle	33.5	44.5	—	22
Wild yak	37	53	10†	
Kiang	42	42	5	11

*Yearlings were combined with adults in several instances because they were almost as large as adults late in the year.
†Figure combines yearlings and young.

Such comparisons for the reserve as a whole are of limited usefulness in evaluating the effect of domestic and wild animals on the rangelands because half of the reserve has little or no livestock. Much wildlife and livestock are concentrated in the *Stipa* zone of Nyima and Shuanghu Counties, an area of 91,520 km² of suitable habitat within the reserve. There were about 0.3–0.4 wild ungulates/km² and 8.7 domestic ungulates/km². Biomass was about 40 kg/km² for wild species and 345 kg/km² for

Table 15.5 Estimated number and biomass (kg/km²) of wild and domestic ungulates in the *Stipa* zone of Shuanghu and Nyima Counties within the Chang Tang Reserve during winter

	Body mass (kg) used in computations	Estimated total number	Number/km²	Biomass (kg/km²)	SEU*/km²
WILD					
Argali	♂ 105, ♀ 68, young 20	250–300	trace	0.19–0.23	trace
Blue sheep	♂ 60, ♀ 40, young 10	4,000–5,000	0.04–0.05	1.75–2.18	0.06–0.08
Chiru	♂ 38, ♀ 26, yrl. 23, young 15	7,000–8,000	0.08–0.09	2.02–2.30	0.09–0.10
Gazelle	♂ and ♀ 15, young 7	9,152–11,895	0.10–0.13	1.32–1.72	0.05–0.07
Yak	♂ 500, ♀ 325, yrl. and young 150	1,500	0.02	6.11	0.11
Kiang	♂ and ♀ 275, yrl. 200, young 50	9,152–11,898	0.10–0.13	24.65–32.05	1.0–1.3
Total		31,104–38,596	0.34–0.42	36.04–44.59	1.32–1.67
DOMESTIC					
Sheep	30	522,851	5.71	171.30	5.71
Goat	25	237,939	2.60	65.00	2.23
Yak	260	33,390	0.36	93.60	2.19
Horse	225	6,692	0.07	15.75	0.66
Total		800,872	8.74	345.65	10.79

Note: The approximate size of this *Stipa* zone is 91,520 km² (excluding 11.6% of lakes, glaciers, and barren peaks). See table 15.4 for population compositions used in the computation; see relevant species chapters for estimates of numbers.

livestock (table 15.5). Thus, wild ungulates composed 4.2% of the total ungulate population and 10.4% of the biomass. Even if wildlife numbers were underestimated by 100%, they would still constitute only a small fraction of the total.

However, certain limited areas may seasonally contain much wildlife, as shown by the Aru Basin. Schaller and Gu (1994) gave numbers and biomass for 1150 km² of habitat in the basin; the biomass figures in table 15.6 have been slightly adjusted from the previously published ones. A total of 1791 wild ungulates of five species (excluding argalis) were observed during a census in August 1990, and an estimated 2100 were present. Domestic animals were twice as numerous in the basin as wild ungulates, but in terms of biomass the converse was true, with 331 kg/km² for wildlife and 169 kg/km² for livestock. Wild yaks made up 73% of the wildlife biomass, followed by kiangs at 17% and chirus at 7%. In the *Stipa* zone, discussed earlier, kiangs contributed the most to biomass, with 70%, and yaks were in second place, with 15% (table 15.5). Harris and Miller (1995) reported on a large seasonal wildlife concentration in Qinghai's Yeniugou during summer 1991. They estimated 7098 wild ungulates in 1051 km². My calculations give a biomass of about 550 kg/km², 40% more

Table 15.6 Ecological density and biomass of ungulates in the Aru Basin (1150 km²) in August 1990

	Number censused	Estimated total number	Number/km²	Biomass (kg/km²)*
WILD				
Argali	0	trace	trace	trace
Blue sheep	121	200	0.17	6.72
Chiru	652	700	0.61	21.80
Gazelle	125	200	0.17	2.25
Yak	681	750	0.65	243.00
Kiang	212	250	0.22	57.37
Total	1791	2100	1.82	331.14
DOMESTIC				
Sheep	—	2750	2.39	71.70
Goat	—	1250	1.09	27.17
Yak	—	300	0.26	67.83
Horse	—	10	trace	1.96
Total		4310	3.74	168.66

Source: Schaller and Gu 1994.
*See table 15.5 for body mass of adults. Most chirus in the sample were males and most blue sheep were female and young.

than in the Aru Basin, and much higher than the 111 kg/km² livestock (table 15.7). The biomass of wild yaks was 52% of the total, kiangs 30%, chirus 7%, and then blue sheep, gazelles, and argalis, the same sequence as in the Aru Basin. This general biomass ranking may reflect actual conditions a century or more ago, and a change in the sequence may point to excessive hunting of a particular species, such as wild yak. The Aru Basin and Yeniugou may thus offer a nostalgic glimpse into the ecological past of the Chang Tang.

In sum, on good rangelands in the reserve, livestock greatly exceeded the wild ungulates in both number (25 : 1) and biomass (9 : 1). Wildlife competed with livestock at most in a minor or local manner. Not coincidentally wildlife density remained highest where nomad and livestock densities remained low.

FOOD HABITS. That food is a limiting resource in the Chang Tang can be deduced from the scarcity or absence of species in desert steppe. Elsewhere 10 wild and domestic species (excluding a few white-lipped deer, cattle, and domestic Bactrian camels at the eastern edge in Qinghai) share a simple habitat with a few plant species whose nutritional level is high only during a few summer months. In such an environment, animals must be opportunistic, and indeed each wild ungulate species forages in several vegetation types, and overlap in diet is considerable. Somewhat different

Table 15.7 Crude density and biomass of ungulates in the Yeniugou, Qinghai (1051 km^2) in summer 1991

	Estimated total number	Number/km^2	Biomass (kg/km^2)*
WILD†			
Argali	245	0.23	11.89
Blue sheep	1200	1.14	34.25
Chiru	2076	1.98	39.50
Gazelle	1511	1.44	15.81
Yak	1223	1.16	283.93
Kiang	843	0.80	165.23
Total	7098	6.75	550.61
DOMESTIC‡			
Sheep	2400	2.28	68.51
Goat	600	0.57	14.27
Cattle, horse, camel	$\leq$100	$\leq$0.10	28.00
Total	3100	2.95	110.78

Source: Based on counts by Harris and Miller 1995.

*Biomass of wild species was calculated on the basis of 3/4 adult female weight, not population composition.

†Excludes a few white-lipped deer.

‡Domestic yaks were grazed in the area only during winter. Adult camels average about 460 kg (Epstein 1969). It is assumed that cattle, horses, and camels each composed a third of 100 large animals.

feeding strategies help reduce competition (see chapter 12). Species tend to select for topography, with, for example, blue sheep near cliffs and chirus on rolling hills and plains. Diet selection is based more on plant type (monocot or dicot) than on plant species. The small gazelles pluck mainly forbs, the medium-sized ungulates are versatile consumers of both graminoids and forbs, and the large yaks and kiangs focus on graminoids. In winter, all species except gazelles feed mostly on grasses and sedges.

The domestic species have been inserted into this well-adapted mammal community, adding species which already have wild counterparts: the domestic yak is the dietary analogue to the wild yak, the horse to the kiang, and the sheep and goat to the blue sheep, argali, and chiru. Livestock forages in the same habitats and on the same plant species as the wild ungulates. Several factors have so far minimized competition. Wildlife density is now low and livestock numbers are as yet not high enough to degrade the rangelands. In winter, when livestock and wild ungulates sometimes concentrate in the same area, graminoids are senescent and tolerate heavy grazing. In summer, wildlife is dispersed, with most chirus even migrating into unpeopled terrain.

When, however, 200 kiangs congregate in an area that has been reserved as winter forage for livestock, the nomads feel, no doubt rightly, that there is competition for limited forage. In some places outside the reserve, villagers have asked the government to reduce blue sheep because they are said to deplete mountain pastures which had been set aside for spring grazing of livestock. Such problems need study and resolution.

It is noteworthy that wild species are generally larger than their closely related domestic counterparts even if all subsist on the same forage (Clutton-Brock 1992). Wild yaks exceed domestic yaks in size, argalis are much larger than domestic sheep, and blue sheep (a goat with sheeplike affinities) are taller and heavier than domestic goats. The argali is not the progenitor of the domestic sheep, and the blue sheep is not the progenitor of the domestic goat, but they are related (see chapter 13). The genetic potential of wild species for improving meat production of domestic species has been recognized by the Chinese government. Hybridization programs between wild and domestic yaks and between argalis and domestic sheep have been initiated. Domestic yaks do poorly in the northern Chang Tang, one reason that nomads have relatively few there, in contrast to wild yaks, which survive superbly on the meager pasturage. Because hybrids may produce more milk and meat and be better adapted to the environment than domesticated breeds, the conservation of the wild relatives is of basic concern to governments (McNeely 1990). The Chang Tang Reserve is an important reservoir of this wild genetic resource.

SEASONALITY OF BREEDING. All ungulate species in the Chang Tang breed seasonally, with rut and birth periods short and fixed, an indication that environmental conditions are rigorous but relatively predictable from year to year (fig. 15.10). The small- to medium-sized wild ungulate species mate in midwinter, a time when the fat reserves that were lost during rut cannot soon be replaced, and yaks and kiangs mate in summer. The timing of the rut is such that the young are born during the short season of plant growth. Chirus and wild yaks give birth early in that season, probably before females have had time to regain much weight after the lean winter, whereas gazelles, blue sheep, and kiangs have their young after nutritious forage has been available for over a month.

Domestic and wild yaks have similar breeding seasons, as do horses and kiangs. But sheep and goats give birth in winter, the former starting in February, when the weather is so severe that mortality of newborns from hypothermia would be high if herdsmen did not dry and shelter such animals. Late gestation and much lactation occur at a time of year when the female has been on a bare subsistence diet for months and has no access to nutritious forage. Compared to the wild species, sheep and goats seem to be poorly adapted to the Chang Tang.

The rut and birth seasons are critical times for ungulates. Domestic and wild species are in no obvious conflict in this respect, except during the mating time of chirus. Because chirus congregate at traditional sites in midwinter for the rut, intensive grazing by livestock at such sites could deplete the food supply, and daily disturbance by nomads could disrupt behavior.

Reserve management

The Tibet Autonomous Region government wrote in its 19 July 1993 administrative permit that legally established the Chang Tang Reserve that "every kind of effective measures must be adopted after the establishment

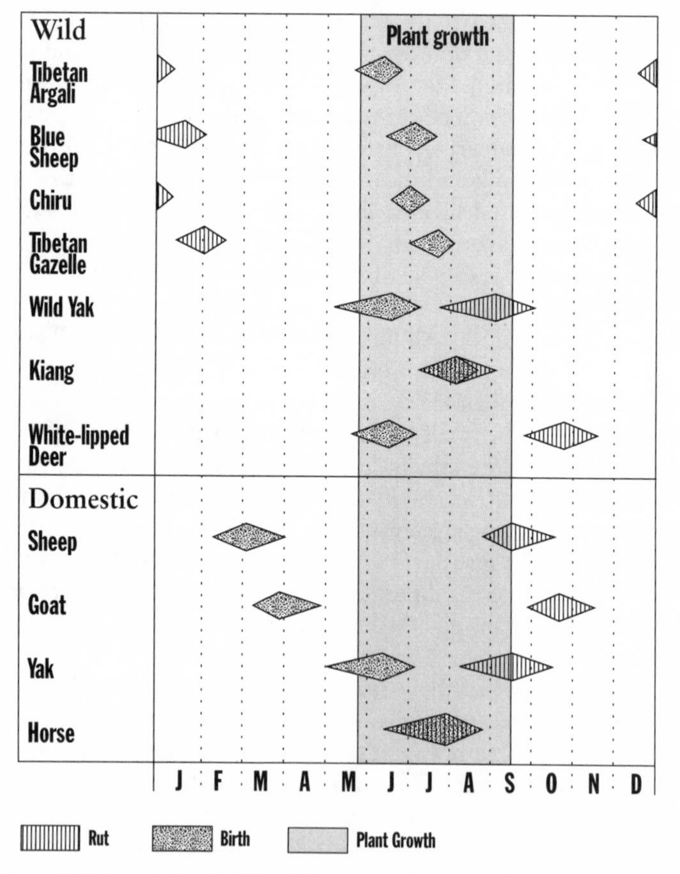

15.10. Rut and birth periods of wild and domestic ungulates in the Chang Tang. The widest part of each bar indicates the approximate peak of activity.

of reserves to protect all species of precious and rare animals that need protection, to make use of their ecological, social, and economic resources with efficiency, to bring reserves up to the national standard. . . ."

The government views the reserve as a multiple-use area in which the needs and aspirations of the nomads with their livestock, as well as those of the wildlife, must be considered in any management effort. This is the correct policy given the large size of the reserve and the many nomads with their large herds of livestock that reside there. The number of people is certain to increase greatly because of a high birthrate and immigration. Since most nomad children remain illiterate, the future offers them mainly a pastoral life. (Tibet has the highest birth and illiteracy rates in China.) Wildlife has been so unsustainably hunted that in the past decades numbers have decreased by as much as 90%. Ensuring the survival of this remnant and encouraging an increase in numbers are basic functions of the reserve. How can the needs of nomads and wildlife be reconciled?

Policy and research requirements

The influx of nomads into the reserve has been so recent that the rangelands remain, on the whole, in good condition. This provides an opportunity to manage an ecosystem before it has been damaged, a situation all too rare in today's world. Stresses on the rangelands still remain low and conservation problems are few, except for hunting, which, however, has been somewhat contained by the Tibet Forest Bureau since the mid-1990s. A management plan must now be designed for the reserve, one that will enable wildlife and nomads with their livestock to use the rangelands sustainably without degradation in the next millennium. A plan has to be innovative, flexible, adapted to addressing local problems, and constantly reviewed and modified to confront uncertainty and reflect changing ideas, methods, and circumstances. Conservation strategies useful today may not be effective in a decade or two. At the same time a plan needs a firm vision, one that can withstand the pressures of compromise when the long-term survival of wildlife is threatened. Any multiple-use area ultimately has to confront the issue of limits—limits on number of people, on number of livestock, on mineral, oil, and other development, and even on number of wild animals. For me to suggest specific limits on livestock and people would be presumptious, except to note that a continuing rapid growth of both will make management of the reserve increasingly difficult. Sustainability demands limits.

I have emphasized the need for flexibility in management because neither the wildlife populations nor the nomads and their livestock are in equilibrium—nor are they ever likely to be for long. Practices of animal selection, grazing rotation, and other forms of livestock management that

15.11. Nomads use domestic yaks to haul supplies or move camp to new grazing grounds. (September 1986)

were developed over the centuries and were efficient and well adapted to the Chang Tang (Goldstein and Beall 1990; Miller and Bedunah 1993) have in recent years been modified and conditions continue to change. About half of the annual calories in a nomad's diet derive from barley (Goldstein and Beall 1989). To obtain barley and other necessities, nomads once trekked south for three to four weeks to trade livestock products for such items (fig. 15.11). Now government trucks haul barley to *xiang*s at subsidized prices, and traders bring consumer goods. Poor families are on welfare, receiving free allotments of barley. When the price of goat cashmere rose on the world market during the early 1990s, nomads began to adjust their herds for more goats and fewer sheep. Since people prefer to eat sheep, a long-lasting change could have unforeseen impacts. In recent years many *xiang*s have fenced one or more pastures for winter grazing or to harvest hay. The government subsidizes 80% of a fence's cost. Some areas of Amdo County consist now of a patchwork of fences, with each household protecting the pastures of its flocks. Some families who lived in tents in 1990 had by 1994 built stone-and-mud huts (fig. 15.12). The owners had to pay for the roof beams but the government delivered the beams for free. Some *xiang*s and even a few households have purchased trucks—in some instances with profits from the sale of chiru wool—and

15.12. In recent years, many nomads have built mud-brick huts as a permanent base. (June 1994)

these are used for trading, hauling household goods between camps, and going on pilgrimage to Lhasa.

The standard of living of the nomads has improved greatly since 1981. But as Goldstein and Beall noted, "by objective measures most of these nomads are still very poor" (1989). In 1981, the nomads they studied had an average of 42.4 head of livestock per person, and by 1988 only 12% of the 57 households had more than 70 head per person. In Nyima and Shuanghu Counties, the figure was 70.5 per person in 1993 (table 15.1), an indication that the nomads in the reserve were comparatively well-off. Although nomads retain their basic way of life and cultural values, both internal changes and outside forces have transformed their existence in the past few decades. What are the trends for the future, especially as they might affect wildlife? Will changes in the economy and government policy lead to overstocking?

I can easily visualize a situation similar to that in the American West over a century ago when the open range was claimed as private land and fenced. In many nomad cultures, such as the Turkana in Kenya (Coppock, Ellis, and Swift 1986), a family exploits a huge area of communal land and may move camp a dozen or more times a year. In contrast, Tibetan nomads generally lack communal land: each household has pastures assigned for its own use, a tradition both in the past and today. Daily herding of sheep and goats in wind and snow is a cold and monotonous task. Will households want to fence all their pastures? Fences would disrupt the movements of wild animals and exclude them from critical winter ranges. Extensive fencing is clearly incompatible with the aims of the reserve.

A Tibetan official in Lhasa told me, "Herdsmen don't know how to use grasslands rationally." Nomads are thought to be inefficient in wool and meat production, and their grazing practices are viewed as maladaptive. Scientific management is the official policy. As Goldstein and Beall (1990) have shown, this viewpoint has sometimes led to unnecessary, and indeed harmful, intervention in the local livestock production system. It is clear that animal husbandry by nomads in various parts of the world is a highly complex biological, social, and economic system which may be locally well adapted (Coppock, Ellis, and Swift 1986; Western and Finch 1986; Miller and Bedunah 1993)—as long as human and livestock densities remain relatively low. But when densities disrupt the equilibrium through overstocking, the results are degradation and erosion. Official ideas concerning the effects of traditional grazing practices in Tibet may in part have been shaped by viewing the ravished rangelands in such heavily populated areas as parts of Qinghai (Cincotta, Zhang, and Zhou 1992), Amdo County, and large tracts of southern Tibet. The last-named region has seriously deteriorated over the centuries, a process hastened by fuel collectors who pull shrubs out by the roots. Critical points here are that livestock management must be evaluated locally on its merits and that policies and practices cannot be extrapolated from one type of terrain or vegetation to another: a management practice that is suitable for the expanses of alpine steppe may have disastrous consequences for alpine meadows on steep hills. Given current market incentives that encourage herd expansion, traditional practices alone may not offer an adequate solution to managing rangelands. But as studies of pastoral societies have concluded (Ellis and Swift 1988; Goldstein and Beall 1990), development policies must build on existing traditional practices and not simply discard them.

Ellis and Swift (1988) identified two forms of stability in grazing systems. In the nonequilibrium system, found mostly in hot deserts, range productivity is more a function of climate than number of livestock. Droughts are common and plant growth is dependent on erratic showers. The vagaries of weather make it difficult to measure the impact of livestock and to monitor resources. In the equilibrium system, the timing of plant production is predictable each year, and the carrying capacity of rangelands is relatively constant. The Chang Tang is of the latter type, although a severe snowstorm may have the same effect as a drought. Range conditions can be closely monitored. But such monitoring requires extensive knowledge, something not yet available. The chemical and physical structure of terrain and soils influences the way herbivores use and affect the habitat, yet little is known about this subject. A detailed vegetation map needs to be prepared, and the nutritional status of forage species and their suitability as livestock and wildlife feed should be investigated, as should the response of these plants to grazing. The food habits of all herbivores, including such small ones as pikas, require more study. Information on wildlife numbers and trends in populations is basic. Intermittent surveys of nomads and their livestock should determine herd size and compositions, changes in herding practices, and the economics of marketing livestock products. Livestock management in a changing economy is obviously a complicated issue, and the wildlife adds a further level of complexity.

The government is primarily concerned with boosting wool and meat production for the market—but with little consideration for the needs and cultural values of the nomads. Because households generally remain at a subsistence level, their first priority must be to produce milk and meat for their own consumption and wool for barter. Any development program that ignores this fact may destabilize the system. Energy conversion efficiency of milk from pasture is 14%, compared to 3% for meat (Western and Finch 1986). A policy that emphasizes meat production could have long-term consequences on the nomads and the rangelands. We also need "a better understanding of the social dimensions of rangeland ecosystems, including social values attached to livestock and livestock management practices, land tenure and community interactions" (Miller and Bedunah 1993) and, it might be added, attitudes toward wildlife. Proposed development plans for the Chang Tang include more fenced pastures, introduction of exotic grasses, and the application of fertilizers to rangelands. The potential impacts of such measures have not been investigated.

As this discussion conveys, any program of livestock development and wildlife conservation must involve the nomads in planning and implementation. It becomes natural science in a social context. Programs should ideally develop at the local level and be based on local data instead of being imposed from the top. However, they have to be organized and monitored at an official level by a management authority with accountability and clearly delineated responsibilities. The need for a multidisciplinary approach is obvious; the one used in Tanzania's Serengeti National Park since the 1950s can serve as a model (Sinclair and Arcese 1995).

The complex problems of biodiversity conservation, environmental protection, economic development, and welfare of a traditional culture in the Chang Tang should be of interest to and the focus of assistance from international development agencies, yet none have so far become involved (Miller and Schaller 1996). At one such agency in Beijing, I was told,

"There are no serious problems there and so our involvement is not necessary." The logic of waiting until an area is degraded before taking action is incomprehensible to me.

The Chang Tang attracted me because of its unique wildlife. I soon realized that the future of this wildlife depended on the nomads, and I noted with apprehension that their changing lifestyles would have an ever-increasing impact on the wild animals. This discussion has reflected such apprehension. My research results are fragmentary, but they do provide a baseline, and more important, they raise issues, ask questions, and pinpoint problems that require thought and action. The reserve is a multiple-use area and this implies that wildlife can, in effect, claim as much right to the land as the nomads and their livestock. This basic premise should always be remembered when contemplating action.

Harvesting of wildlife

Hunting of all species except the wolf is currently forbidden in the reserve. Such a protectionist policy does "not acknowledge reality of wildlife use by communities in developing countries," according to Allen and Edwards (1995). But it is precisely because of this reality that protection is at present essential. The hunters in the Chang Tang kill according to economic rules, not biological rules. Unregulated exploitation has decimated one of the world's great wildlife populations and, given a free reign, hunters would have further depleted the resource and possibly exterminated certain species. Populations need time to recover. However, there is now widespread realization that a wildlife resource can be managed on a long-term basis only if local people have a positive attitude toward it and if they derive benefits from it (West and Brechin 1991).

Most nomads in the reserve are recent immigrants without a deep-rooted spirit of stewardship for the land. An incentive is required to stimulate the people to protect wildlife and rangelands. One can argue that the nomads already have certain privileges associated with living in a reserve, such as pastures in good condition, and that being there entails obligations and restrictions. However, the fact remains that wild ungulates will be viewed increasingly as livestock competitors and will continue to be regarded as potential food and income sources. The challenge is to devise a means of benefiting people directly or indirectly from the presence of wildlife and to raise funds for reserve management. Governments usually claim lack of funds to explain laxness in caring for natural resources. Unless the Chang Tang Reserve ultimately funds itself as much as possible, any management program is unlikely to persist for long.

Tourism is sometimes viewed as a panacea for natural areas in that it may assist the local economy and stimulate conservation. If properly managed, tourism can indeed bring benefits. But the reserve is too remote, too high, and with too little variety to entice tourists in even moderate numbers. As Harris (1993) has shown for Qinghai's Yeniugou, only a few specialty tours would come annually to view wild yaks and chirus. While such tours would help to raise awareness of the natural treasures hidden in the Chang Tang, their contribution to the local economy would be small.

A sustainable harvest of certain species could benefit the reserve and all nomads living in it—but only after the wild ungulate populations have recovered. Residents should be given the responsibility of maintaining the wildlife, with benefits linked directly to number of animals (Harris 1993). Such responsibility would include a prohibition on all casual, subsistence, and commercial hunting for immediate private gain and the expulsion of outsiders who enter the reserve to hunt. To prevent outsiders from settling in the reserve to obtain the special benefits there, further immigration should be banned. The reserve offers two luxury products of international interest: wildlife trophies and chiru wool.

TROPHY HUNTS. Trophy hunting by foreigners on the Tibetan Plateau began in Qinghai, Xinjiang, and Gansu during the 1980s (see chapters 4 and 5). Tibet is now also considering such hunts. Local governments find trophy hunting an attractive way of making money because foreign intrusion is brief, logistics are simple, and profits are great. Usually trophy hunting benefits the hunter and the government, not the animals and local people. A more acceptable approach has been applied with some success in Qinghai's Dulan International Hunting Reserve, located in the Burhan Budai Shan just east of the Yeniugou. Between 1985 and 1990, 51 foreign hunters killed 59 wild ungulates, 82% of them blue sheep and the rest gazelles and white-lipped deer, for a gross income of $330,000. Of this, 55% went to the government agencies and 45% to reserve management, including salaries for a local guard force and elementary education for nomads (Liu 1993).

Cai, Liu, and O'Gara (1989) and Harris (1993) suggested a similar approach to conserving the Yeniugou. The principal species of interest to hunters there are wild yaks, Tibetan argalis, and chirus. All three are considered to be endangered by China, and the yaks and chirus are fully protected (Class I); provincial authorities cannot issue a hunting permit but the central government may. All three have been listed in Appendix I of CITES since the 1970s, meaning that most countries prohibit the import of trophies. In the United States, the Endangered Species Act also prevents hunters from bringing such trophies into the country. In spite of such complete protection, two North American hunters, Donald Cox and James Conklin, accompanied by biologist Richard Mitchell, received permission from Xinjiang to shoot two chirus in 1987 (Schapiro 1992), and a Spaniard shot one in 1992. In a well-protected area, an annual kill of 2% of a population for trophies can readily be sustained, and the potential income from yaks, argalis, and chirus is great (Harris 1993). Should the Yeniugou become a hunting reserve?

The Yeniugou, only 4000 km^2 in size, is the finest and most accessible wildlife area in Qinghai. To make it a hunting reserve rather than a nature reserve conveys that endangered species have only monetary value when in fact national pride, ethics, aesthetics, and religious sentiment are also part of a people's value system. Some areas are so unique and special that they should be maintained solely for their own sake, and the wildlife in them should not have to depend on market forces for survival. Surely a few percent of the funds taken by the government from hunts in the Dulan International Hunting Reserve and elsewhere can be devoted to maintaining the Yeniugou as a nature reserve. Hunting and tourism in the same small area are not compatible, because animals become so stressed after being pursued that they do not allow close approach for viewing and photography (Hutchins and Geist 1987). The Yeniugou has admittedly various problems that will make the management of a nature reserve difficult, principally a continuing influx of Tibetan and Mongolian nomads who, unlike the nomads in Tibet, lack allocated rangelands in that area (Harris 1993), as well as much illegal hunting.

I have discussed the Yeniugou because Harris (1993) has already considered the pros and cons of trophy hunting there and I disagree with him. Should animals be killed for trophies in the Chang Tang Reserve, which is so large that it could be zoned, with a small portion designated as a hunting area? Blue sheep and gazelles are readily accessible outside the reserve, argalis are now too rare to permit hunts anywhere in Tibet, and kiangs are not coveted as trophies. Wild yaks and chirus would be in demand, but both are protected by international convention. To permit wealthy foreigners to kill animals for pleasure soon after the nomads were prohibited from hunting them for subsistence might not promote a positive local attitude toward conservation even if a community derives some material benefits. Trophy hunting should not be allowed at present.

SUSTAINABLE HARVEST OF WILD UNGULATES. Exploitation for individual gain has been wholly unsustainable in the reserve, and the best hope for the survival of large wildlife populations is ultimately a regulated harvest that benefits all nomads, not just a small percentage as the illegal kill does now. The meat of several wild ungulate species, such as kiang and blue sheep, could some day contribute to the subsistence of nomads again. But it is the chiru that has the greatest economic potential because of the international luxury market for its wool. It may seem ironic that I worked for years to protect the chirus and suppress the wool trade, yet now suggest that a harvest might some day resume. Having observed the uncontrolled slaughter, I believe that the chiru's valuable wool could help to save the species from extinction. A combination of chiru and livestock management could also lead to a more sustainable use of rangelands, with nomads careful to maintain good chiru habitat: after all, a chiru is more valuable than a sheep.

As the saiga antelope and vicuña have shown, a species can recover after declining to very low numbers and become an important wildlife resource again. The migratory saiga is found on the steppes of Kazakhstan and the Kalymkia region of Russia. During the late nineteenth and early twentieth centuries the saigas were killed so indiscriminately for horns, meat, and hides that they were reduced to a few small, scattered populations. One Russian customs office alone recorded the export of 3,950,000 pairs of saiga horns to China for use in traditional medicines. The Soviet Union prohibited hunting in 1921 and the saigas soon increased and expanded their range until by the end of the 1950s there were nearly 3 million (Teer 1991). Commercial hunting was resumed in 1951, but this time the saiga was managed sustainably by state-controlled cooperatives which guarded the animals, monitored them, and harvested them, with the cull conducted by professional teams in autumn before the rut (Milner-Gulland 1994). The harvest reached 500,000 in some years (Reed 1995). Effective management ceased with the dissolution of the Soviet Union.

The vicuña, a camelid, had problems similar to those facing the chiru at present. It inhabits the grasslands of the Andes at 4000–5000 m, where it once numbered perhaps one million animals. Relentlessly slaughtered for meat and wool, the vicuña declined until by the mid-1960s only about 10,000 were left (Rabinovich, Hernández, and Cajal 1985). The vicuña's wool has an average diameter of 12 microns (Otte and Hofmann 1981), second only to the chiru's in fineness, and it is much in demand on the international market. Then, well guarded by frequent patrols, the population increased to an estimated 84,000 by 1981. An economic incentive now had to be found to compensate the local people for allowing vicuñas and livestock to share the communally owned pastures. In 1987, CITES moved several vicuña populations from Appendix I to Appendix II, which permitted the export of vicuña cloth woven in Peru. During annual roundups, vicuñas are corralled, sheared like sheep, and released. The local people became involved in and received part of all benefits from the management and harvest activities. A 1994 CITES ruling permitted a limited export of raw wool rather than only cloth. Poaching increased immediately because it is impossible to determine the source of the wool on the overseas market. Tibet might ultimately wish to establish its own weaving industry for chiru wool.

The saiga and vicuña each represent a major conservation success story that could be emulated with the chiru. The Tibet Forest Bureau has discussed the possibility of captive breeding. To establish and maintain a viable breeding program would take years and require a huge investment of funds—and it will ultimately fail unless all the difficult problems of captive management are resolved. Decades of effort to rear musk oxen for their wool have not been self-supporting because of various unanticipated costs such as that of fencing, supplemental feeding, and problems in harvesting wool, and because of a decline in reproduction and health of captive herds (White, Tiplady, and Groves 1989; Groves 1995). Furthermore, such a program would serve no role in the preservation of chirus in the wild. Shearing the wool from live wild animals, as with vicuñas, is another option. But attempts to capture the fleet and fragile chirus in large numbers would be difficult and cause mortality either during the roundup or afterward because the shorn animals would lack protection from the cold. However, a harvest of animals for meat, horns, and hides could succeed with a proper management program. Saigas and vicuñas differ from chirus in various aspects. For example, saigas can sustain a much higher harvest rate than chirus because they are more fecund, breeding at a younger age and often bearing twins, and vicuñas are easier to manage because they are sedentary, with most animals in small, territorial herds. However, the saiga and vicuña programs can serve as models and provide lessons on a complex management issue.

No chiru harvest can be considered until much more information about the species is available. What is the size of each migratory population? What age and sex should be harvested? What percentage of the population can be safely removed? Natural mortality of young is high and it varies from year to year, requiring annual monitoring as a basis for calculating the harvest. Population models and computer simulation models have been developed for various species, including vicuñas and saigas. Such models help in evaluating data, identifying critical population processes, and predicting how a population will react under different harvest strategies based on number, age, and sex of animals culled (Rabinovich, Hernández, and Cajal 1985; Franklin and Fritz 1991; Milner-Gulland 1994; van Rooyen 1994), and a similar effort is needed for chirus. What is the most effective method of harvesting? Trained field biologists are essential for such tasks. The saigas and vicuñas had recovered quite well within 20 years, and the chirus have a similar potential. Planning for that goal should begin now.

Conclusion

My vision for tomorrow is the past when humans, livestock, and wild animals lived in the vast steppes of the Chang Tang in ecological harmony. The beauty of these steppes and peaks will persist, but without wildlife they will be empty and the Tibetans will have lost part of their natural and cultural heritage.

As Goldstein and Beall wrote: "The nomads of Pala like their way of life and want to maintain it in the years ahead, choosing to incorporate or ignore new items as they see fit. They want nothing more than to be allowed to pursue the life of their ancestors, and flourish or fail as the gods and their own abilities dictate. Although there are problems and issues yet to be resolved, for now, and for the foreseeable future, the nomadic pastoral way of life is alive and well on the Changtang—and all of us are richer for it" (1990). We should strive to provide the same for the chirus, kiangs, wild yaks, wolves, and others that inhabit these unbounded uplands.

Guidelines for Conservation Action in the Chang Tang Reserve

THE PRIMARY objective in managing the Chang Tang Reserve is to maintain an undamaged ecosystem with vigorous populations of all native animal and plant species. To achieve this objective, it will be necessary to provide the nomads with economic benefits by permitting livestock grazing and ultimately by offering a sustainable harvest of chirus and possibly other wild ungulate species. Until a management plan is available, an interim program of conservation action is needed to protect the resources. Because most of the eleven points listed below involve policy guidelines, they could be implemented quickly and at little cost.

1. A management committee for the Chang Tang Reserve should be established to formulate policy, initiate and monitor programs, and seek funding. Such a committee would consist of scientists, managers, and officials from the Tibet Forestry Department, the Animal Husbandry Department, the Tibet Academy of Social Sciences, and other relevant organizations, as well as representatives from the nomad communities.

2. Nomads have left the northern half of the reserve (north of about 33°30′ N) virtually unoccupied because it offers poor grazing. This marginal area cannot contribute much to livestock production, and no further attempts to settle it should be made. Human intrusion into this part of the reserve should be prohibited, except for brief trips on special permit, to give the wild animals a safe and undisturbed haven. The calving grounds of chirus are there and the survival of wild yaks depends on such large uninhabited tracts.

3. The Aru Basin had and perhaps still has the finest seasonal concentration of wildlife in the reserve, and as such it has valuable potential for research and tourism. But an influx of nomads and hunters began to degrade the area in 1991. The basin should be designated as a special zone in which livestock grazing is prohibited or restricted to a few seasonal households with a limit on number of domestic animals. Hunting should be permanently banned. Similar special zones might also be necessary in other areas that are of importance to wildlife, such as key mating grounds of chirus.

4. Fences are a threat to wildlife. An occasional fenced pasture of a few

hectares poses little problem, but long fences, as at Garco, are not compatible with the aims of the reserve. A policy on fencing is necessary.

5. Human population growth will make sustainable management of resources in the reserve increasingly difficult. To limit the rate of growth, no nomad household should be permitted to migrate into the reserve and settle there.

6. Illegal commercial hunting has been the main cause for the decline of wildlife. Enforcement of wildlife laws should continue and the patrol effort should be increased, with particular emphasis on apprehending traders and other middlemen. Only after wildlife populations have recovered should a sustainable harvest under strict controls be considered. The wildlife resources, including any harvest programs, should ideally be co-managed by a local community and the government, with regulations and quotas established jointly for a particular area. The current policy of killing all wolves is not in the spirit of conservation. Wolves that habitually prey on livestock may be eliminated, but to others the gift of life should be extended.

7. Methods that help nomads increase economic returns from their livestock products should be sought to offset in part restrictions on livestock numbers. Such initiatives could include a system that enables households to sell wool and other products at a higher price directly to markets and the development of a home industry of weaving items for international sale.

8. A conservation education program for the inhabitants of the reserve is essential to convey the spiritual and economic benefits of good resource management. Communal discussions led by monks, wildlife biologists, social anthropologists, and others would be of value, as would printed materials.

9. Extractive activities, such as for gold and oil, will increasingly intrude on the reserve. Firm regulations concerning development must be established, regulations that limit construction of roads and buildings, enforce removal of debris and litter, and in other ways ensure that the reserve retains its beauty.

10. A long-term study of the ecology and behavior of wild ungulates and large carnivores should be initiated, at first with emphasis on conducting censuses of each species, tracing movements, and monitoring populations to record numbers, fecundity, and mortality. Potential competition in food habits between wild and domestic ungulates, particularly between livestock and kiangs, also needs immediate investigation.

11. The reserve should be designated a UNESCO World Heritage Site under the World Cultural and Natural Heritage Convention because of its "outstanding universal value." The Convention provides aid for the

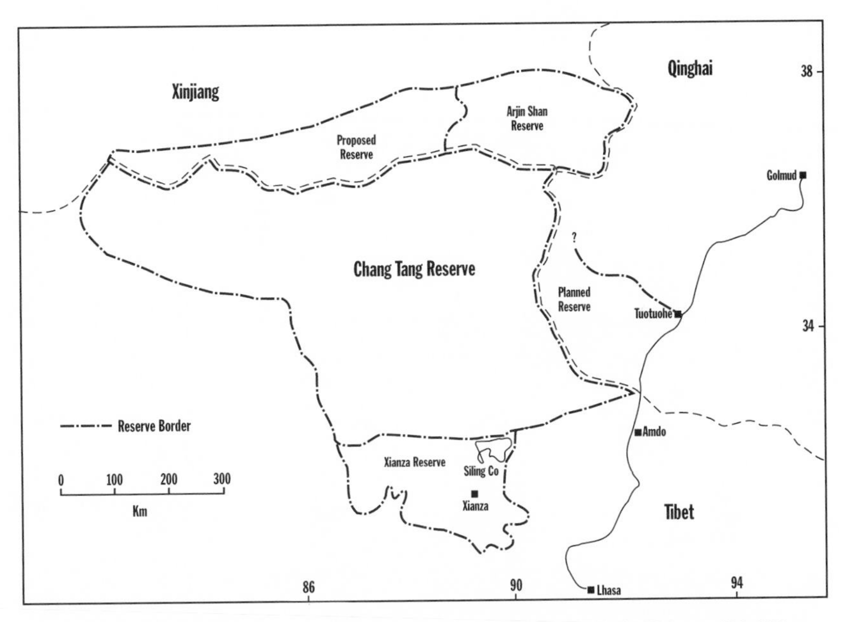

Current and proposed reserves that would offer protection to a total of about 500,000 km^2 of the northwestern part of the Tibetan Plateau. Such protection would enable both the wildlife and the nomads with their livestock to continue their traditional lives.

protection of sites. The reserve should also become a biosphere reserve within the Man and Biosphere Programme, which emphasizes the integration of traditional patterns of land use with protection of natural ecosystems.

Much of this eleven-point program applies equally to other reserves and proposed protected areas that are contiguous with the Chang Tang Reserve (see the map). The Arjin Shan (45,000 km^2) and Xianza (40,000 km^2) Reserves already exist. Xinjiang is considering a westward extension (50,000 km^2) of the Arjin Shan Reserve. In the near future, the southwest corner of Qinghai will be transferred to Tibet. Once the transfer is completed, the Tibet Forestry Department plans to add about 60,000 km^2 to the eastern margin of the Chang Tang Reserve. Thus, around 500,000 km^2 (192,500 square miles)—an area twice the size of Great Britain—would be devoted to wildlife conservation, an extraordinary initiative by China.

Epilogue

I CANNOT PREDICT the end of my involvement with the Chang Tang. The last data in this book were collected in 1994. But in June and July 1997, I was once again in these uplands, surveying wildlife in an area that I had not seen before. Together with Liu Wulin and Zhang Hong of the Tibet Forestry Department and Lü Zhi of Peking University, I traveled cross-country through a region that is known as Koko Xili in southwest Qinghai, adjoining Tibet's Chang Tang Reserve. The region is remote and high, usually between 4700 and 5000 m in elevation, and it has a number of large lakes, among them Ulan Ul Hu, Xijir Ulan Hu, and Hoh Xil Hu (see figures 1.4 and 3.2). Much of the Koko Xili is barren or covered sparsely with cushion plants and the coarse sedge *Carex moorcroftii*, too barren to support nomads, but south of about 35°N is a fine steppe. Part of this region will probably be transferred from Qinghai to Tibet before the end of the century and incorporated into the Chang Tang Reserve. This epilogue presents a summary of our journey.

The purpose of the survey was twofold. One goal was to ascertain the status of wildlife. Reports had reached us that nomads, gold prospectors, and motorized, armed gangs of Tibetan and Chinese poachers from as far away as the towns of Xining and Golmud in Qinghai and of Nagqu and Chamdo in Tibet had decimated the wildlife, especially the chiru because of its valuable wool. A shipment of 500 kg of chiru wool is reputed to have been sold illegally by China to an Italian dealer within the past two years, and the consignment is said to have consisted mainly of Qinghai animals. The second purpose was to trace the migration of chiru females to their calving grounds in the desolate northern part of Koko Xili. We suspected that the calving grounds of Tibet's East Chang Tang population were somewhere there and that the main Qinghai population also migrated to that area (see chapter 3).

The weather in the Chang Tang is seldom benign. But that June heavy storms lashed us so often that the ground became a morass from melting snow, and we ultimately had to walk a week to the community of Tuotuohe to seek help for our mired vehicles. More disconcerting was the scarcity of wildlife. During the survey, which covered about 1000 km, we tallied only nine wolves, eight Tibetan brown bears, 239 Tibetan gazelles, and

296 kiang of which 177 were in one concentration. Of just 120 wild yaks observed, 70% were lone bulls or bulls in small bachelor herds. In contrast to parts of the Chang Tang Reserve, such as the Aru Basin, reproduction in these yaks was adequate: one female herd with 36–38 animals included at least 5 calves, the tracks of a herd of 61 showed 15 calves, and a lone female with a newborn was observed on 4 July. The region has little suitable habitat for blue sheep, and only five animals were seen. No Tibetan argalis were encountered. At one site with several old skulls a nomad told us, "There were many. They have been hunted out."

We tallied a total of about 3400 chirus, some possibly counted more than once during an attempt to follow one migratory population. Of the total, 10.6% were yearling and adult males, most of them widely scattered singly and in small herds. Males were particularly common near the Tibetan border southwest of Xijir Ulan Hu in mid-June. The animals were probably part of the East Chang Tang population, judging by the eastward movement of males we observed in 1994. Females from that population also appeared to have moved through the area, as suggested by the many tracks and droppings, heading north and northwest toward Lixioidan Co and the Kunlun Shan.

Between 12 and 24 June, we followed part of the migratory route of the Qinghai population which winters in the south around the upper Tuotuohe valley, as we noted in October 1993. The females, both yearling and adult, traveled along a narrow route from the northern rim of the Ulan Ul Hu Basin past the eastern side of Xijir Ulan Hu over some low ranges and along broad valleys northeast to the eastern end of Hoh Xil Hu. Some animals tarried there, perhaps to give birth or to await the melting of heavy snows that covered the range to the east. However, most females entered this range, moving apparently toward the southern part of the Zonag Co Basin. Many were heavily pregnant, and one animal, killed by a wolf, had a full-term fetus. (A lone female with about a two-day-old young was observed on 30 June). These migrating females were often in large herds, some numbering 100–200 animals and one about 500, and several herds often traveled near each other. There were 14.8 yearlings to 100 adult females in a sample of 780. We traced the migration about 125 km, at most half of their total distance, but failed to reach the calving grounds because of the bad weather. Populations other than the two mentioned here no doubt persist, but they are small and their movements probably local. The three thousand females we observed represent a sizable proportion of those that survive in Qinghai. Unless these remnants are vigorously protected, the annual migrations will soon cease, and part of the chiru's life will remain a mystery.

Between 6 and 12 July we walked more than 75 km southeast out of the Koko Xili area to the community of Tuotuohe. Our route traversed alpine steppe, a superb grazing land lushly covered with grasses and sedges, dominant among them *Stipa purpurea*, and with many forb species, particularly leguminous ones such as purple-flowered *Oxytropis*. Nomads have settled the area only during the past four decades. The population remains sparse and the steppe is in good condition, yet acrimony over pasture rights has led to armed conflict, most recently in 1996. One aspect of our walk was dispiriting: we looked in vain for a vista with wildlife. Wild yaks are gone, their bountiful meat always in demand, and argalis persist at most as a tiny remnant. We observed a total of 9 gazelles, 11 chiru males, and a lone kiang. I would not expect outsiders, with their focus on quick profits, to have empathy for the land and its wildlife. But local Tibetans have also been involved in the slaughter. Some in Tuotuohe "have become rich" from killing chirus, commented a resident. Nomads have hunted wildlife for centuries, but certain traditions from the past are not necessarily an appropriate part of today's culture, especially if the motive is unrestrained profit. Was there a need to kill so much, to kill almost everything within several thousand square kilometers? Have the people become so estranged from their religion and from the animals that once gave them sustenance that they now treat the future of wildlife with indifference?

While walking the empty steppe, I also realized that we know as yet too little about this unique high-altitude ecosystem. There is vague talk of managing and modifying the steppe but without any plan to probe its secrets, to learn by looking closely into how the deep-rooted *Carex*, white-flowered cushions of *Androsace*, snow finches, pikas, Tibetan brown bears, and broad-mouthed yaks function together to create fertility and diversity in this seemingly barren land. We cannot save biodiversity, including the interdependent natural processes, by casually degrading the grasslands and decimating the herbivores, as if we knew which species are vital to maintaining the natural community. Clumps of *Stipa* grass, although modest and ephemeral in appearance, may have lived for many decades, even centuries, their biomass mostly concealed underground as roots. The graminoids evolved with the wild herbivores which in turn have supported a variety of predators. Together they can perhaps teach us to manage the steppe in such a way that livestock too can remain as a sustainable part of the system. The Chang Tang as a whole fortunately remains intact, functionally complete, its component parts still in dynamic stability, an unusual situation in today's world.

Humans have been part of the Chang Tang for thousands of years and they left us great herds of wildlife and rich pastures. Some of this natural wealth has been squandered, as our survey in Qinghai illustrated. When I looked at that steppe, I saw only what had vanished. But I also knew that even my pessimism reflected a dogged hope for the future, a pragmatic vision that much of the vast Chang Tang can be saved. And that is what the government of the Tibet Autonomous Region is now attempting. To bequeath the Chang Tang far into the next millennium will require a never-ending moral vigilance, a passion to understand the ecology, and a deep commitment to a harmonious coexistence between the nomads with their livestock and the wildlife. Without such dedication there will ultimately be a desert where only howling winds break a deadly silence.

Appendix A
Common and Scientific Names of Wild Mammal Species Mentioned in Text

Order Primata	
Mountain gorilla	*Gorilla gorilla beringei*
Order Lagomorpha	
Tibetan woolly hare	*Lepus oiostolus*
Black-lipped pika	*Ochotona curzoniae*
Order Rodentia	
Himalayan marmot	*Marmota himalayana*
Prairie dog	*Cynomys ludovicianus*
Order Carnivora	
Wolf	*Canis lupus*
Dhole	*Cuon alpinus*
Red fox	*Vulpes vulpes*
Sand fox	*Vulpes ferrilata*
Brown bear	*Ursus arctos*
Asiatic black bear	*Ursus thibetanus*
Giant panda	*Ailuropoda melanoleuca*
Red panda	*Ailurus fulgens*
Steppe polecat	*Mustela eversmanni*
Black-footed ferret	*Mustela nigripes*
Eurasian badger	*Meles meles*
Manul (Pallas's cat)	*Felis manul*
Puma (mountain lion)	*Felis concolor*
Lynx	*Felis (Lynx) lynx*
Snow leopard	*Uncia (Panthera) uncia*
Tiger	*Panthera tigris*
Order Proboscidea	
African elephant	*Loxodonta africana*

Order Perissodactyla

Common name	Scientific name
Kiang (Tibetan wild ass)	*Equus kiang*
Asiatic wild ass	*Equus hemionus*
Plains zebra	*Equus burchelli*

Order Artiodactyla

Common name	Scientific name
Family Suidae	
Wild pig	*Sus scrofa*
Family Camelidae	
Dromedary (one-humped camel)	*Camelus dromedarius*
Bactrian (two-humped) camel	*Camelus (bactrianus) ferus*
Vicuña	*Vicugna vicugna*
Family Moschidae	
Musk deer	*Moschus spp.*
Family Cervidae	
Tufted deer	*Elaphodus cephalophus*
Muntjac (barking deer)	*Muntiacus spp.*
Axis deer	*Cervus (Axis) axis*
Sambar	*Cervus unicolor*
Red deer and elk	*Cervus elaphus*
Swamp deer (barasingha)	*Cervus duvauceli*
White-lipped deer	*Cervus albirostris*
Père David's deer	*Elaphurus davidianus*
Caribou	*Rangifer arcticus*
Roe deer	*Capreolus capreolus*
Family Antilocapridae	
Pronghorn	*Antilocapra americana*
Family Bovidae	
American bison	*Bison bison*
European bison	*Bison bonasus*
Yak	*Bos grunniens*
Gaur	*Bos gaurus*
African buffalo	*Syncerus caffer*
Bushbuck	*Tragelaphus scriptus*
Nyala	*Tragelaphus angasi*
Eland	*Taurotragus oryx*
Nilgai	*Boselaphus tragocamelus*
Four-horned antelope	*Tetracerus quadricornis*
Black-fronted duiker	*Cephalophus nigrifrons*
Maxwell's duiker	*Cephalophus maxwelli*
Common duiker	*Cephalophus grimmia*
Impala	*Aepyceros melampus*
Kirk's dik-dik	*Madoqua kirki*
Steenbok	*Ramphicerus campestris*
Oribi	*Ourebia ourebi*
Klipspringer	*Oreotragus oreotragus*

Beira	*Dorcatragus megalotis*
Gerenuk	*Litocranius walleri*
Springbok	*Antidorcas marsupialis*
Dibatag	*Ammodorcas clarkei*
Blackbuck antelope	*Antilope cervicapra*
Dorcas gazelle	*Gazella dorcas*
Thomson's gazelle	*Gazella thomsoni*
Grant's gazelle	*Gazella granti*
Goitered gazelle	*Gazella subgutturosa*
Speke's gazelle	*Gazella spekei*
Tibetan gazelle	*Procapra picticaudata*
Przewalski's gazelle	*Procapra przewalskii*
Mongolian gazelle	*Procapra (Prodorcas) gutturosa*
Saiga	*Saiga tatarica*
Rhebok	*Pelea capreolus*
Kob	*Kobus kob*
Puku	*Kobus vardoni*
Lechwe	*Kobus leche*
Bohor reedbuck	*Redunca redunca*
Arabian oryx	*Oryx leucoryx*
Gemsbok	*Oryx gazella*
Sable	*Hippotragus niger*
Roan	*Hippotragus equinus*
Blesbok	*Damaliscus dorcas*
Topi	*Damaliscus lunatus*
Jackson's hartebeest	*Alcelaphus buselaphus*
Common wildebeest	*Connochaetes taurinus*
Black wildebeest	*Connochaetes gnou*
Mountain goat	*Oreamnos americanus*
Chamois	*Rupicapra rupicapra*
Goral	*Naemorhedus goral*
Japanese serow	*Nemorhaedus (Capricornis) crispus*
Serow	*Nemorhaedus (Capricornis) sumatraensis*
Takin	*Budorcas taxicolor*
Musk ox	*Ovibos moschatus*
Himalayan tahr	*Hemitragus jemlahicus*
Wild goat	*Capra aegagrus*
Asiatic ibex	*Capra ibex*
Markhor	*Capra falconeri*
Dagestan tur	*Capra cylindricornis*
Blue sheep (bharal)	*Pseudois nayaur*
Urial and mouflon	*Ovis orientalis*
Argali	*Ovis ammon*
Bighorn sheep	*Ovis canadensis*
Aoudad	*Ammotragus lervia*
Chiru (Tibetan antelope)	*Pantholops hodgsoni*

Appendix B
Bird and Reptile Species Observed in the Chang Tang Reserve

Over 500 bird species have been recorded for Tibet alone (Vaurie 1972; Zheng et al. 1983), but the northern Chang Tang harbors few species. I recorded birds casually in the Chang Tang Reserve and tallied 37 species, a list that probably could be doubled in length with the addition of scarce migrants and birds from somewhat lower elevations bordering the reserve. For example, Achuff and Petocz (1988) recorded 58 species in the Arjin Shan Reserve. And just southwest of the reserve, in the Domar and Zapug areas, I noted common mergansers *(Mergus merganser)*, great crested grebes *(Podiceps cristatus)*, common pochards *(Aythya ferina)*, and black-necked cranes *(Grus nigricollis)*, to name just four. The most abundant birds in the reserve were horned larks, rufous-necked snow finches, and Blanford's snow finches. In winter, after the departure of migrants, I sometimes noted only a dozen or so bird species in a month's work. Winter residents are marked with an asterisk (*).

The reserve lacks amphibians, but there is one reptile, a small lizard which can be found as high as 5200 m.

Birds

Bar-headed goose	*Anser indicus*
Ruddy shelduck	*Tadorna ferruginea*
Tufted duck	*Aythya fuligula*
Common teal	*Anas crecca*
*Lammergeyer	*Gypaetus barbatus*
*Himalayan griffon	*Gyps himalayensis*
*Upland hawk	*Buteo hemilasius*
*Eurasian kestrel	*Falco tinnunculus*
*Saker falcon	*Falco cherrug*
*Tibetan snowcock	*Tetraogallus tibetanus*
Little ringed plover	*Charadrius dubius*
Mongolian plover	*Charadrius mongolus*
Common redshank	*Tringa totanus*
Common greenshank	*Tringa nebularia*
Common sandpiper	*Tringa hypoleucos*
Brown-headed gull	*Larus brunnicephalus*
Common tern	*Sterna hirundo*
*Tibetan sandgrouse	*Syrrhaptes tibetanus*

*Blue hill pigeon	*Columba rupestris*
Hoopoe	*Upupa epops*
Hume's short-toed lark	*Calandrella acutirostris*
Crested lark	*Galerida cristata*
*Horned lark	*Eremophila alpestris*
White wagtail	*Motacilla alba*
Brown accentor	*Prunella fulvescens*
*Hume's ground jay	*Pseudopodoces humilis*
*Red-billed chough	*Pyrrhocorax pyrrhocorax*
*Raven	*Corvus corax*
Plumbeous redstart	*Rhyacornis fuliginosus*
White-winged redstart	*Phoenicurus erythrogaster*
Hodgson's redstart	*Phoenicurus hodgsoni*
Desert wheatear	*Oenanthe deserti*
*Brandt's mountain finch	*Leucosticte brandti*
*Tibet snow finch	*Montifringilla adamsi*
*Rufous-necked snow finch	*Montifringilla ruficollis*
*Blanford's snow finch	*Montifringilla blanfordi*
Common rosefinch	*Carpodacus erythrinus*

Reptiles

—	*Phrynocephalus theobaldi*

References

Achuff, P., and R. Petocz. 1988. *Preliminary resource inventory of the Arjin Mountains Nature Reserve, Xinjiang, People's Republic of China.* Gland, Switzerland: World Wide Fund for Nature.

Ackerman, B., F. Lindzey, and T. Hemker. 1984. Cougar food habits in southern Utah. *J. Wildl. Manage.* 48:147–155.

Alexander, B. 1987. A beast for all seasons. *Int. Wildl.* 17(3):44–51.

Allen, C. 1982. *A mountain in Tibet.* London: André Deutsch.

Allen, C., and S. Edwards. 1995. The sustainable-use debate: Observations from IUCN. *Oryx* 19:92–98.

Allen, G. 1939. Zoological results of the second Dolan expedition to western China and eastern Tibet, 1934–1936. Part III, Mammals. *Proc. Academy Nat. Sciences Philadelphia* 90:261–294.

———. 1940. *Natural history of central Asia.* Vol. 2, *The mammals of China and Mongolia.* New York: American Museum of Natural History.

Amato, G. 1994. A systematic approach for identifying evolutionarily significant units for conservation: The dilemma of subspecies. Ph.D. diss., Yale University.

Amato, G., D. Wharton, Z. Zainuddin, and J. Powell. 1995. Assessment of conservation units for the Sumatran rhinoceros (*Dicerorhinus sumatrensis*). *Zoo Biology* 14:395–402.

An, Z. 1982. Palaeoliths and microliths from Shenja and Shuanghu, northern Tibet. *Current Anthropology* 23:493–499.

Andrews, R. 1932. *Natural history of central Asia.* Vol. 1, *The new conquest of central Asia.* New York: American Museum of Natural History.

———. 1933. The Mongolian wild ass. *Nat. Hist.* 33:3–16.

Annenkov, B. 1990. The snow leopard *(Uncia uncia)* in the Dzungarsky Ala Tau. *Int. Pedigree Book of Snow Leopards* 6:21–24.

Atkins, D. 1994. Initial report on captive wild camel breeding program and census surveys of wild camel populations of Trans Altai 'A' Reserve. Mongolia/UNDP-GEF Biodiversity Project. Typescript.

———. 1995. Project report on the second winter season enclosure of captive wild camels for breeding and winter census survey of wild camel population, Great Gobi National Park 'A' Zone. Mongolia/UNDP-GEF Biodiversity Project. Typescript.

Avise, J., and C. Aquadro. 1982. A comparative summary of genetic distances in the vertebrates. *Evolutionary Biology* 15:151–185.

Bagla, P. 1995. "Sustainable" tigers? *BBC Wildlife* 15(5):55.

Bailey, F. 1911. Notes on game animals near Gyantse and in the Chumbi Valley. *J. Bombay Nat. Hist. Soc.* 20:1029–1032.

———. 1914. Exploration on the Tsangpo or upper Brahmaputra. *Geogr. J.* 44: 341–354.

———. 1945. *China—Tibet—Assam.* London: Jonathan Cape.

Ballard, J., and M. Kreitman. 1995. Is mitochondrial DNA a strictly neutral marker? *Trends in Ecology and Evolution* 12:485–488.

Bannikov, A. 1957. Distribution géographique et biologie du cheval sauvage et du chameau de Mongolie (*Equus przewalskii* et *Camelus bactrianus*). *Mammalia* 21: 152–160.

———. 1976. Wild camels of the Gobi. *Wildlife* 18:398–403.

Bannikov, A., L. Zhirnov, L. Lebedeva, and A. Fandeev. 1967. *Biology of the saiga.* Jerusalem: Israel Program for Scientific Translations.

Bartz, F. 1935. Das Tierleben Tibets und des Himalaya-Gebirges. *Wissenschaftliche Veröffentlichungen des Museums für Länderkunde zu Leipzig* 3:115–177.

Belovsky, G., and O. Schmitz. 1994. Plant defenses and optimal foraging by mammalian herbivores. *J. Mammal.* 75:816–832.

Berger, J. 1981. The role of risks in mammalian combat: Zebra and onager fights. *Z. Tierpsychologie* 56:297–304.

———. 1986. *Wild horses of the Great Basin.* Chicago: University of Chicago Press.

———. 1990. Persistence of different-sized populations: An empirical assessment of rapid extinctions in bighorn sheep. *Cons. Biol.* 4:91–98.

Berry, S. 1990. *A stranger in Tibet.* London: Collins.

Blanford, W. 1888–1891. *The fauna of British India: Mammalia.* London: Taylor and Francis.

Blank, D. 1992. Peculiarities of social and reproductive behavior of *Gazella subgutturosa* in Iliisky Valley. *Antelope Specialist Group Gnusletter* 11(3):10–11.

Bonnemaire, J. 1976. Le yak domestique et son hybridation. In *Le yak*, pp. 46–76. Ethnozootechnie, no. 15. Paris: Sociéte d'Ethnozootechnie.

Bonvalot, G. 1892. *Across Tibet.* New York: Cassell.

Boutton, T., L. Tieszen, and S. Imbamba. 1988. Biomass dynamics of grassland vegetation in Kenya. *Afr. J. Ecol.* 26:89–101.

Bower, H. 1894. *Diary of a journey across Tibet.* New York: Macmillan.

Braden, K. 1982. The geographic distribution of snow leopards in the USSR: Maps of areas of snow leopard habitation in the USSR. *Int. Pedigree Book of Snow Leopards* 3:25–39.

Brantingham, P., J. Olsen, and G. Schaller. In press. Paleolithic sites and late Pleistocene climates in northwest Tibet, China. *J. Field Archaeology.*

Bristow, M. 1996. Not dead yet. *BBC Wildlife* 14(5):68–74.

Bunch, T., R. Mitchell, and A. Maciulis. 1990. G-banded chromosomes of the Gansu argali *(Ovis ammon jubata)* and their implications in the evolution of the *Ovis* karyotype. *J. Heredity* 81:227–230.

Bunch, T., and C. Nadler. 1980. Giemsa-band patterns of the tahr and chromosomal evolution of the tribe Caprini. *J. Heredity* 71:110–116.

Bunch, T., C. Nadler, and L. Simmons. 1978. G-band patterns, hemoglobin, and transferrin types of the bharal. *J. Heredity* 69:316–320.

Bunnell, F., and M. Gillingham. 1985. Foraging behavior: Dynamics of dining out. In *Bioenergetics of wild herbivores*, ed. R. Hudson and R. White, pp. 53–79. Boca Raton, Fla.: CRC Press.

Burdsall, R., and A. Emmons. 1935. *Men against the clouds.* New York: Harper.

Burrard, G. 1925. *Big game hunting in the Himalayas and Tibet.* London: H. Jenkins.

Butler, J., P. Achuff, and L. Johnston. 1986. *Arjin Mountains Nature Reserve, Xinjiang, People's Republic of China.* IUCN/WWF Report. Gland, Switzerland: IUCN/WWF.

Caccone, A., G. Amato, and J. Powell. 1987. Intraspecific DNA divergence in *Drosophila:* A study on parthenogenetic *D. mercatorum. Mol. Biol. Evol.* 4:343–350.

Cai, G. 1982. Notes on birds and mammals in the region of sources of the Yangtze River. *Acta Biologica Plateau Sinica* 1:135–149. (In Chinese.)

———. 1988. Notes on white-lipped deer *(Cervus albirostris)* in China. *Acta Theriologica Sinica* 8:7–12. (In Chinese.)

Cai, G., Y. Liu, and B. O'Gara. 1989. Observations of large mammals in the Qaidam Basin and its peripheral mountainous areas in the People's Republic of China. *Can. J. Zool.* 68:2021–2024.

Cai, L., and G. Wiener. 1995. *The yak.* Bangkok, Thailand: Food and Agriculture Organization of the UN.

Caughley, G. 1970. *Cervus elaphus* in southern Tibet. *J. Mammal.* 51:611–614.

The central Asian expedition of Captain Roborovsky and Lieut. Kozloff. 1896. *Geogr. J.* 8:161–173.

Chang, C., N. Chen, M. Coward, et al. 1986. Preliminary conclusions of the Royal Society and Academia Sinica 1985 geotraverse of Tibet. *Nature* 323:501–507.

Chang, D. 1981. The vegetation zonation of the Tibetan Plateau. *Mountain Research and Development* 1:29–48.

Chang, D., and H. Gauch. 1986. Multivariate analysis of plant communities and environmental factors in Ngari, Tibet. *Ecology* 67:1568–1575.

Chen, L., E. Reiter, and Z. Feng. 1985. The atmospheric heat source over the Tibetan Plateau: May–August 1979. *Monthly Weather Review* 113:1771–1790.

Chen, W. 1981. Natural environment of the Pliocene basin in Gyirong, Xizang. In *Geological and ecological studies of Qinghai-Xizang Plateau*, ed. D. Liu, vol. 1, pp. 343–352. Beijing: Science Press.

Cheng, H., Z. Ni, S. Sun, X. Yu, and D. Chen. 1981. Regional differentiation of agriculture on the Qinghai-Xizang Plateau. In *Geological and ecological studies of Qinghai-Xizang Plateau*, ed. D. Liu, vol. 2, pp. 2021–2026. Beijing: Science Press.

Cheng, J. 1984. The geographical distribution of wild Bactrian camel in Gansu. *Acta Theriologica Sinica* 4:186. (In Chinese.)

Chestin, I. 1996. Variability in skulls of central Asian brown bears. *J. Wildlife Research* (Poland) 1:70–74.

Child, G., and J. Le Riche. 1969. Recent springbok treks (mass movements) in south-western Botswana. *Mammalia* 33:499–504.

Chundawat, R., and G. Rawat. 1994. Food habits of snow leopard in Ladakh, India. In *Proceedings of the Seventh International Snow Leopard Symposium*, ed. J. Fox and J. Du, pp. 127–132. Seattle: International Snow Leopard Trust.

Cincotta, R., P. Van Soest, J. Robertson, C. Beall, and M. Goldstein. 1991. Foraging ecology of livestock on the Tibetan Changtang: A comparison of three adjacent grazing areas. *Arctic and Alpine Research* 23:149–161.

Cincotta, R., Y. Zhang, and X. Zhou. 1992. Transhumant alpine pastoralism in northeastern Qinghai Province: An evaluation of livestock population response during China's agrarian reform. *Nomadic Peoples* 30:3–25.

Clark, L. 1954. *The marching wind.* New York: Funk and Wagnall.

Clarke, G. 1987. China's reforms of Tibet, and their effects on pastoralism. *Institute of Development Studies* (University of Sussex, England), pp. 63–131.

Clutton-Brock, J. 1992. The process of domestication. *Mammal Review* 22:79–85.

The conservation atlas of China. 1990. Beijing: Science Press.

Coppock, D., J. Ellis, and D. Swift. 1986. Livestock feeding ecology and resource utilization in a nomadic pastoral ecosystem. *J. Applied Ecology* 23:573–583.

Corbet, G. 1980. *The mammals of the Palaearctic region: A taxonomic review.* Ithaca: Cornell University Press.

Crandall, L. 1964. *The management of wild mammals in captivity.* Chicago: University of Chicago Press.

Dang, H. 1967. The snow leopard and its prey. *The Cheetal* (India) 10:72–84.

Dash, Y., A. Szaniawski, G. Child, and P. Hunkeler. 1977. Observations on some large mammals of the Transaltai, Djungarian, and Shargin Gobi, Mongolia. *Terre et Vie* 31:587–596.

Deasy, H. 1901. *In Tibet and Chinese Turkestan.* New York: Longmans Green.

Demidoff, E. 1900. *After wild sheep in the Altai and Mongolia.* London: Rowland Ward.

Derbyshire, E., Y. Shi, J. Li, B. Zheng, S. Li, and J. Wang. 1991. Quaternary glaciation of Tibet: The geological evidence. *Quaternary Science Reviews* 10:485–510.

Dolan, B. 1939. Zoological results of the second Dolan expedition to western China and eastern Tibet, 1934–1936. *Proc. Academy Nat. Sciences Philadelphia* 90:159–184.

Dolan, J. 1988. A deer of many lands—A guide to the subspecies of the red deer *Cervus elaphus* L. *Zoonooz* (San Diego Zoo) 62(10):4–34.

Dolan, J., and L. Killmar. 1988. The shou, *Cervus elaphus wallichi* Cuvier, 1825: A rare and little known cervid, with remarks on three additional Asiatic elaphines. *Zool. Garten* 58:84–96.

Dollman, J., and J. Burlace, eds. 1922. *Rowland Ward's records of big game.* London: Rowland Ward.

Donner, F. 1985. Overland from China. *Foreign Service J.*, Apr.:38–40.

East, R. 1984. Rainfall, soil nutrient status, and biomass of large African savanna mammals. *Afr. J. Ecol.* 22:245–270.

Economic birds and mammals in Qinghai. 1983. Qinghai: People's Publishing House. (In Chinese.)

Einarsen, A. 1948. *The pronghorn antelope and its management.* Washington, D.C.: Wildlife Management Institute.

Eisenberg, J. 1981. *The mammalian radiations.* Chicago: University of Chicago Press.

Eisenberg, J., and M. Lockhart. 1972. An ecological reconnaissance of Wilpattu National Park, Ceylon. *Smithsonian Contributions to Zoology* 101:1–118.
Ekvall, R. 1964. *Fields on the hoof.* New York: Holt, Rinehart, and Winston.
Ellis, J., and D. Swift. 1988. Stability of African pastoral ecosystems: Alternate paradigms and implications for development. *J. Range Management* 41:450–459.
Emmons, L. 1987. Comparative feeding ecology of felids in a neotropical rainforest. *Behavioral Ecology and Sociobiology* 20:271–283.
Engelmann, C. 1938. Über die Grossäuger Szetschwans, Sikongs und Osttibets. *Z. Säugetierkunde* 13:1–76.
Epstein, H. 1969. *Domestic animals of China.* Farnham Royal: Commonwealth Agricultural Bureaux.
Estes, R. 1972. The role of the vomeronasal organ in mammalian reproduction. *Mammalia* 36:315–341.
———. 1976. The significance of breeding synchrony in the wildebeest. *East Afr. Wildl. J.* 14:135–152.
———. 1991. *The behavior guide to African mammals.* Berkeley and Los Angeles: University of California Press.
Expedition for rare animal species in Sichuan. 1977. Chengdu: Sichuan Forest Bureau. (In Chinese.)
Fan, Y. 1981. Chemical characteristics of Xizang lakes. In *Geological and ecological studies of Qinghai-Xizang Plateau,* ed. D. Liu, vol. 2, pp. 1705–1711. Beijing: Science Press.
Fang, J., and K. Yoda. 1991. Climate and vegetation in China. V, Effect of climatic factors on the upper limit of distribution of evergreen broadleaf forest. *Ecol. Res.* 6:113–125.
Feh, C., T. Boldsukh, and C. Tourenq. 1994. Are family groups in equids a response to cooperative hunting by predators? The case of Mongolian kulans (*Equus hemionus luteus* Matschie). *Rev. Ecol.* 49:11–20.
Feng, Z. 1991a. On the status and conservation of wildlife resources in the Karakorum-Kunlun mountain region, China. *Chinese J. Arid Land Res.* 4:65–74.
———. 1991b. Wild animal resources in the Hoh Xil region. *Chinese J. Arid Land Res.* 4:247–253.
Feng, Z., G. Cai, and C. Zheng. 1986. *The mammals of Xizang.* Beijing: Science Press. (In Chinese.)
Field, C. 1972. The food habits of wild ungulates in Uganda by analyses of stomach contents. *E. Afr. Wildlife J.* 10:17–42.
Flerov, K. 1952. *Fauna of USSR.* Vol. 1, no. 2, *Musk deer and deer.* Washington: Israel Program for Scientific Translations and National Science Foundation.
Floyd, T., L. Mech, and P. Jordan. 1978. Relating wolf scat content to prey consumed. *J. Wildl. Manage.* 42:528–532.
Fox, J. 1994. Snow leopard conservation in the wild—A comprehensive perspective on a low density and highly fragmented population. In *Proceedings of the Seventh International Snow Leopard Symposium,* ed. J. Fox and J. Du, pp. 3–15. Seattle: International Snow Leopard Trust.

Fox, J., C. Nurbu, and R. Chundawat. 1991a. The mountain ungulates of Ladakh, India. *Biol. Cons.* 58:167–190.

———. 1991b. Tibetan argali *(Ovis ammon hodgsoni)* establish a new population. *Mammalia* 55:448–452.

Fox, J., S. Sinha, R. Chundawat, and P. Das. 1988. A field survey of snow leopard presence and habitat use in northwestern India. In *Proceedings of the Fifth International Snow Leopard Symposium*, ed. H. Freeman, pp. 99–111. Seattle and Dehra Dun: International Snow Leopard Trust and Wildlife Institute of India.

———. 1991. Status of the snow leopard *Panthera uncia* in northwest India. *Biol. Cons.* 55:283–298.

Franklin, W., and M. Fritz. 1991. Sustained harvesting of the Patagonia Guanaco: Is it possible or too late? In *Neotropical wildlife use and conservation*, ed. J. Robinson and K. Redford, pp. 317–336. Chicago: University of Chicago Press.

Frenzel, B. 1968. The Pleistocene vegetation of northern Eurasia. *Science* 161:637–649.

Fryxell, J. 1987. Food limitation and demography of a migratory antelope, the white-eared kob. *Oecologia* (Berlin) 72:83–91.

———. 1995. Aggregation and migration by grazing ungulates in relation to resources and predators. In *Serengeti II*, ed. A. Sinclair and P. Arcese, pp. 257–273. Chicago: University of Chicago Press.

Fryxell, J., J. Greever, and A. Sinclair. 1988. Why are migratory ungulates so abundant? *Am. Nat.* 131:781–798.

Fryxell, J., and A. Sinclair. 1988. Seasonal migration by white-eared kob in relation to resources. *Afr. J. Ecol.* 26:17–31.

Gai, P. 1985. Microlithic industries in China. In *Palaeoanthropology and Palaeolithic archaeology in the People's Republic of China*, ed. R. Wu and J. Olsen, pp. 225–241. New York: Academic Press.

Ganhar, J. 1979. *The wildlife of Ladakh.* Srinagar: Haramukh.

Gao, X., and J. Gu. 1989. The distribution and status of the Equidae in China. *Acta Theriologica Sinica* 9:269–274. (In Chinese.)

Gao, X., K. Xu, J. Yao, and Z. Jia. 1996. The population structure of goitered gazelle in Xinjiang. *Acta Theriologica Sinica* 16:14–18. (In Chinese.)

Gao, Y. 1987. *Fauna Sinica, Mammalia.* Vol. 8, *Carnivora.* Beijing: Science Press. (In Chinese.)

Gasse, F., J. Fontes, E. Campo, and K. Wei. 1996. Holocene environmental changes in Bangong Co (western Tibet). Part 4, Discussion and conclusions. *Palaeogeography, Palaeoclimatology, Palaeoecology* 120:79–92.

Gatesy, J., G. Amato, E. Vrba, G. Schaller, and R. DeSalle. 1997. A cladistic analysis of mitochondrial DNA from the Bovidae. *Molecular Phylogenetics and Evolution* 7:303–319.

Gatesy, J., R. DeSalle, and W. Wheeler. 1993. Alignment-ambiguous nucleotide sites and the exclusion of systematic data. *Molecular Phylogenetics and Evolution* 2:152–157.

Gatesy, J., D. Yelon, R. DeSalle, and E. Vrba. 1992. Phylogeny of the Bovidae

(Artiodactyla, Mammalia), based on mitochondrial ribosomal DNA sequences. *Mol. Biol. Evol.* 9:433–446.

Gauthier-Pilters, H., and A. Dagg. 1981. *The camel.* Chicago: University of Chicago Press.

Gee, E. 1967. Occurrence of the Nayan or great Tibetan sheep, *Ovis ammon hodgsoni,* in Bhutan. *J. Bombay Nat. Hist. Soc.* 64:553.

Geist, V. 1971. *Mountain sheep.* Chicago: University of Chicago Press.

———. 1991. On the taxonomy of giant sheep (*Ovis ammon* Linnaeus, 1766). *Can. J. Zool.* 69:706–723.

———. 1992. Endangered species and the law. *Nature* 357:274–276.

Gentry, A. 1968. The extinct bovid genus *Qurliqnoria* Bohlin. *J. Mammal.* 49:769.

———. 1992. The subfamilies and tribes of the family Bovidae. *Mammal Review* 22:1–32.

Georgiadis, N., and S. McNaughton. 1988. Interactions between grazers and a cyanogenic grass, *Cynodon plectostachyus. Oikos* 51:343–350.

Gill, R., L. Carpenter, R. Bartmann, D. Baker, and G. Schooveld. 1983. Fecal analysis to estimate mule deer diets. *J. Wildl. Manage.* 47:902–915.

Goldstein, M., and C. Beall. 1989. The impact of China's reform policy on the nomads of western Tibet. *Asian Survey* 29:619–641.

———. 1990. *Nomads of western Tibet.* Berkeley and Los Angeles: University of California Press.

Grasslands and grassland sciences in northern China. 1992. Washington, D.C.: National Academy Press.

Green, M. 1986. The distribution, status, and conservation of Himalayan musk deer *Moschus chrysogaster. Biol. Cons.* 35:347–375.

Grenard, F. [1903] 1974. *Tibet: The country and its inhabitants.* Reprint, Delhi: Cosmo Publications.

Grenot, C. 1991. Wildlife management: Ecophysiological characteristics of large Saharan mammals and their effects on the ecosystem. In *Mammals in the Palaearctic desert: Status and trends in the Sahara-Gobian region,* ed. J. McNeely and V. Neronov, pp. 103–135. Moscow: Russian Academy of Sciences.

Groves, C. 1967. On the gazelles of the genus *Procapra* Hodgson, 1846. *Z. Säugetierkunde* 32:144–149.

———. 1969. On the smaller gazelles of the genus *Gazella* de Blainville, 1816. *Z. Säugetierkunde* 34:38–60.

———. 1974. *Horses, asses, and zebras in the wild.* Hollywood, Fla.: Curtis Books.

———. 1978. The taxonomic status of the dwarf blue sheep (Artiodactyla; Bovidae). *Säugetierkundliche Mitteilungen* 26:177–183.

———. 1981. Systematic relationships in the Bovini (Artiodactyla; Bovidae). *Z. Zoologische Systematik und Evolutionsforschung* 19:264–278.

———. 1985. An introduction to the gazelles. *Chinkara* 1:4–16.

Groves, C., and P. Grubb. 1987. Relationships of living deer. In *Biology and management of the Cervidae,* ed. C. Wemmer, pp. 21–59. Washington: Smithsonian Institution Press.

Groves, C., and V. Mazák. 1967. On some taxonomic problems of Asiatic wild

asses; with the description of a new subspecies (Perissodactyla; Equidae). *Z. Säugetierkunde* 32:321–384.
Groves, P. 1995. *Muskox husbandry.* Biological Papers of the University of Alaska, Special Report no. 9. Fairbanks: University of Alaska.
Groves, P., and G. Shields. 1996. Phylogenetics of the Caprinae based on cytochrome b sequence. *Molecular Phylogenetics and Evolution* 5:467–476.
Gu, J. 1990. The ungulates in the Arjin and east Kunlun Mountains. *Chinese J. Arid Land Res.* 3:97–104. (In Chinese.)
Gu, J., C. Fu, and Z. Gu. 1984. Large hoofed animals of Arjin Shan and eastern Kunlun. Academia Sinica, Xinjiang. Typescript.
Gu, J., and X. Gao. 1985. The wild two-humped camel in the Lop-Nor region. In *Contemporary mammalogy in China and Japan,* ed. T. Kawamichi, pp. 117–120. Osaka: Mammal Society of Japan.
Gu, J., X. Gao, and J. Zhou. 1991. Status and distribution of wild two-humped camel in Xinjiang. In *Studies on the animals in Xinjiang,* pp. 1–9. Beijing: Scientific Publishing House. (In Chinese.)
Guthrie, R. 1990. *Frozen fauna of the mammoth steppe.* Chicago: University of Chicago Press.
Gyllensten, U., and H. Erlich. 1988. Generation of single-stranded DNA by the polymerase chain reaction and its application to direct sequencing of the HLA-DQA locus. *Proc. Nat. Acad. Sci.* 85:7652–7656.
Habibi, K., C. Thouless, and N. Lindsay. 1993. Comparative behaviour of sand and mountain gazelles. *J. Zool. Lond.* 229:41–53.
Hare, J. 1995. Status and distribution of wild Bactrian camels, *Camelus bactrianus ferus,* in the Xinjiang Uygur Autonomous Region and Gansu Province in the People's Republic of China. UN Environment Programme. Typescript.
———. 1996. Footprints on the wind. *BBC Wildlife* 14(3):24–30.
———. 1997. The wild Bactrian camel *Camelus bactrianus ferus* in China: The need for urgent action. *Oryx* 31:45–48.
Harris, R. 1991. Conservation prospects for musk deer and other wildlife in southern Qinghai, China. *Mountain Research and Development* 11:353–358.
———. 1993. Wildlife conservation in Yeniugou, Qinghai Province, China. Ph.D. diss., University of Montana, Missoula.
Harris, R., and G. Cai. 1993. Autumn home range of musk deer in Baizha Forest, Tibetan Plateau. *J. Bombay Nat. Hist. Soc.* 90:430–436.
Harris R., and D. Miller. 1995. Overlap in summer habitats and diets of Tibetan Plateau ungulates. *Mammalia* 59:197–212.
Harrison, T., P. Copeland, W. Kidd, and Y. An. 1992. Raising Tibet. *Science* 255:1663–1670.
Hayes, M. 1987. Veterinary notes for horse owners. London: Stanley Paul.
Hedin, S. 1898. *Through Asia.* Vol. 1. London: Methuen.
———. 1903. *Central Asia and Tibet.* 2 vols. London: Hurst and Blackett.
———. 1909. *Trans-himalaya.* 2 vols. New York: Macmillan.
———. [1922] 1991. *Southern Tibet.* Vols. 3 and 4. Reprint, Delhi: B. R. Publ. Corp.
Helle, T. 1980. Abundance of warble fly *(Oedemgenga tarandi)* larvae in semi-

domestic reindeer (*Rangifer tarandus*) in Finland. *Rep. Kevo Subarctic Res. Station* 16:1–6.

Helle, T., and H. Tarvainen. 1984. Effects of insect harassment on weight gain and survival in reindeer calves. *Rangifer* 4(1):24–27.

Hou, S. 1991. *An outline of Tibetan archeology*. Lhasa: People's Publishing House of Tibet. (In Chinese.)

Heptner, V., A. Nasimovic, and A. Bannikov. 1966. *Die Säugetiere der Sowjetunion*. Vol. 1. Jena: Gustav Fischer Verlag.

Houston, D. 1988. Digestive efficiency and hunting behavior in cats, dogs, and vultures. *J. Zool., London,* 216:603–605.

Hövermann, J., and W. Wang, eds. 1987. *Reports on the northeastern part of the Qinghai-Xizang (Tibet) Plateau*. Beijing: Science Press.

Huang, C., and Y. Liang. 1981. Based upon palynological study to discuss the natural environment of the central and southern Qinghai-Xizang Plateau of Holocene. In *Geological and ecological studies of Qinghai-Xizang Plateau,* ed. D. Liu, vol. 1, pp. 215–224. Beijing: Science Press.

Huang, R. 1987. Vegetation in the northeastern part of the Qinghai-Xizang Plateau. In *Reports on the northeastern part of the Qinghai-Xizang (Tibet) Plateau,* ed. J. Hövermann and W. Wang, pp. 438–489. Beijing: Science Press.

Huang, X., Z. Wang, J. Wu, and L. Liu. 1986. The breeding biological characteristics of *Marmota himalayana* in Reshuitan and Wulannaotan, Haiyan County, Qinghai Province. *Acta Theriologica Sinica* 6:307–311. (In Chinese.)

Huc, E., and J. Gabet. [1850] 1987. *Travels in Tartary, Thibet, and China, 1844–1846*. Reprint, New York: Dover.

Hudson, R., and R. White, eds. 1985. *Bioenergetics of wild herbivores*. Boca Raton, Fla.: CRC Press.

Huntly, N., and O. Reichman. 1994. Effect of subterranean mammalian herbivores on vegetation. *J. Mammal.* 75:852–859.

Hutchins, M., and V. Geist. 1987. Behavioural considerations in the management of mountain-dwelling ungulates. *Mountain Research and Development* 7:135–144.

Jackson, R. 1979a. Aboriginal hunting in West Nepal with reference to musk deer *(Moschus moschiferus)* and snow leopard (*Panthera uncia*). *Biol. Cons.* 16:63–72.

———. 1979b. Snow leopards in Nepal. *Oryx* 15:191–195.

———. 1991. Snow leopards and other wildlife in the Qomolangma Nature Preserve in Tibet. *Snow Line* (International Snow Leopard Trust) 9(1):9–12.

———. 1996. Home range, movements, and habitat use of snow leopard *(Uncia uncia)* in Nepal. Ph.D. thesis, University of London.

Jackson, R., and G. Ahlborn. 1984. A preliminary habitat suitability model for the snow leopard, *Panthera uncia,* in West Nepal. *Int. Pedigree Book of Snow Leopards* 4:43–52.

———. 1988. Observations on the ecology of snow leopard in West Nepal. In *Proceedings of the Fifth International Snow Leopard Symposium,* ed. H. Freeman, pp. 65–87. Seattle: International Snow Leopard Trust.

———. 1989. Snow leopards *(Panthera uncia)* in Nepal—Home range and movements. *National Geographic Res.* 5:161–175.

Jackson, R., Z. Wang, X. Lu, and Y. Chen. 1994. Snow leopards in the Qomolangma Nature Preserve of the Tibet Autonomous Region. In *Proceedings of the Seventh International Snow Leopard Symposium*, ed. J. Fox and J. Du, pp. 85–95. Seattle: International Snow Leopard Trust.

Jaczewski, Z. 1986. Einige Angaben über den Weisslippenhirsch (*Cervus albirostris* Przewalski 1883). *Z. Jagdwissenschaft* 32:75–83.

Jarman, P. 1983. Mating system and sexual dimorphism in large, terrestrial, mammalian herbivores. *Biol. Review* 58:485–520.

Jarman, P., and A. Sinclair. 1979. Feeding strategy and the pattern of resource partitioning in ungulates. In *Serengeti, dynamics of an ecosystem*, ed. A. Sinclair and M. Norton-Griffiths, pp. 130–163. Chicago: University of Chicago Press.

Jiang, Z., Z. Feng, Z. Wang, L. Chen, P. Chai, and Y. Li. 1994. Saving the Przewalski's gazelle. *Species* 23:59–60.

Jiang, Z., and W. Xia. 1983. Utilization of the food resources by Plateau pika. *Acta Theriologica Sinica* 5:251–262. (In Chinese.)

———. 1987. The niches of yaks, Tibetan sheep, and Plateau pikas in the alpine meadow ecosystem. *Acta Biologica Plateau Sinica* 6:115–146. (In Chinese.)

Johnson, K., G. Schaller, and J. Hu. 1988. Comparative behavior of red and giant pandas in the Wolong Reserve, China. *J. Mammal.* 69:552–564.

Jones, M. 1993. Longevity of ungulates in captivity. *Int. Zoo Yearbook* 32:159–169.

Kaiser, M., and A. Gebauer. 1993. Bemerkungen zur Biologie und zur Haltung des Schwarzlippen-Pfeifhasen, *Ochotona curzoniae* (Hodgson, 1858). *Milu* (Berlin) 7: 474–504.

Kaji, K. 1985. An expedition to the source of the Yellow River in China. *Honyurui Kagaku* 51:1–9. (In Japanese.)

Kaji, K., N. Ohtaishi, S. Miura, T. Koizumi, K. Tokida, and J. Wu. 1993. Distribution and status of white-lipped deer and associated ungulate fauna in the Tibetan Plateau. In *Deer of China*, ed. N. Ohtaishi and H. Sheng, pp. 147–158. Amsterdam: Elsevier.

Kaji, K., N. Ohtaishi, S. Miura, and J. Wu. 1989. Distribution and status of white-lipped deer *(Cervus albirostris)* in the Qinghai-Xizang (Tibet) Plateau, China. *Mammal Review* 19:35–44.

Karou: A Neolithic site in Tibet. 1985. Beijing: Cultural Relics Publishing House. (In Chinese.)

Kelsall, J. 1968. *Migratory barren-ground caribou of Canada.* Canadian Wildlife Service Monograph no. 3. Ottawa: Canadian Wildlife Service.

———. 1969. Structural adaptations of moose and deer to snow. *J. Mammal.* 50: 302–310.

Kinloch, A. 1892. *Large game shooting in Thibet, the Himalayas, Northern and Central India.* Bombay: Thacker, Spink.

Kishimoto, R. 1989. Early mother and kid behavior of a typical "follower," Japanese serow *Capricornis crispus. Mammalia* 53:165–174.

Kitchen, D., and O'Gara, B. 1982. Pronghorn. In *Wild mammals of North America*, ed. J. Chapman and G. Feldhamer, pp. 960–971. Baltimore: Johns Hopkins University Press.

Klingel, H., and U. Klingel. 1966. Tooth development and age determination in the plains zebra (*Equus quagga boehmi* Matschie). *Zool. Garten* 33:34–54.

Koerth, B., L. Krysl, B. Sowell, and F. Bryant. 1984. Estimating seasonal diet quality of pronghorn antelope from fecal analysis. *J. Range Manage.* 37:560–563.

Köhler-Rollefson, I. 1991. *Camelus dromedarius. Mammalian Species* 375:1–8.

Koizumi, T., N. Ohtaishi, K. Kaji, Y. Yu, and K. Tokida. 1993. Conservation of white-lipped deer in China. In *Deer of China*, ed. N. Ohtaishi and H. Sheng, pp. 309–318. Amsterdam: Elsevier.

Koshkarev, E. 1989. *Snow leopard in Kirgizia: Population, ecology, and conservation.* Frunze: Academy of Sciences of Kirgizia. (In Russian.)

Kozloff [Kozlov], P. 1908. Through eastern Tibet and Kam. *Geogr. J.* 31:402–415, 522–534, 649–661.

Kozlov, P. 1899. Report. In *Proceedings of the Imperial Russian Geographic Society expedition to central Asia, 1893–1895, led by V. I. Roborovskiy*, p. 143. St. Petersburg: Imperial Russian Geographic Society. (In Russian.)

Kuhle, M. 1987a. The problem of Pleistocene inland glaciation of the northeastern Qinghai-Xizang Plateau. In *Reports on the northeastern part of the Qinghai-Xizang (Tibet) Plateau*, ed. J. Hövermann and W. Wang, pp. 250–315. Beijing: Science Press.

———. 1987b. Die Wiege der Eiszeit. *Geo.*, Feb.:80–94.

Kuroki, K. 1987. Animals at the headwaters area of the Yangtze River—The ecology of the chiru. In *The distant headwaters of the Yangtze River—The results of the research team to the Qinghai-Xizang Plateau*, ed. M. Matasumoto and M. Matsubara, pp. 84–107. Tokyo: Nihon Hoso Shuppan Kyokai. (In Japanese.)

Kurtén, B. 1968. *Pleistocene mammals of Europe.* London: Weidenfeld and Nicolson.

Lachungpa, U. 1994. Tibetan wild ass or kiang not extinct in Sikkim. *Sikkim Science Society News Letter* 3(3):7–9.

Landor, H. 1899. *In the forbidden land.* 2 vols. New York: Harper.

Larrick, J., and K. Burck. 1986. Tibet's all-purpose beast of burden. *Natural History* 95(1):56–65.

Larsen, J. 1987. Bounty of the barrenlands. *Garden* 11(6):16–22.

Lawson, B., and R. Johnson. 1982. Mountain sheep. In *Wild mammals of North America*, ed. J. Chapman and G. Feldhamer, pp. 1036–1055. Baltimore: Johns Hopkins University Press.

Leche, W. 1904. *Scientific results of a journey in central Asia, 1899–1902.* Vol. 6, part 1, *Zoologie.* Stockholm: Lithographic Institute of the General Staff of the Swedish Army.

Leopold, A. 1949. *A sand county almanac.* Oxford: Oxford University Press.

Leuthold, W. 1977. *African ungulates.* Berlin: Springer Verlag.

Lhalungpa, L. 1983. *Tibet, the sacred realm.* New York: Aperture.

———, trans. 1985. *The life of Milarepa.* Boston: Shambala.

Li, B. 1985. The zonation of vegetation in the Mount Namjagbarwa region. *Mountain Research* 3:291–298. (In Chinese.)

Li, B., J. Zhang, J. Wang, and W. Chen. 1985. The alpine cushion vegetation of Xizang. *Acta Botanica Sinica* 27:311–317. (In Chinese.)

Li, M. 1981. Soil formation and distribution on Qiangtang Plateau. In *Geological and ecological studies of Qinghai-Xizang Plateau*, ed. D. Liu, vol. 2, pp. 1877–1896. Beijing: Science Press.

Li, Y. 1993. A discussion of the Upper Paleolithic industries of China. *Acta Anthropologica Sinica* 12:214–223. (In Chinese.)

Liao, Y. 1985. The geographical distribution of ounces in Qinghai Province. *Acta Theriologica Sinica* 5:183–188. (In Chinese.)

Liao, Y., and B. Tan. 1988. A preliminary study on the geographical distribution of snow leopards in China. In *Proceedings of the Fifth International Snow Leopard Symposium*, ed. H. Freeman, pp. 51–63. Seattle: International Snow Leopard Trust.

Littledale, S. 1894. Field-notes on the wild camel of Lob-Nor. *Proc. Zoöl. Soc. London*, pp. 446–448.

———. 1896. A journey across Tibet, from north to south and west to Ladak. *Geogr. J.* 7:453–483.

Liu, W., and B. Yin, eds. 1993. *Precious and rare wildlife and its protection in Tibet.* Beijing: China Forestry Publishing House. (In Chinese.)

Liu, Y. 1993. *International hunting and the involvement of local people, Dulan, Qinghai, People's Republic of China.* M.S. thesis, University of Montana.

Liu, Z., and X. Wu. 1981. Climatic classification of Qinghai-Xizang Plateau. In *Geological and ecological studies of Qinghai-Xizang Plateau*, ed. D. Liu, vol. 2, pp. 1575–1580. Beijing: Science Press.

Lorenz, K. 1961. Phylogenetische Anpassung und adaptive Modifikation des Verhaltens. *Z. Tierpsychol.* 18:139–187.

Lovari, S., and M. Apollonio. 1994. On the rutting behaviour of the Himalayan goral *Nemorhaedus goral* (Hardwicke, 1825). *J. Ethol.* 12:25–34.

Lovari, S., and M. Locati. 1994. Site features of territorial dung-marking in mainland serow. *Mammalia* 58:153–156.

Lu, X., R. Jackson, and Z. Wang. 1994. Herd characteristics and habitat use of a blue sheep population in the Qomolangma Nature Preserve. In *Proceedings of the Seventh International Snow Leopard Symposium*, ed. J. Fox and J. Du, pp. 97–103. Seattle: International Snow Leopard Trust.

Lu, Z., K. Li, and H. Ju. 1993. Distribution, ecological types, and utilization of wild yak in China. *Chinese J. Zoology* 28(4):41–45. (In Chinese.)

Luo, N., and J. Gu. 1991. The blue sheep resource in the western Arjin Mountains, with reference to game utilization. In *Studies on the animals in Xinjiang*, pp. 16–20. Beijing: Scientific Publishing House. (In Chinese.)

Lushchekina, A., V. Neronov, G. Ogureeva, and A. Sokolova. 1986. Distribution, ecology, protection, and rational utilization of the Mongolian gazelle (*Procapra gutturosa* Pallas 1777) in the Mongolian People's Republic. *Bull. Moscow Biol. Studies of the Investigation of Nature* 91(5):73–82. (In Russian.)

Lydekker, R. 1898. *Wild oxen, sheep, and goats of all lands.* London: Rowland Ward.

Ma, L. 1991. *Glimpses of northern Tibet.* Beijing: Chinese Literature Press.

Macintyre, D. 1891. *Hindu-Koh.* Edinburgh: Blackwood and Sons.

Mallon, D. 1984. The snow leopard, *Panthera uncia*, in Mongolia. *Int. Pedigree Book of Snow Leopards* 4:3–9.

———. 1991. Status and conservation of large mammals in Ladakh. *Biol. Cons.* 56: 101–109.

Martinka, C. 1967. Mortality of northern Montana pronghorns in a severe winter. *J. Wildl. Manage.* 31:159–164.

Mayr, E., and P. Ashlock. 1991. *Principles of systematic zoology*. New York: McGraw-Hill.

McNaughton, S. 1979. Grassland-herbivore dynamics. In *Serengeti, dynamics of an ecosystem*, ed. A. Sinclair and M. Norton-Griffiths, pp. 46–81. Chicago: University of Chicago Press.

———. 1983. Serengeti grassland ecology: The role of composite environmental factors and contingency in community organization. *Ecological Monographs* 53: 291–320.

———. 1985. Ecology of a grazing ecosystem: The Serengeti. *Ecological Monographs* 55:259–294.

———. 1988. Mineral nutrition and spatial concentrations of African ungulates. *Nature* 334(6180):343–345.

———. 1990. Mineral nutrition and seasonal movement of African migratory ungulates. *Nature* 345(6276):613–615.

McNeely, J. 1990. How wild relatives of livestock contribute to a balanced environment. *IUCN/SSC Asian Wild Cattle Specialist Group Newsletter* 3:10–15.

Meagher, M., and M. Meyer. 1994. On the origin of brucellosis in bison of Yellowstone National Park: A review. *Cons. Biol.* 8:645–653.

Mech, L. 1970. *The wolf.* Garden City, N.Y.: Natural History Press.

Migot, A. 1957. *Tibetan marches.* Harmondsworth: Penguin Books.

Miller, D. 1990. Grasslands of the Tibetan Plateau. *Rangelands* 12:159–163.

———. 1992. Wild yaks of Kunlun. *Himal* (Nepal) 5(3):35–36.

———. 1994. The wool that makes the carpets. *Nepalese-Tibetan Carpet*, Jan.:48–56.

Miller, D., and D. Bedunah. 1993. High-elevation rangeland in the Himalaya and Tibetan Plateau: Issues, perspectives, and strategies for livestock development and resource conservation. *Proceedings of the Seventeenth International Grassland Congress*, pp. 1785–1790. Palmerston North: New Zealand Grassland Association.

Miller, D., R. Harris, and G. Cai. 1994. Wild yaks and their conservation on the Tibetan Plateau. In *Proceedings of the 1st International Congress on Yak*, ed. R. Zhang, J. Han, and J. Wu, pp. 27–34. Lanzhou: Gansu Agricultural University.

Miller, D., and R. Jackson. 1994. Livestock and snow leopards: Making room for competing users on the Tibetan Plateau. In *Proceedings of the Seventh International Snow Leopard Symposium*, ed. J. Fox and J. Du, pp. 315–328. Seattle: International Snow Leopard Trust.

Miller, D., and G. Schaller. 1996. Rangelands of the Chang Tang Wildlife Reserve in Tibet. *Rangelands* 18:91–96.

———. 1998. Rangelands dynamics in the Chang Tang Wildlife Reserve, Tibet. In *Karakoram-Hindukush-Himalaya: Dynamics of change*, ed. I. Stellrecht. Köln: Rüdiger Köppe.

Milner-Gulland, E. 1994. A population model for the management of the saiga antelope. *J. Applied Ecology* 31:25–39.

Miura, S., K. Kaji, N. Ohtaishi, T. Koizumi, K. Tokida, and J. Wu. 1993. Social organization and mating behavior of white-lipped deer in the Qinghai-Xizang Plateau, China. In *Deer of China*, ed. N. Ohtaishi and H. Sheng, pp. 220–234. Amsterdam: Elsevier.

Miura, S., N. Ohtaishi, K. Kaji, J. Wu, and S. Zheng. 1989. The threatened white-lipped deer *Cervus albirostris*, Gyaring Lake, Qinghai Province, China, and its conservation. *Biol. Cons.* 47:237–244.

Miyamoto, M., S. Tanhauser, and P. Laipis. 1989. Systematic relationships in the artiodactyl tribe Bovini (family Bovidae), as determined from mitochondrial DNA sequences. *Syst. Zool.* 38:342–349.

Miyashita, M., and K. Nagase. 1981. Breeding the Mongolian gazelle *Procapra gutturosa* at Osaka Zoo. *Int. Zoo. Yearbook* 21:158–162.

Moehlman, P. 1985. The odd-toed ungulates: Order Perrisodactyla. In *Social odors in mammals*, ed. R. Brown and D. Macdonald, pp. 531–549. Oxford: Clarendon Press.

Moen, A. 1973. *Wildlife ecology*. San Francisco: W. H. Freeman.

Molnar, P. 1989. The geologic evolution of the Tibetan Plateau. *Am. Scient.* 77: 350–360.

Mukasa-Mugerwa, E. 1981. *The camel: A bibliographical review*. Addis Ababa: International Livestock Centre for Africa.

Murray, M. 1995. Specific nutrient requirements and migration of wildebeest. In *Serengeti II*, ed. A. Sinclair and P. Arcese, pp. 231–256. Chicago: University of Chicago Press.

Nasonov, N. 1923. Geographic distribution of wild sheep of the Old World. *Russian Academy of Sciences* (St. Petersburg), pp. 94–110. (In Russian.)

Nelson, G., and N. Platnick. 1981. *Systematics and biogeography: Cladistics and vicariance*. New York: Columbia University Press.

Neumann-Denzau, G. 1991. In pursuit of a mirage. *BBC Wildlife* 9(8):546–553.

Ni, Z. 1980. *An enumeration of the vascular plants of Xizang*. Lhasa: Scientific and Technical Communications of the Tibet Autonomous Region. (In Chinese.)

No ancient ice sheet on Tibetan Plateau. 1990. *China's Tibet* 1(4):48.

Novacek, M., A. Wyss, and M. McKenna. 1988. The major groups of eutherian mammals. In *The phylogeny and classification of the tetrapods*, ed. M. Benton, vol. 2, pp. 31–71. Oxford: Clarendon Press.

Novoa, C. 1970. Reproduction in Camelidae. *J. Reprod. Fert.* 22:3–20.

Ohtaishi, N., and Y. Gao. 1990. A review of the distribution of all species of deer (Tragulidae, Moschidae, and Cervidae) in China. *Mammal Review* 20:125–144.

Oli, M. 1994a. Ghost in the snow. *BBC Wildlife* 12(8):30–35.

———. 1994b. Snow leopards and a local human population in a protected area: A case study from the Nepalese Himalaya. In *Proceedings of the Seventh International Snow Leopard Symposium*, ed. J. Fox and J. Du, pp. 51–64. Seattle: International Snow Leopard Trust.

———. 1994c. Snow leopards and blue sheep in Nepal: Densities and predator: prey ratio. *J. Mammal.* 75:998–1004.

———. 1996. Seasonal patterns in habitat use of blue sheep *Pseudois nayaur* (Artiodactyla, Bovidae) in Nepal. *Mammalia* 60:187–193.

Oli, M., and M. E. Rogers. 1996. Seasonal pattern in group size and population composition of blue sheep in Manang, Nepal. *J. Wildl. Manage.* 60:797–801.

Oli, M., I. Taylor, and M. Rogers. 1993. Diet of snow leopard (*Panthera uncia*) in the Annapurna Conservation Area, Nepal. *J. Zool., London,* 231:365–370.

———. 1994. Snow leopard *Panthera uncia* predation of livestock: An assessment of local perceptions in the Annapurna Conservation Area, Nepal. *Biol. Cons.* 68: 63–68.

Olschak, B., A. Gansser, and E. Bührer. 1987. *Himalaya.* New York: Facts on File.

Olsen, S. 1988. Records of early Asian camel domestication from dated artwork. *Camel Newsletter* 3:6–9.

———. 1990. Fossil ancestry of the yak, its cultural significance and domestication in Tibet. *Proc. Academy Nat. Sciences Philadelphia* 142:73–100.

Otte, K., and R. Hofmann. 1981. The debate about the vicuna population in Pampa Galeras Reserve. In *Problems in management of locally abundant wild mammals,* ed. P. Jewell and S. Holt, pp. 259–275. New York: Academic Press.

Owen-Smith, N., and S. Cooper. 1989. Nutritional ecology of a browsing ruminant, the kudu (*Tragelaphus strepsiceros*), through the seasonal cycle. *J. Zool., London,* 219:29–43.

Phelps, E. [1900] 1983. Yak shooting in Tibet. In *A century of natural history,* ed. J. Daniel, pp. 152–159. Reprint, Bombay: Bombay Natural History Society.

Piao, R. 1994. An assessment of *Asinus kiang* population density with the Fourier series. In *A collection of Tibet forestry papers,* ed. W. Zhen and W. Liu, pp. 186–188. Lhasa: People's Publishing House. (In Chinese.)

Piao, R., and W. Liu. 1994. Research on the status of the Tibetan gazelle population. In *A collection of Tibet forestry papers,* ed. W. Zhen and W. Liu, pp. 189–194. Lhasa: People's Publishing House of Tibet. (In Chinese.)

Pielou, E. 1991. *After the ice age.* Chicago: University of Chicago Press.

Pilgrim, G. 1939. *Memoirs of the Geological Survey: Palaeontologia Indica,* n.s., vol. 26, memoir no. 1. Delhi: Geological Survey of India.

Plumptre, A. 1995. The chemical composition of montane plants and its influence on the diet of the large mammalian herbivores in the Parc National des Volcans, Rwanda. *J. Zool., London,* 235:323–337.

Pocock, R. 1910. On the specialized cutaneous glands of ruminants. *Proc. Zool. Soc. London,* pp. 840–986.

———. 1918. On some external characters of ruminant artiodactyla, part II. *Annals and Magazine of Natural History* 9(2):125–144.

Pohle, C. 1974. Haltung und Zucht der Saiga-Antilope *(Saiga tatarica)* im Tierpark Berlin. *Zool. Garten* 44:387–409.

———. 1991. Zur neueren Geschichte der Kiang-Haltung in Tiergärten ausserhalb Chinas. *Zool. Garten* 61:263–266.

Pousargues, E. de. 1898. Étude sur les ruminants de l'Asie centrale. *Mémoires de la Société Zoologique de France* 11:126–224.

Prejevalsky [Przewalski], N. 1876. *Mongolia, the Tangut country, and the solitudes of northern Tibet.* 2 vols. London: Sampson, Low, Marston, Searle, and Rivington.

———. 1879. *From Kulja across the Tian Shan to Lob-Nor.* London: Sampson, Low, Marston, Searle, and Rivington.

Prschewalski [Przewalski], N. 1884. *Reisen in Tibet am oberen Lauf des Gelben Flusses in den Jahren 1879 bis 1880.* Jena: Hermann Costernoble.

Pu, R. 1993. Numbers and distribution of blue sheep in the west part of Tibet. In *A collection of Tibet forestry papers*, ed. W. Zhen and W. Liu, pp. 201–205. Lhasa: People's Publishing House of Tibet. (In Chinese.)

Rabinovich, J., M. Hernández, and J. Cajal. 1985. A simulation model for the management of vicuña populations. *Ecological Modelling* 30:275–295.

Rangelands of Tibet. 1992. Beijing: Science Publishing Co. (In Chinese.)

Rasool, G. 1989. Wildlife in the wilderness. *Natura* (WWF Pakistan Newsletter), Mar.:3–6.

Rawling, C. 1905. *The great plateau.* London: Edward Arnold.

Raymo, M., and W. Ruddiman. 1992. Tectonic forcing of late Cenozoic climate. *Nature* 359:117–129.

Reed, H. 1995. Saiga antelope threatened by massive trade. *TRAFFIC USA* 14(2): 1–2.

Rees, D. 1995. Observation of mixed bharal/ibex groups. *Caprinae News* (Species Survival Commission) 8/9:12.

Reiter, E. 1981. The Tibet connection. *Nat. Hist.* 90(9):65–71.

Reiter, E., and D. Gao. 1982. Heating of the Tibet Plateau and movements of the South Asian High during spring. *Monthly Weather Review* 110:1694–1711.

Research on flora and fauna in the Ali Prefecture. 1979. Beijing: Scientific Publishing House. (In Chinese.)

Rice, C. 1995. On the origin of sexual displays in caprids. *Z. Säugetierkunde* 60: 53–62.

Roborovsky, V., and P. Kozloff. 1896. The central Asian expedition of Captain Roborovsky and Lieut. Kozloff. *Geogr. J.* 8:161–173.

Rockhill, W. 1891. *The land of the lamas.* New York: Century.

———. 1894. *Diary of a journey through Mongolia and Tibet in 1891 and 1892.* Washington, D.C.: Smithsonian Institution.

———. 1895. Big game of Mongolia and Tibet. In *Hunting in many lands*, ed. T. Roosevelt and G. Grinnell, pp. 255–277. New York: Forest and Stream Publishing.

Ronaldshay, Earl of. 1902. *Sport and politics under an eastern sky.* Edinburgh: Blackwood and Sons.

Roosevelt, T., and K. Roosevelt. 1926. *East of the sun and west of the moon.* New York: Charles Scribner's.

Ruddiman, W., and J. Kutzbach. 1991. Plateau uplift and climatic change. *Sci. Am.* 264:66–75.

Ryder, O., and L. Chemnick. 1990. Chromosomal and molecular evolution in Asiatic wild asses. *Genetica* 83:67–72.

Schäfer, E. 1936. Über das osttibetische Argalischaf (*Ovis ammon* subsp?). *Zool. Garten* 8:253–258.

———. 1937a. Über das Zwergblauschaf (*Pseudois* spec. nov.) and das Grossblauschaf (*Pseudois nahoor* Hdgs) in Tibet. *Zool. Garten* 9:263–278.

———. 1937b. *Unbekanntes Tibet.* Berlin: Paul Parey.
———. 1937c. Der wilde Yak, *Bos (Poëphagus) grunniens mutus* Prez. *Zool. Garten* 9:27–34.
———. 1937d. Zur Kenntnis des Kiang (*Equus kiang* Moorcroft). *Zool. Garten* 9: 122–139.
Schaller, G. 1967. The deer and the tiger. Chicago: University of Chicago Press.
———. 1976. Mountain mammals in Pakistan. *Oryx* 13:351–356.
———. 1977a. Aggressive behaviour of domestic yak. *J. Bombay Nat. Hist. Soc.* 73: 385–389.
———. 1977b. *Mountain monarchs: Wild sheep and goats of the Himalaya.* Chicago: University of Chicago Press.
———. 1980. *Stones of Silence.* New York: Viking Press.
———. 1995. Tracking the Gobi's last wild bears and camels. *Int. Wildlife* 25(1): 18–23.
———. 1997. *Tibet's Hidden Wilderness.* New York: Abrams.
Schaller, G., and B. Gu. 1994. Comparative ecology of ungulates in the Aru Basin of northwest Tibet. *National Geographic Research and Exploration* 10:266–293.
Schaller, G., and A. Hamer. 1978. Rutting behavior of Père David's deer, *Elaphurus davidianus. Zool. Garten* 48:1–15.
Schaller, G., J. Hu, W. Pan, and J. Zhu. 1985. *The giant pandas of Wolong.* Chicago: University of Chicago Press.
Schaller, G., H. Li, Talipu, H. Lu, J. Ren, M. Qiu, and H. Wang. 1987. Status of large mammals in the Taxkorgan Reserve, Xinjiang, China. *Biol. Conserv.* 42: 53–71.
Schaller, G., H. Li, Talipu, J. Ren, and M. Qiu. 1988. The snow leopard in Xinjiang, China. *Oryx* 22:197–204.
Schaller, G., and W. Liu. 1996. Distribution, status, and conservation of wild yak *Bos grunniens. Biol. Cons.* 76:1–8.
Schaller, G., W. Liu, and X. Wang. 1996. Status of Tibet red deer. *Oryx* 30:269–274.
Schaller, G., and J. Ren. 1988. Effects of a snow storm on Tibetan antelope. *J. Mammal.* 69:631–634.
Schaller, G., J. Ren, and M. Qiu. 1988. Status of snow leopard in Qinghai and Gansu Provinces, China. *Biol. Conserv.* 45:179–194.
———. 1991. Observations on the Tibetan antelope *(Pantholops hodgsoni). Applied Animal Behaviour Sci.* 29:361–378.
Schaller, G., Q. Teng, K. Johnson, X. Wang, H. Shen, and J. Hu. 1989. The feeding ecology of giant pandas and Asiatic black bears in the Tangjiahe Reserve, China. In *Carnivore behavior, ecology, and evolution,* ed. J. Gittleman, pp. 212–241. Ithaca: Cornell University Press.
Schaller, G., Q. Teng, W. Pan, Z. Qin, X. Wang, J. Hu, and H. Shen. 1986. Feeding behavior of Sichuan takin (*Budorcas taxicolor*). *Mammalia* 50:311–322.
Schaller, G., J. Tserendeleg, and G. Amarsanaa. 1994. Observations on snow leopards in Mongolia. In *Proceedings of the Seventh International Snow Leopard Symposium,* ed. J. Fox and J. Du, pp. 33–46. Seattle: International Snow Leopard Trust.

Schaller, G., R. Tulgat, and B. Navantsatsvalt. 1993. Observations on the Gobi brown bear in Mongolia. In *Bears of Russia and adjacent countries—State of populations*, vol. 2, pp. 110–125. Moscow: Ministry of Environmental Protection.

Schapiro, M. 1992. Moonlighting. *Outside*, Nov.:23–24.

Schemnitz, A., ed. 1980. *Wildlife management techniques manual.* Washington, D.C.: Wildlife Society.

Schweinfurth, U. 1957. Die horizontale und vertikale Verbreitung der Vegetation im Himalaya. Bonn: F. Dümmlers Verlag.

Shah, N. 1994. *Status survey of southern kiang* (Equus kiang polyodon) *in Sikkim.* Baroda, Gujarat: Faculty of Science, Maharaja Sayajirao University.

Shaw, R. [1871] 1984. *Visits to High Tartary, Yarkand, and Kashgar.* Reprint, Oxford: Oxford University Press.

Sheffield, W., B. Fall, and B. Brown. 1983. *The nilgai antelope in Texas.* College Station: Texas Agricultural Experiment Station.

Shen, R., E. Reiter, and J. Bresch. 1986. Numerical simulation of the development of vortices over the Qinghai-Xizang (Tibet) Plateau. *Meteorol. Atmos. Phys.* 35: 70–95.

Sheng, H. 1986. Recent information on studies of international endangered mammals. *Chinese Wildlife* 5:1–4. (In Chinese.)

———, ed. 1992. *The deer in China.* Shanghai: East China Normal University Press. (In Chinese.)

Sheng, H., and H. Lu. 1982. Distribution, habits, and resource status of the tufted deer *(Elaphodus cephalophus). Acta Zoologica Sinica* 28:307–311. (In Chinese.)

Sheng, H., and N. Ohtaishi. 1993. The status of deer in China. In *Deer of China*, ed. N. Ohtaishi and H. Sheng, pp. 1–10. Amsterdam: Elsevier.

Simpson, G. 1945. The principles of classification and a classification of mammals. *Bull. Amer. Mus. Nat. Hist.* 85:1–350.

Sinclair, A. 1975. The resource limitation of trophic levels in tropical grassland ecosystems. *J. Anim. Ecol.* 44:497–520.

———. 1977. *The African buffalo.* Chicago: University of Chicago Press.

———. 1979. The eruption of the ruminants. In *Serengeti, dynamics of an ecosystem*, ed. A. Sinclair and M. Norton-Griffiths, pp. 82–103. Chicago: University of Chicago Press.

———. 1983. The adaptations of African ungulates and their effects on community function. In *Tropical savannas*, ed. F. Bourlière, pp. 401–426. Amsterdam: Elsevier.

Sinclair, A., and P. Arcese, eds. 1995. *Serengeti II.* Chicago: University of Chicago Press.

Smith, A. 1988. Patterns of pika (genus *Ochotona*) life history variation. In *Evolution of life histories of mammals*, ed. M. Boyce, pp. 233–256. New Haven: Yale University Press.

Smith, A., H. Smith, X. Wang, X. Yin, and J. Liang. 1986. Social behavior of the steppe-dwelling black-lipped pika. *National Geographic Research* 2:57–74.

Sokolov, V. 1974. Saiga tatarica. *Mammalian Species* 38:1–4.

Soma, H., T. Kiyokawa, K. Matayoshi, I. Tarumoto, M. Miyashita, and K. Nagase. 1979. The chromosomes of the Mongolian gazelle *(Procapra gutturosa)*, a rare species of antelopes. *Proc. of the Japan Academy* 55:6–9.

Sopin, A. 1982. Intraspecific relationships in *Ovis ammon* (Artiodactyla, Bovidae). *Zoologicheskii Zhurnal* 61:1882–1892. (In Russian.)

Soulé, M., and B. Wilcox, eds. 1980. *Conservation biology.* Sunderland, Mass.: Sinauer.

Status review, critical evaluation, and recommendations on proposed threatened status for argali (Ovis ammon). 1991. Evergreen, Colo.: Domestic Technology International.

Stein, R. 1972. *Tibetan civilization.* Stanford: Stanford University Press.

Sterndale, R. 1884. *Natural history of the mammalia of India and Ceylon.* Calcutta: Thacker, Spink.

Stockley, C. 1928. *Big game shooting in the Indian Empire.* London: Constable.

Suchbat, C., Ganzoring, N. Dawaa, and Z. Borgil. 1989. Ergebnisse der Bestandserfassung der Kropfantilope (*Procapra gutturosa* Pallas 1777) auf dem Territorium des Suche-Bator-und Ost-Aimaks der MVR im Juli, August 1981. *Erforschung Biologischer Ressourcen der Mongolischen Volksrepublic* (Halle-Wittenberg) 3:61–65.

Suo, T. 1964. *Economic animals of China.* Beijing: Scientific Publishing House. (In Chinese.)

Sutcliffe, A. 1985. *On the track of ice age mammals.* Cambridge: Harvard University Press.

Swofford, D. 1993. *PAUP: Phylogenetic analysis using parsimony, Version 3.1.1.* Champaign: Illinois Natural History Survey.

Takatsuki, S., N. Ohtaishi, K. Kaji, Y. Han, and J. Wu. 1988. A note on fecal and rumen contents of white-lipped deer on eastern Qinghai-Tibet Plateau. *J. Mammal Soc. Japan* 13:133–137.

Talwar, R., and Chundawat, R. 1995. A report of the survey of Tibetan antelope, *Pantholops hodgsoni,* in the Ladakh region of Jammu and Kashmir. Typescript.

Teer, J. 1991. Conservation status of the saiga antelope in Kalymkia and Kazakhstan. *Species* 17:35–38.

Teichman, E. 1922. *Travels of a consular officer in eastern Tibet.* Cambridge: At the University Press.

Thompson, L., E. Mosley-Thompson, M. Davis, J. Bolzan, J. Dai, T. Yao, N. Gundestrup, X. Wu, L. Klein, and Z. Xie. 1989. Holocene–Late Pleistocene climatic ice core records from Qinghai-Tibetan Plateau. *Science* 2462: 74–77.

Tibet: Environment and development issues 1992. 1992. Dharamsala, India: Department of Information and International Relations.

Tong, B. 1981. Some features of permafrost on the Qinghai-Xizang Plateau and the factors influencing them. In *Geological and ecological studies of Qinghai-Xizang Plateau,* ed. D. Liu, pp. 1781–1795. Beijing: Science Press.

Tsalkin, V. 1951. Mountain sheep of Europe and Asia. *Moscow Nat. Hist. Soc.* 27: 255–265. (In Russian.)

Tulgat, R. 1995. Fluctuations in wild camel (*Camelus ferus* Prz.) numbers and reproductivity of the herds. In *Natural conditions and biological resources in Great Gobi National Park,* ed. J. Badamkhand, pp. 87–90. Ulaanbaatar: Ministry of Nature and Environment. (In Mongolian.)

Tulgat, R., and G. Schaller. 1992. Status and distribution of wild Bactrian camels *Camelus bactrianus ferus*. *Biol. Cons.* 62:11–19.

Valdez, R. 1983. *Wild sheep and wild sheep hunters of the Old World.* Mesilla, N.M.: Wild Sheep and Goat International.

van Rooyen, A. 1994. Harvesting strategies for impala using computer simulation. *S. Afr. J. Wildl. Res.* 4:82–88.

Van Soest, P. 1982. *Nutritional ecology of the ruminant.* Corvallis, Oreg.: O and B Books.

Vane-Wright, R., C. Humphries, and P. Williams. 1991. What to protect?—systematics and the agony of choice. *Biol. Cons.* 55:235–254.

Vaurie, C. 1972. *Tibet and its birds.* London: Witherby.

von Roy, E. 1958. Einige Bemerkungen zur Systematik der Paarhufer. *Säugetierkundliche Mitteilungen* 6:150–153.

Vrba, E. 1987. Ecology in relation to speciation rates: Some case histories of Miocene-Recent mammal clades. *Evol. Ecol.* 1:283–300.

———. 1995. The fossil record of African antelopes (Mammalia, Bovidae) in relation to human evolution and paleoclimate. In *Paleoclimate and evolution with emphasis on human origins*, ed. E. Vrba, G. Denton, T. Partridge, and L. Burckle, pp. 385–424. New Haven: Yale University Press.

Vrba, E., and G. Schaller. Forthcoming. Phylogeny of Bovidae (Mammalia) based on behavior, glands, and skull morphology. In *Ruminant Artiodactyla: Past, present, and future.* New Haven: Yale University Press.

Waddell, L. 1905. *Lhasa and its mysteries.* New York: Dutton.

Wadia, D. 1966. *Geology of India.* London: Macmillan.

Wallace, H. 1913. *The big game of central and western China.* London: John Murray.

Walsh, N., S. Fancy, T. McCabe, and L. Pank. 1992. Habitat use by the Porcupine caribou herd during predicted insect harassment. *J. Wildl. Manage.* 56:465–473.

Walsh, P., D. Metzger, and R. Higuchi. 1991. Chelex 100 as a medium for simple extraction of DNA for PCR-based typing from forensic material. *BioTechniques* 10:506–513.

Walter, H., and E. Box. 1983a. The deserts of central Asia. In *Temperate deserts and semi-deserts*, ed. N. West, pp. 193–236. Amsterdam: Elsevier.

———. 1983b. The Pamir—An ecologically well-studied high-mountain desert biome. In *Temperate deserts and semi-deserts*, ed. N. West, pp. 237–269. New York: Elsevier.

Walther, F. 1979. Das Verhalten der Hornträger (Bovidae). *Handbuch der Zoologie* 8(10):1–184.

Walther, F., E. Mungall, and G. Grau. 1983. *Gazelles and their relatives.* Park Ridge, N.J.: Noyes.

Wang, F., B. Li, and Q. Zhang. 1981. The Pliocene and Quaternary environment on the Qinghai-Xizang Plateau. In *Geological and ecological studies of Qinghai-Xizang Plateau*, ed. D. Liu, pp. 231–235. Beijing: Science Press.

Wang, H., and H. Sheng. 1988. Studies on population densities, conservation, and exploitation of forest musk deer *(Moschus berezovskii)* in the northwest of the Sichuan Basin. *Acta Theriologica Sinica* 8:241–249. (In Chinese.)

Wang, J. 1981. On the fundamental characteristic of the steppe vegetation in Xizang Plateau. In *Geological and ecological studies of Qinghai-Xizang Plateau*, ed. D. Liu, pp. 1929–1936. Beijing: Science Press.

Wang, P. 1984. Progress in late Cenozoic palaeoclimatology of China: A brief review. In *The evolution of the East Asian environment*, ed. R. Whyte, vol. 1, pp. 165–187. Hong Kong: University of Hong Kong.

Wang, X., and R. Hoffmann. 1987. *Pseudois nayaur* and *Pseudois schaeferi*. *Mammalian Species* 278:1–6.

Wang, X., J. Li, and K. Song. 1988. The identification report about the Qilian Mountain sheep's subspecies, Gansu Province. Unpublished report from Ministry of Forestry, Beijing, to U.S. Department of the Interior.

Wang, X., and A. Smith. 1988. On the natural winter mortality of the Plateau pika *(Ochotona curzoniae)*. *Acta Theriologica Sinica* 8:152–156. (In Chinese.)

Wang, X. M., and G. Schaller. 1996. Status of large mammals in western Inner Mongolia, China. *J. East China Normal University, Natural Science*, Special Issue, Dec.:93–104.

Wang, Z., and S. Wang. 1986. Distribution and recent status of the Felidae in China. In *Cats of the world*, ed. S. Miller and D. Everett, pp. 201–209. Washington, D.C.: National Wildlife Federation.

Ward, A. 1923. Game animals of Kashmir and adjacent hill provinces. *J. Bombay Nat. Hist. Soc.* 28:335–344.

———. 1924. The mammals and birds of Kashmir and the adjacent hill provinces. *J. Bombay Nat. Hist. Soc.* 29:879–887.

Weaver, J. 1993. Refining the equation for interpreting prey occurrence in gray wolf scats. *J. Wildl. Manage.* 57:534–538.

Wegge, P. 1979. Aspects of the population ecology of blue sheep in Nepal. *J. Asian Ecology* 1:10–20.

———. 1989. Wild caprids and their predators in Khunjerab National Park, northern Pakistan. p. 41. In *World Conference on Mountain Ungulates*, Abstracts. Camerino, Italy: Università degli Studi di Camerino.

Wei, W., W. Zhou, N. Fan, and D. Biggins. 1996. Activity rhythm and home range of alpine weasel. *Acta Theriologica Sinica* 16:35–42.

Wellby, M. 1898. *Through unknown Tibet.* Philadelphia: J. B. Lippincott.

Wemmer, C., and J. Murtaugh. 1980. Olfactory aspects of rutting behavior in the Bactrian camel *(Camelus bactrianus ferus)*. In *Chemical signals*, ed. D. Müller-Schwarze, pp. 107–124. New York: Plenum.

West, P., and S. Brechin, eds. 1991. *Resident peoples and national parks: Social dilemmas and strategies in international conservation.* Tucson: University of Arizona Press.

Western, D. 1975. Water availability and its influence on the structure and dynamics of a savannah large mammal community. *E. Afr. Wildl. J.* 13:265–286.

Western, D., and V. Finch. 1986. Cattle and pastoralism: Survival and production in arid lands. *Human Ecology* 14:77–94.

Wheeler, W., and D. Gladstein. 1994. *MALIGN, Version 2.1.* New York: American Museum of Natural History.

Whicker, A., and J. Detling. 1988. Ecological consequences of prairie dog disturbances. *Bioscience* 38:778–785.

White, R., B. Tiplady, and P. Groves. 1989. Qiviut production from muskoxen. In *Wildlife production systems,* ed. R. Hudson, K. Drew, and L. Baskin, pp. 387–400. Cambridge: Cambridge University Press.

Wild yaks slaughtered in Amdo. 1992. *Tibetan Environment and Development News* (Washington, D.C.) 6:1.

Wilson, E. [1913] 1986. *A naturalist in western China.* Reprint, London: Cadogan Books.

Wilson, P. 1981. Ecology and habitat utilisation of blue sheep *Pseudois nayaur* in Nepal. *Biol. Cons.* 21:55–74.

———. 1984. Aspects of reproductive behaviour of bharal *(Pseudois nayaur)* in Nepal. *Z. Säugetierkunde* 49:36–42.

———. 1985. The status of *Pseudois nayaur* and *Ovis* populations in Nepal. In *Wild sheep,* ed. M. Hoefs, pp. 172–178. Whitehorse, Yukon: Northern Wild Sheep and Goat Council.

Wong, H. 1993. Distress message from a nature reserve. Typescript.

Wu, Y., C. Yuan, J. Hu, J. Peng, and P. Tao. 1990. A biological study of dwarf blue sheep. *Acta Theriologica Sinica* 10:185–188. (In Chinese.)

Wu, X., and Z. Lin. 1981. A preliminary analysis of climatic change during the historical time of Qinghai-Xizang Plateau. In *Geological and ecological studies of Qinghai-Xizang Plateau,* ed. D. Liu, vol. 2, pp. 1581–1587. Beijing: Science Press.

Xu, R. 1981. Vegetational changes in the past and the uplift of Qinghai-Xizang Plateau. In *Geological and ecological studies of Qinghai-Xizang Plateau,* ed. D. Liu, vol. 1, pp. 139–143. Beijing: Science Press.

Xu, S. 1981. The evolution of the palaeogeographic environments in the Tanggula Mountains in the Pliocene-Quaternary. In *Geological and ecological studies of Qinghai-Xizang Plateau,* ed. D. Liu, vol. 1, pp. 247–253. Beijing: Science Press.

Le yak. 1976. Ethnozootechnie, no. 15. Paris: Sociéte d'Ethnozootechnie.

Yang, Q. 1994. Further study on the geographical distribution and conservation of snow leopard in Qinghai, China. In *Proceedings of the Seventh International Snow Leopard Symposium,* ed. J. Fox and J. Du, pp. 73–77. Seattle: International Snow Leopard Trust.

Yang, Y., B. Li, Z. Yong, et al. 1983. *Geomorphology of Xizang (Tibet).* Beijing: Science Press. (In Chinese.)

Young, J. 1935. Hunting notes. In *Men against the clouds,* ed. N. Burdsall and A. Emmons, pp. 239–258. New York: Harper and Brothers.

Yu, Y., S. Miura, J. Pen, and N. Ohtaishi. 1993. Parturition and neonatal behavior of white-lipped deer. In *Deer of China,* ed. N. Ohtaishi and H. Sheng, pp. 235–241. Amsterdam: Elsevier.

Zeuner, F. 1963. *A history of domesticated animals.* London: Hutchinson.

Zhang, C. 1984. Account of a survey of wildlife of the Qinghai-Xizang Plateau. *Chinese Wildl.* 1:43–46. (In Chinese.)

Zhang, J., J. Wang, W. Chen, and B. Li. 1981. On the vegetation zoning in the Qinghai-Xizang Plateau. In *Geological and ecological studies of Qinghai-Xizang Plateau,* ed. D. Liu, vol. 2, pp. 1919–1928. Beijing: Science Press.

Zhang, R. 1989. *China: The yak.* Gansu: Science and Technology Publishers. (In Chinese.)

Zhang, R., J. Han, and J. Wu, eds. 1994. *Proceedings of the 1st International Congress on Yak.* Lanzhou: Gansu Agricultural University.

Zhen, J., and S. Zhu. 1990. Some ecological information on argali *(Ovis ammon hodgsoni)* in the Burhanbuda Mountain of Qinghai Province. *Acta Theriologica Sinica* 10:304–307. (In Chinese.)

Zhen, W., and W. Liu, eds. 1994. *A collection of forestry papers in Tibet.* Lhasa: People's Publishing House of Tibet. (In Chinese.)

Zheng, B., and J. Li. 1981. Quaternary glaciation of the Qinghai-Xizang Plateau. In *Geological and ecological studies of Qinghai-Xizang Plateau,* ed. D. Liu, pp. 1631–1640. Beijing: Science Press.

Zheng, S., and N. Pi. 1979. Research on the ecology of musk deer. *Acta Zoologica Sinica* 25:176–186. (In Chinese.)

Zheng, S., J. Wu, and Y. Han. 1989. Preliminary investigation on the food habits and reproduction of the white-lipped deer. *Acta Theriologica Sinica* 9:123–129. (In Chinese.)

Zheng, S., Y. Yu, Y. Han, and J. Wu. 1989. Studies of the ungulate community at Yanchiwan Nature Reserve. *Acta Theriologica Sinica* 9:130–136. (In Chinese.)

Zheng, Z., D. Li, Z. Wang, et al. 1983. *The avifauna of Xizang.* Beijing: Science Press. (In Chinese.)

Zhirnov, L., and V. Ilyinsky. 1986. *The Great Gobi National Park—Refuge for rare animals of the central Asian deserts.* Moscow: Centre for International Projects.

Author Index

Subject Index

The letter *t* following a locator refers to a table; *f* refers to a figure.

Om mani padme hum

(the Tibetan invocation for the salvation of all living creatures)